Greenhouse horticulture

Greenhouse horticulture

Technology for optimal crop production

Cecilia Stanghellini

Bert van 't Ooster

Ep Heuvelink

BRILL | WAGENINGEN ACADEMIC

EAN: 9789004697034
e-EAN: 9789004697041
ISBN: 978-90-04-69703-4
e-ISBN: 978-90-04-69704-1
DOI: 10.1163/9789004697041

First published, 2019
2nd edition, 2024

© Koninklijke Brill NV
The Netherlands, 2024

Preface

Good greenhouse management requires an understanding of the processes that determine crop production. These include how microclimate affects crops, and how the greenhouse structure, the technology installed and its management, together with conditions outside the greenhouse, affect microclimate.

The course 'Greenhouse technology' we teach at Wageningen University is a unique combination of crop science, physics and engineering. These are different scientific disciplines and often the much-needed communication between them is hindered by not speaking each other's 'language'. It is our experience that 'Plant Science' students taking the course are often uneasy about 'technical' topics whereas 'Biosystems Engineering' students struggle with crop physiology.

So, we have long felt the need for an up-to-date book that covers all the topics on the course. And we aren't alone: a question that we often hear when teaching dedicated on-site courses worldwide or giving lectures at the annual Wageningen Summer School on Greenhouse Horticulture, is: 'Can you recommend a book that covers the course content?' Obtaining the Wageningen Excellent Education Award for our course twice in a row was a further incentive to write one ourselves.

There are several books available. But, as far as we know, none of them focus on an integrated approach to crop physiology and the technical aspects of greenhouse cultivation and climate management. The main purpose of this book is to fill that gap, combining the analysis of the relationship between crop production and ambient climate with an explanation of the processes that determine the climate in a protected environment. This book outlines the methods and gives several examples of how to make optimal choices about technology and its operation, by weighing the costs and benefits.

In spite of the word 'technology', this was never meant to be a book specifically about high-tech greenhouses. Usually, fewer than half the students taking 'Greenhouse Technology' are Dutch, and we hope that this textbook will be appreciated by university students and professionals around the world. We feel confident about achieving our objective.

Writing a book always involves being selective: not all topics within the wide-ranging field of greenhouse crop technology receive (equal) attention. For instance, we do not address labour management or robotics, and climate is approached with steady-state equations. The aim is to help readers to understand so that they are able to deal with disparate issues, rather than giving them 'rules' to address specific problems.

We are grateful to Tijs Kierkels, a freelance journalist, who produced written documents based on our lectures as a first step towards book chapters. We have been extremely lucky to be able to rely on the huge greenhouse expertise of our colleagues. In particular, Esteban Baeza, Sjaak Bakker, Jouke Campen, Silke Hemming, Frank Kempkes, Leo Marcelis, Rachel Van Ooteghem and Wim Voogt have read and improved earlier versions of several chapters. Luca Incrocci (University of Pisa, Italy) and Juan Ignacio Montero (IRTA, Spain) and a number of anonymous reviewers have also read and improved a number of sections. Dai Jianfeng (Signify, China) spotted some errors while taking care of the Chinese translation and was so kind to share his findings with us. Of course, any inaccuracy that may remain is our fault and not theirs.

Finally, we wish to thank Wageningen Academic Publishers, and Mike Jacobs in particular, for their enthusiasm and support in making this book possible.

Cecilia Stanghellini, Bert van 't Ooster and *Ep Heuvelink*

Table of contents

Preface   6

Abbreviations   12

Chapter 1
Introduction   14
    1.1 Greenhouses: why?   15
    1.2 What is a greenhouse?   19
    1.3 Different conditions, different greenhouses   22
    1.4 Greenhouse technology: why?   23

Chapter 2
Crop as a production machine   26
    2.1 Development and growth   27
    2.2 Growth analysis   32
    2.3 Concluding remarks   53

Chapter 3
Radiation, greenhouse cover and temperature   54
    3.1 Solar radiation   55
    3.2 Thermal radiation: the laws of Stefan-Boltzmann and Wien   65
    3.3 Heat transfer – conduction, convection and thermal storage   67
    3.4 Energy balance and greenhouse temperature   69
    3.5 Concluding remarks   73

Chapter 4
Properties of humid air and physics of air treatment   74
    4.1 Properties of humid air   75
    4.2 Physics of air treatment   89
    4.3 Concluding remarks   98

Chapter 5
Ventilation and mass balance   100
    5.1 Introduction   101
    5.2 Ventilation performance   102
    5.3 The temperature in a ventilated greenhouse   107
    5.4 Mass balance   108
    5.5 Concluding remarks: [semi]-closed greenhouses   111

Chapter 6
Crop transpiration and humidity in greenhouses   114
    6.1 Transpiration from a leaf surface   115
    6.2 Transpiration of a crop   119
    6.3 Calculating transpiration of a greenhouse crop   121
    6.4 Humidity in greenhouses   124
    6.5 Concluding remarks   130

Chapter 7
Crop response to environmental factors   132
    7.1 Light   133
    7.2 Temperature   146
    7.3 Carbon dioxide (CO_2)   151
    7.4 Humidity   155
    7.5 Drought and salinity   157
    7.6 Concluding remarks   162

Chapter 8
Heating in climate-controlled greenhouses 164
  8.1 Heat demand of a greenhouse 165
  8.2 Heating systems of greenhouses 176
  8.3 Transport units servicing compartments, different greenhouses 179
  8.4 The energy centre of the greenhouse 180
  8.5 Energy sources 181
  8.6 Industrial residual heat 188
  8.7 Heat pump 189
  8.8 The role of buffer tanks 189
  8.9 Economic feasibility 190
  8.10 Concluding remarks 192

Chapter 9
Cooling and dehumidification 194
  9.1 Evaporative cooling 196
  9.2 Air conditioning through heat pumps 201
  9.3 Other cooling and dehumidification methods 207
  9.4 Increasing the solar collector function of a greenhouse 208
  9.5 Concluding remarks 211

Chapter 10
Supplementary lighting 212
  10.1 Introduction 213
  10.2 Light measurement units and variables used 214
  10.3 The amount of light at a certain point 216
  10.4 Major lamp types and their characteristics 220
  10.5 Sources of electricity 225
  10.6 Concluding remarks 225

Chapter 11
Carbon dioxide supply 226
  11.1 Introduction 227
  11.2 Units 227
  11.3 CO_2 sources 228
  11.4 CO_2 supply strategy 232
  11.5 CO_2 distribution 234
  11.6 CO_2 sensors 237
  11.7 Concluding remarks 237

Chapter 12
Managing the shoot environment 238
    12.1 Introduction 239
    12.2 The limiting factor 240
    12.3 Managing climate: source/sink balance 244
    12.4 Optimizing a single climate factor 246
    12.5 Concluding remarks 253

Chapter 13
Root zone management:
how to limit emissions 256
    13.1 Precision irrigation 257
    13.2 Closed irrigation systems 266
    13.3 Saline irrigation water 268
    13.4 Optimal management of irrigation 274
    13.5 Water use efficiency of growing systems 276
    13.6 Concluding remarks 277

Chapter 14
Vertical farms 278
    14.1 Introduction 279
    14.2 Light and energy 280
    14.3 Resource requirement 281
    14.4 Environmental footprint 282
    14.5 Future perspective 284

Chapter 15
The future 286
    15.1 The future of greenhouses 287
    15.2 The greenhouses of the future 288

References 290

About the authors 310

List of boxes 312

Abbreviations

AR	anti-reflection
AS	artificial solar light
B	blue (light)
CAM	Crassulacean acid metabolism
CFD	computational fluid dynamics
CHP	combined heat and power
COP	coefficient of performance
COP_c	coefficient of performance for cooling
COP_h	coefficient of performance for heating
DFT	deep flow technique
DIF	difference between day and night temperature
DLI	daily light integral
DMC	dry matter content
DoS	degree of saturation
DROP	drop-in temperature
DVR	development rate
EC	electrical conductivity
ETFE	ethyl tetra fluor ethylene
EVA	ethylene-vinyl-acetate
FR	far-red (light)
FvCB	Farquhar-von Caemmerer-Berry
GR	growth rate
HET0	high transpiration
HI	harvest index
HPS	high pressure sodium
HU	heat units
IR	infrared
LAI	leaf area index
LAR	leaf area ratio
LED	light emitting diode
LET0	low transpiration
LUE	light use efficiency
LWR	leaf weight ratio

NAR	net assimilation rate
NDIR	non-dispersive infrared
NFB	net financial benefit
NFT	nutrient film technique
NIR	near infrared
PAR	photosynthetically active radiation
PC	polycarbonate
PCR	photosynthetic carbon reduction
PE	polyethylene
PER	primary energy ratio
PFAL	plant factories with artificial lighting
PMMA	poly methyl methacrylate
PPF	photosynthetic photon flux
PPFD	photosynthetic photon flux density
R	red (light)
RGR	relative growth rate
RH	relative humidity
RTD	resistance temperature detectors
RUE	radiation use efficiency
SAS	shade-avoidance syndrome
SLA	specific leaf area
STES	seasonal thermal energy storage
TIR	thermal infrared
UV	ultraviolet
VMD	volume median diameter
VPD	vapour pressure deficit
WUE	water use efficiency

Chapter 1
Introduction

Photo: Biagio Dimauro, Regione Sicilia

It is approximately 12,000 years since agriculture was first practised in different regions of the world. In spite of the huge impact this has had on our welfare, until recently the progress in agriculture has been 'limited' to management of the root zone environment (ploughing, seeding, irrigation, fertilisation). The 'chemical' revolution of the 20th century, besides changing the whole concept of fertilisation, added (chemical) crop protection to the agricultural toolbox. Nevertheless, most farmers can still only hope for good harvests and worry about early frosts or late snow, heavy rain or prolonged droughts, not to mention pests that are ever more resistant and regulations that reduce the arsenal of chemical weapons with which to fight them.

1.1 Greenhouses: why?

The usefulness of covers against such hazards has long been appreciated: the Romans, for instance, grew exotic plants in shelters covered with the mica they used for windows, something which Martialis 93AD (Epigrammata, VIII, 14) praised for its transparency. Things became easier when glass-making technology enabled the production of large enough plates of clear glass for the south-facing, generous windows of the Renaissance orangeries. It is a fact, however, that only very special and exotic plants were deemed worthy of such care up to the first half of the 20th century. By then grapes were grown in stove-heated, glass-roofed buildings in Holland, the very first attempt at commercial protected horticulture (Figure 1.1). That grapes were not destined to be the crop of the future is attested by the fact that of the 9,500 ha glasshouses operating now in the Netherlands, there are barely any that contain grapes. The plastic revolution allowed for much cheaper crop shelters, with the result that protected horticulture around the world has grown to an estimated area of anything between 500,000 to 2 million ha (depending on what is counted as a greenhouse) commercial vegetable production in permanent structures (greenhouses), of which only 40,000 are glass-covered (glasshouses) (Rabobank, 2018a). Protected cultivation, controlled environment agriculture (CEA) and greenhouse horticulture are often used as synonyms, all three terms referring to crop production systems that allow for manipulation of the aerial environment.

Figure 1.1. Wall greenhouses (left: a modern reconstruction) and orangeries built in gardens of European monasteries and castles allowed for cultivation out of season and the growing of exotic plants, the first steps in the development of commercial greenhouse horticulture. Flat glass (left and right, foreground) was applied to obtain earlier yields and to produce seedlings for field transplant. When flat glass was put on a construction one could walk under, the first Venlo-type greenhouses were born (right, background) (photo left: historische druivenkwekerij Sonnehoeck; photo right: archive Greenhouse Horticulture, Wageningen University & Research).

In fact, because crop growth and development depend on several factors of the aerial environment (Figure 1.2), we now know that the combination of clever technology and a confined environment would enable full control of crop growth and production, although full control may well be financially unviable. Nevertheless, a greenhouse grower has many more tools for managing a crop than a field grower could ever dream of, the only problem being the all-too-frequent lack of 'instructions' for using such tools.

Although protected cultivation developed mainly in relatively cold regions, with the purpose of raising the temperature so as to cultivate exotic and/or out-of-season crops, we now know that there are many more reasons to put crops under cover. In rainy sub-tropical regions, for example, shelters are used to prevent damage from heavy rain, even though the temperature and humidity in the open air would be better for crop growth than within the shelter. Similarly, net houses are used in desert regions to limit crop evaporation and prevent leaf burning, despite the reduced photosynthesis rates in the shade.

Reliable statistics for the surface area of protected crop cultivation (Table 1.1) are quite difficult to find, both because some statistics include temporary structures and some do not (and the lack of a clear definition for what is temporary), and because the constant increase in area worldwide quickly make any statistics obsolete. The main driving factors for this increasing surface area may be an increasing international demand for high-quality horticultural products (including out-of-season vegetables and foreign products) coupled with improved transportation and post-harvest management, which have allowed production to take place in regions far from the main markets.

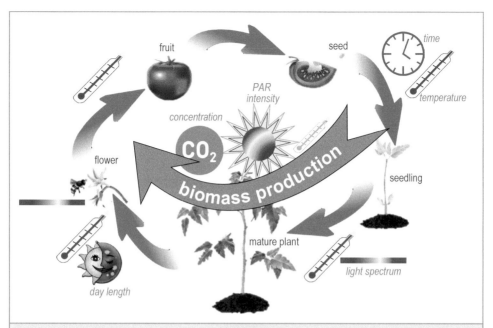

Figure 1.2. The life cycle of a plant and the (aerial) factors that determine it. As there must be growth (biomass) before any development (transition to the next life stage) can take place, photosynthetic active radiation (PAR) and carbon dioxide concentration (the determinants of photosynthesis) are the primary factors for crop production. Temperature is the most obvious determinant of development, whereas growth (photosynthesis) depends rather weakly on it (provided it is within a reasonable range). We are only now starting to understand the role light spectrum (wavelength composition) plays in several processes, including the behaviour of pollinators (and of pests and their predators).

Table 1.1. Estimated area (ha) of greenhouses for vegetable production worldwide, allocated to the major climate zones (please refer to Figure 1.8).Chinese solar greenhouses (Box 3.8) and greenhouses for ornamentals are not included, for lack of information (Hickman, 2018). In rich countries (such as the Netherlands), the total acreage for ornamental production can be similar to the vegetable acreage. In equatorial countries (both in Africa and America), ornamentals are produced (usually at high elevation) for export, and their acreage can be several times over the acreage for vegetable production.

	Sub-arctic to temperate	Mediterranean to sub-tropic	Tropic to equatorial	Total
Asia + Oceania	2,300	173,400	100	175,800
Europe	47,000	131,000		178,000
Americas	2,400	6,800	7,300	16,500
Africa		30,200	100	30,300
Near East		74,800		74,800
Total	51,700	416,200	7,500	475,400

Rabobank (2000) identified several 'critical success factors' that determine the viability of a greenhouse business in various places:

- geography: distance to the main market(s);
- climate;
- cost of production means/resources;
- internal market: ability of the home market to absorb production in case of disruption of export flows;
- cost of manpower;
- access to finance: investment climate;
- infrastructure: reliability of transport, energy/electricity supply, irrigation water, etc.;
- access to knowledge: adequate labour and extension services;
- socio-political environment: its stability determines the investment horizon.

Indeed, every relevant greenhouse area worldwide can be traced to favourable conditions in the majority of these categories.

What is certain is that the increase of protected cultivation areas is particularly prevalent in otherwise marginal agricultural land, thanks to relatively high returns coupled with high efficiency in the use of (scarce) resources. It is a fact that conversion of land from open field cultivation to protected cultivation enormously increases the output and earnings per unit of soil surface. The example of Almeria (SE Spain; Figure 1.3) with approximately 30,500 hectares (Cajamar, 2017) of plastic greenhouses is notable. Protected crop cultivation was responsible for the astonishing economic growth of this province over the 20 years preceding the 2007 economic crisis (Caja Rural Intermediterránea, 2005).

Protected crop cultivation is a small fraction of the total agricultural land but, as a capital-, labour- and (often) energy-intensive cropping system, it may play an important role in the local, or even the national, economy. For instance, in the Netherlands glasshouse cultivation covered in 2016 0.7% of the overall agricultural land, but accounted for 20% of the annual gross income from agriculture (CBS, 2018), with annual crop revenue exceeding often 1 million € ha^{-1} (KWIN, 2017). Protected crops are generally concentrated in relatively small areas (Figure 1.4), often close to urban areas with evident implications from an environmental point of view.

The production technology worldwide is currently undergoing a modernisation process to face the increasing competition arising from the globalisation of both production and marketing. Moreover, an increase in the awareness of environmental pollution provoked by agriculture, the demand for healthy foods and the shortage of resources such as water are pushing growers towards the application of more sustainable growing techniques. For instance, the greenhouse area using biological agents for pest control in Almeria increased from 3% in 2006 to 28% of the total area in 2007 (Bielza *et al.*, 2008) in response to stricter chemical controls in the main export markets.

Figure 1.3. Plastic greenhouses (the white areas) in the plain of Campo de Dalia, to the west of the city of Almeria (SE Spain). Images: courtesy of Earth Sciences and Image Analysis Laboratory, NASA Johnson Space Center (http://eol.jsc.nasa.gov) ISS008-E-14686, inset: ISS004-E-13199.

Figure 1.4. Left: glasshouses covering nearly each square meter soil in the Dutch Westland (photo: archive Greenhouse Horticulture, Wageningen University & Research). Right: plastic greenhouses not far from Kuala Lumpur, Malaysia (photo: Jouke Campen, Greenhouse Horticulture, Wageningen University & Research).

1.2 What is a greenhouse?

Whenever a physical shelter is used to somehow modify the environment around a crop we speak of 'protected cultivation'. That is, cultivation carried out through some kind of permanent or temporary shelter covering the entire crop with the aim of enhancing its productivity. As

Figure 1.5-1.7 indicate, there is a very broad range of systems that fit this description, therefore further detailing (Box 1.1) may be useful. Here we draw from the definitions applied by EFSA (2010: pp. 13-18), to which the reader is referred for a greater selection of pictures.

'Greenhouse' denotes a structure that is tall enough to enter, is permanent (multi-annual), and has some means for controlling the environment, such as ventilation openings that can be somehow managed. We speak of high-tech greenhouses when there are more advanced, computer-controlled actuators for environmental control, such as (but not necessarily all of them): ventilation (either through computer-controlled openings or fans), heating, carbon dioxide supply, misting, cooling and de-humidification, movable screens for shading and

Figure 1.5. Extreme examples of protected cultivation: low, temporary tunnels in Southern Italy (left) and a high-end glasshouse in the Netherlands (right) with controlled ventilation, air conditioning and supplementary lighting. (photos: Cecilia Stanghellini)

Figure 1.6. High, partial shelter, typically used for strawberries (Belgium, left) and table grapes, with insect/bird nets on the front (Portugal, right). (photo: left: Paolo Battistel, Ceres s.r.l.; photo right: Frank Kempkes, Greenhouse Horticulture, Wageningen University & Research)

Figure 1.7. Tomatoes in a net house in Arica (Chile, left) and lettuce in a multi-tunnel (Italy, right) (photos: Cecilia Stanghellini).

Box 1.1. Various types of crop protection and what they do.

Low tunnel	Passive increase of temperature, tunnel removed well before harvest.
High, partial shelter	Protection against rain and birds, passive increase of temperature.
Shade/net house	Used in high-radiation, arid environments. Prevention of leaf burn, higher humidity in the shoot environment.
Walk-in tunnel	Passive increase of temperature.
Low-tech greenhouse/multi-tunnel	Passive increase of temperature, very limited control of temperature through regulation (usually by hand) of the openings and seasonal whitewash.
High-tech green/glasshouse	Computer-driven control of temperature through ventilation and heating, control of humidity through ventilation and fogging, control of carbon dioxide concentration through artificial supply and ventilation, control of light level possible through shading and supplemental light.
Plant factory/vertical farm	Fully removed from external disturbances, full control of all production factors.

energy saving, and artificial light. The word glasshouse is often used for high-tech greenhouses, as glasshouses are usually equipped with high-tech actuators. However, since the amount of technology installed has little to do with the material for covering, in this book we will reserve the term glasshouse for glass-covered greenhouses.

Finally, plant factories (or vertical farms) are systems completely separated from the external climate, in which production can be fully controlled. The obvious downside of this format is that artificial light is used to make up for the lack of sunlight. Graamans *et al.* (2018) have shown this can be an economical proposition only when a significant market gain is possible or when the value of other (scarce) resources (ground, water) offsets the electricity costs required for lighting.

1.3 Different conditions, different greenhouses

It must be clear by now that there are several factors behind the variety of greenhouse systems we have covered. The most obvious is climate: in some locations a grower may need to shade against leaf burn, while elsewhere there is little that would grow without heating. To some extent, the desired crop also plays a role: sweet peppers, for instance, have very different temperature requirements than blueberries. However, as we have seen above, with the right amount of technology it is possible to grow everything everywhere. The real issue becomes the ability to make a profit.

The final factor to determine the most appropriate greenhouse system in a certain location is the bottom line: the market value of what the greenhouse produces puts a limit on the production costs that may be allowed, which in turn puts a limit on the amount of technology and resources a grower can afford.

The amount of technology typically installed in greenhouses increases with latitude, which is both a consequence of the climate and of the affluence of the local market (with few exceptions). In Figure 1.8 we provide a summary of the challenges faced by growers operating typical greenhouses in the three main climatic regions of the world.

There are, however, some trends that may cause relatively rapid changes in the greenhouse outlook worldwide. In cold regions, greenhouse management will need to adapt, due to the

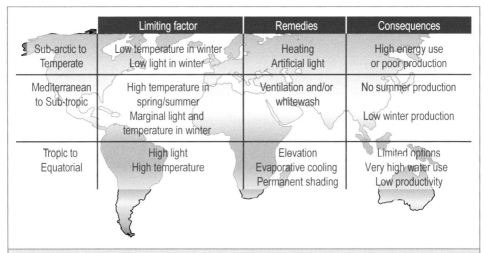

	Limiting factor	Remedies	Consequences
Sub-arctic to Temperate	Low temperature in winter Low light in winter	Heating Artificial light	High energy use or poor production
Mediterranean to Sub-tropic	High temperature in spring/summer Marginal light and temperature in winter	Ventilation and/or whitewash	No summer production Low winter production
Tropic to Equatorial	High light High temperature	Elevation Evaporative cooling Permanent shading	Limited options Very high water use Low productivity

Figure 1.8. Main factors limiting the productivity of local greenhouses in the main climatic regions of the world, what is often done (remedies) to mitigate climate within these greenhouses and what are the consequences. The main limiting factors are common to a given climate zone, although there is obviously a gradient in the severity within a zone.

increasing disapproval of heating from the burning of fossil fuels. On the other hand, increasing disposable income and awareness of food security in the temperate-to-warm regions will cause a shift to demand-driven production, which will need to be achieved through a wider use of technology in the production system than is now the case.

1.4 Greenhouse technology: why?

What is profitable in the Netherlands may not be as wise in Mexico and yet a whole body of accepted wisdom worldwide derives from the 'rules' developed by Dutch growers (the 'best in the world') years ago. In addition, conditions may and most likely will change. There may be a shift in what consumers want growers to produce, as well as what growers are able to deliver at a profit. What is cost-effective today may not be tomorrow and vice versa, meaning the rules growers apply today may not be right ones in the future. To fully exploit its potential, protected horticulture will need to move from its present body of rules to a base of knowledge. Understanding crop production and the processes that determine it (at climate, greenhouse and crop level) makes it possible for growers and professionals to recognise the consequences of any changes they may consider and to guide informed decisions, as shown by Figure 1.9.

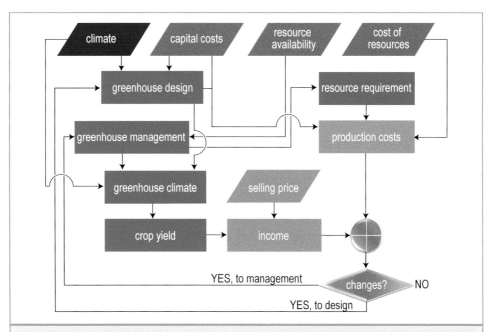

Figure 1.9. Decision scheme (the 'adaptive greenhouse') to identify sensible improvements in greenhouse design and/or management (after Vanthoor, 2011). Environmental constraints (such as restrictions on the use of fossil fuels, for instance) limit the availability of resources. Other resources may be constrained as well: water, for instance, or skilled labour.

In this book we combine understanding of crop production in relation to microclimate with knowledge of the processes that determine microclimate within a shelter: the blue rectangles in Figure 1.9. The processes addressed in this book (crop growth, energy balance, mass exchange) apply equally to any kind of shelter, so in spite of the word 'technology', this is not a book only about high-tech greenhouses. What technology does affect is the ability of the grower to modify the environment within the shelter in order to increase crop productivity. Therefore, most examples in this book will refer to greenhouses, that is, shelters fitted with some form of technology that a grower may need to utilise.

Chapter 2
Crop as a production machine

Photo: Paolo Battistel Ceres s.r.l.

In order to understand and predict the effect of climatic factors and crop management (e.g. planting density or fruit pruning) on crop growth and yield, a thorough understanding of plant behaviour is needed. In this chapter we will focus on crop yield and its underlying processes. The principles of plant and crop growth analysis will be explained, the differences between development and growth will be defined, and processes such as light interception, photosynthesis, respiration, as well as biomass partitioning will be discussed. For more detailed information on the behaviour of plants, including topics only briefly discussed in this chapter such as the physiology of flowering and plant hormones, we refer the reader to the book Plant Physiology in Greenhouses (Heuvelink and Kierkels, 2015).

2.1 Development and growth

Development and growth are often used interchangeably, however for a clear analysis we need to draw a clear distinction. Growth refers to an increase in size: length, area or weight. The main factors that influence growth are light and CO_2.

Development, on the other hand, is a different process, linked to the initiation of new organs on the plant (e.g. leaves or flowers). It is the transition from one phase in the life cycle to another; for example, from a seed to a young vegetative plant. The related development step is referred to as 'germination'. The next phase transition is from a juvenile to an adult plant that has the ability to flower. Note that growth also takes place during this developmental stage. In some cases development takes place without growth, however, often increase in length, volume, weight, etc. is a prerequisite to reaching the next phase. After all, the young plant must have a certain photosynthetic capacity (leaf area) or sugar content to support reproduction. However, growth and development are not directly linked to each other: tomato plants that have grown for 50 days and already have the first flower open can be either big or small (same developmental stage, but clearly different growth rates). Development can be defined as 'the systematic movement along the genetically programmed sequence of events during the life cycle' (Figure 2.1), of which the main influencing factor is temperature. Often a phase transition includes the formation of new organs.

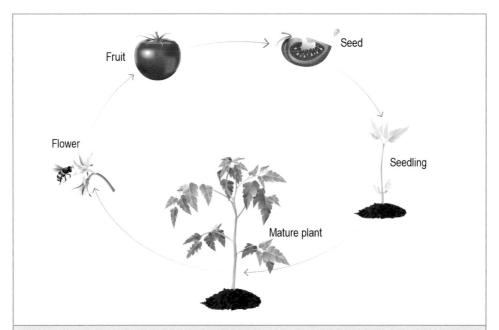

Figure 2.1. Development of a tomato plant. The life cycle starts with a seed, that germinates. After emergence (plant appears above the substrate (soil)) vegetative growth (root, stem, leaves) takes place. The plant becomes generative, a flower truss is initiated. After this flower truss 3 leaves are initiated and again a flower truss, this is repeated many times for indeterminate cultivars. A flower at anthesis (fully open) gets pollinated, this results in fruit set, and the fruit develops into a harvest ripe fruit. This fruit contains seeds and the life cycle starts again (image from Let's Talk Science).

A developmental stage is a characteristic point in the life cycle, e.g. the appearance of the first flower. The rate of development is the inverse of the time needed between two events, e.g. between anthesis and breaker stage (this is the point at which the green colour of a tomato fruit 'breaks' and develops into red). When it takes 40 days to transition from one developmental stage to the next, the development rate is 1/time = 1/40 = 0.025 per day.

As stated previously, development primarily depends on temperature; this relationship is an optimum response (e.g. time to flowering in chrysanthemum (Larsen and Persson, 1999)). However, the development rate (DVR) increases linearly with temperature (e.g. leaf initiation rate in Karlsson *et al.*, 1991; Figure 2.2), often over a large range of suboptimal temperatures. In that case a fixed temperature sum (a certain number of heat units) is needed to complete the development, which in this example means the initiation of a new leaf. This temperature sum (in degree days) or heat units is always the same and depends on the crop or cultivar. The minimum temperature is also relevant: at temperatures lower than this minimum, no development takes place.

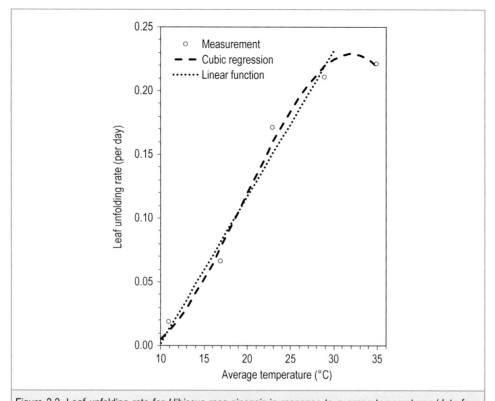

Figure 2.2. Leaf unfolding rate for *Hibiscus rosa-sinensis* in response to average temperatures (data from Karlsson *et al.*, 1991).

In a formula (assuming the temperature *T* remains constant):

$$HU = d \cdot (T - T_{min}) \tag{2.1}$$

HU: heat units (or degree days; °C d); *T*: average daily temperature (°C); T_{min}: minimum temperature required or base temperature (°C); *d*: number of days.

This formula can be rewritten as:

$$1/d = T/HU - T_{min}/HU \tag{2.2}$$

which means that from the linear relationship between development rate (1 d^{-1}) and temperature (*T*), we can deduce *HU* and T_{min} (Box 2.1). *HU* is the inverse of the slope of the linear relationship and $-T_{min}/HU$ is the intercept.

Box 2.1. Example 1: the time from anthesis to a harvest-ready tomato.

The following data have been observed (DVR = development rate; e.g. 0.012987 = 1/77).

Temperature	Days	DVR
17	77	0.012987
20	63	0.015873
23	52	0.019231

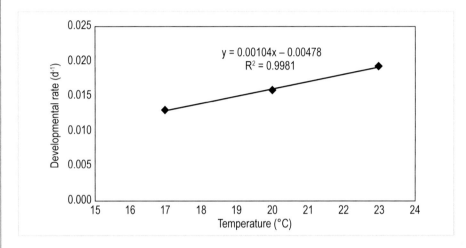

Based on the linear relationship between developmental rate and temperature (shown above) and Equation 2.2, $HU =$ 1/0.00104 = 961 °C d and T_{min} = 0.00478 · 961= 4.6 °C.

For calculating the developmental stage (Box 2.2), and taking an actual temperature for each day, rather than assuming a constant temperature, Equation 2.3 applies:

$$DVS_t = \sum_{i-1}^{t} (T_i - T_{min})/HU \qquad (2.3)$$

where DVS_t is developmental stage at time t; T_i is actual daily temperature (°C); T_{min}: minimum temperature required (°C); HU: heat units (°C d), and $T_i - T_{min}$ is the effective temperature (°C).

Similar to the above for the suboptimal temperature range, heat units can be calculated for the supra-optimal temperature range using a T_{max} (maximum temperature above which no development takes place).

Figure 2.3 shows that both the tomato fruit development rate (from anthesis to harvest) and the plant development rate (number of trusses appearing per day) are (almost) linearly related to temperature in the range from 17 till 24 °C. At 22 °C, the plant development rate

Box 2.2. Example 2: radish, time from emergence to harvest.

Assume that radish requires a minimum temperature of 5 °C and needs a temperature sum of 500 °C d (degree days) from emergence to harvest. It can now be calculated how many days cultivation would take at 10 or 15 °C.

A constant temperature of 10 °C each day contributes 10-5 = 5 degree days to the temperature sum (effective temperature is therefore 5 °C). It takes 500/(10-5) = 100 days until a harvest-ready product.

At a constant temperature of 15 °C this is 500/(15-5) = 50 days. Therefore, a temperature increase from 10 to 15 °C leads to a 50% reduction in cultivation time.

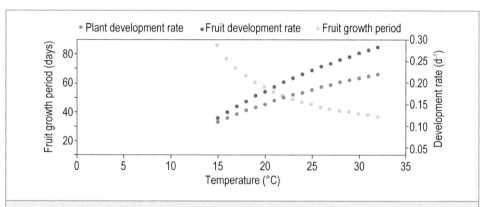

Figure 2.3. Relationship between fruit growth period, fruit development rate (×10), plant development rate (leaf appearance rate) and temperature in tomatoes (data from De Koning, 1994).

is 0.16 d⁻¹. This means that a new truss appears every $1/0.16 \approx 6$ days. At 22 °C, the fruit growth period (time between anthesis and harvest-ready fruit) is about 50 days. This means a fruit development rate of $1/50 = 0.02$ per day. De Koning (2000) showed that salinity or fruit load (number of fruits on the plant) barely influenced the fruit growth period. While cultivars did differ in fruit growth periods, their temperature response was the same (no interaction between temperature and cultivar). This author reported the same for plant development rate, measured by the leaf appearance rate: differences of about 10% between cultivars, but the same temperature response for different cultivars (De Koning, 1994: p36).

The development rate is equally related to day and night temperatures and, therefore, to the 24 h mean temperature, as observed for tomato, cucumber, chrysanthemum, and Easter lily (references in De Koning, 1994: p39). However, as shown by the slightly curved response

(Figure 2.3), significant differences between day and night temperatures may cause lower developmental rates than would be expected from the 24 h mean temperature.

While growers, as well as researchers, usually measure and control air temperature, it is the plant or organ temperature that is relevant for development and growth. Organ temperature is often assumed to be equal to air temperature, but large deviations may occur. For example, Savvides *et al.* (2013) report that meristem temperatures ($T_{meristem}$; determines rate of leaf initiation) substantially deviate from air temperature T_{air} in a species-specific manner under moderate environments. This deviation ranged between -2.6 and 3.8 °C in tomatoes and between -4.1 and 3.0 °C in cucumbers. The lower $T_{meristem}$ observed in cucumbers was linked with higher transpiration of the bud foliage sheltering the meristem when compared with tomato plants.

2.2 Growth analysis

Growth analysis is a quantitative method of analysing differences in plant and crop growth caused by e.g. genotype, climate, nutrition, etc. It focuses on leaf area, light interception, photosynthesis, biomass increase and partitioning of assimilates. Unless explicitly stated otherwise, in this paragraph weight or mass refer to dry weight or dry mass. Growth analysis was the first step towards modelling plant growth and its concepts are still very useful. A typical pattern of crop growth over time is a sigmoid: at first an exponential phase, then a linear part and finally a slowdown in growth (e.g. Lee *et al.*, 2003). To compare growth in the linear part (e.g. between years), the difference between the slopes (growth rates, which is constant in the linear part) is a good indicator. In the exponential phase it is not the growth rate, but rather the relative growth rate (see below) that is constant. In the exponential phase (early crop phase), plants are growing without competition from their neighbours.

2.2.1 Young plants (early crop phase)

Plant growth analysis is an explanatory, holistic and integrative approach to interpreting plant form and function. It uses simple primary data in the form of weights, areas, volumes and contents of plant components to investigate processes within and involving the whole plant (Hunt *et al.*, 2002).

Plant (absolute) growth rate (GR; g d⁻¹) is calculated as the difference in weight (g) between two points in time divided by the difference in time (d):

$$GR = (W_2 - W_1) / (t_2 - t_1) = \Delta W / \Delta t \tag{2.4}$$

W_2 is weight at time 2 t_2 and W_1 is weight at time 1 t_1. When the time interval becomes infinitely small, we find

$$GR = dW / dt \tag{2.5}$$

which means that the growth rate is the derivative of the function of plant weight against time.

In an early stage of plant growth, the GR is not a good parameter for comparison, because it is constantly changing (there is no linear relationship between weight and time). Young plants grow in an exponential way and exponential growth can be calculated with the following equation:

$$W_t = W_0 \cdot e^{RGR \cdot t} \tag{2.6}$$

where W_t is plant weight (g) at time t; W_0 is initial plant weight (g) at time zero; e is the base of natural logarithms; RGR is relative growth rate (a constant; d^{-1}) and t is time since W_0 (d).

Exponential growth means that there is a linear relationship between the natural logarithm of weight and time. We can calculate the slope of this linear relationship in the same way as the growth rate above. This slope is called the relative growth rate (RGR; d^{-1}):

$$RGR = (\ln W_2 - \ln W_1) / (t_2 - t_1) = \Delta \ln W / \Delta t \tag{2.7}$$

When the time interval becomes infinitely small, we find

$$RGR = d\ln W / dt. \tag{2.8}$$

$d\ln W/dt$ represents the derivative of $\ln W$ with respect to time. According to mathematics, the derivative of $\ln W$ is $1/W$; hence

$$RGR = 1/W \cdot dW/dt \tag{2.9}$$

dW/dt is in Equation 2.9 since W itself is a function of t. The RGR (g g^{-1} d^{-1} or d^{-1}) is the increase in plant weight per unit of weight and per unit of time; or one could say it is the growth rate dW/dt per unit of plant weight. The connection between the formula for *RGR* (Equation 2.7) and the exponential growth function (Equation 2.6) is:

$$\ln W_t = \ln W_0 + \ln (e^{RGR \cdot t}) = \ln W_0 + RGR \cdot t.$$

Hence, $RGR = (\ln W_t - \ln W_0)/t$

RGR separated in two underlying components

RGR is described by Equation 2.9 (where W is plant weight) and the outcome of the formula does not change when at the same time we divide and multiply by leaf area A, thereby introducing leaf area in both the numerator and the denominator:

$$RGR = 1/W \cdot dW/dt = 1/A \cdot dW/dt \cdot A/W = NAR \cdot LAR \qquad (2.10)$$

where A represents leaf area (cm^2), NAR is net assimilation rate (g cm^{-2} d^{-1}) and LAR is leaf area ratio (cm^2 g^{-1}).

$$NAR = 1/A \cdot dW/dt \qquad (2.11)$$

$$LAR = A/W \qquad (2.12)$$

NAR reflects the plant growth rate per unit of leaf area and is therefore closely related to leaf photosynthesis rate minus respiration. Leaf area is an important driving variable for plant growth, as only light intercepted by green leaf area results in photosynthesis and growth. RGR depends not only on the net assimilation rate (NAR), but also on how much of the plant's total biomass is included in the leaf area. The leaf area ratio (LAR) is the ratio between leaf area and total plant weight.

LAR separated in two underlying components

As was done in Equation 2.10, we can use Equation 2.12 to separate LAR into its underlying components by introducing leaf weight (W_{leaf}; in grams) in both the numerator and the denominator:

$$LAR = A/W = A/W_{leaf} \cdot W_{leaf}/W = SLA \cdot LWR \qquad (2.13)$$

where SLA is specific leaf area (cm^2 g^{-1}) and LWR is leaf weight ratio (g g^{-1}).

$$SLA = A/W_{leaf} \qquad (2.14)$$

$$LWR = W_{leaf}/W \qquad (2.15)$$

If the leaves are thin, SLA is high, meaning a large leaf area per unit of leaf weight. For thick leaves, SLA has a low value. Leaf weight ratio (LWR) is the ratio between leaf weight and total plant weight and it reflects the partitioning of assimilates to the leaves.

$$RGR = NAR \cdot LAR; LAR = SLA \cdot LWR; hence RGR = NAR \cdot SLA \cdot LWR \qquad (2.16)$$

In other words, relative growth rate is the product of 3 components: a photosynthetic component (NAR), a component reflecting leaf thickness (SLA), and a partitioning component (LWR).

The two latter components together reflect changes in plant morphology. RGR is higher when the leaf photosynthesis rate (NAR) is higher, when leaves are thinner (higher SLA) or when a larger fraction of total plant growth is partitioned to the leaves (higher LWR).

Equation 2.16 enables us to explain the differences in young plant growth between climate conditions and crop species. For example, Shibuya *et al.* (2016) use plant growth analysis to explain differences in RGR of cucumber seedlings grown under light with different red to far-red ratios.

The question of why young tomato plants grow so much faster than young carnation plants is answered by Bruggink (1992) using plant growth analysis (Figure 2.4). For young tomato plants the RGR is about five times higher than that of young carnation plants (Figure 2.4A and B). Furthermore, there is little difference between the species with regard to their NAR (Figure 2.4C and 2.4D). Therefore, the rate of photosynthesis per unit leaf area does not explain the observed significant differences in RGR. The explanation is found in the morphology: the LAR of carnation plants (Figure 2.4E) is about five times lower than that of tomato plants (Figure 2.4F) and it does not respond to light, whereas the LAR of tomato plants increases at low light levels. The much higher LAR of tomatoes results from a much higher SLA, whereas the LWR is more or less the same for both crops (SLA and LWR are not shown in the graphs; LAR = SLA · LWR). Tomato plants have thin leaves (high SLA) and carnation plants have much thicker leaves, which accounts for the much higher RGR of tomato plants. For young plants it is favourable to produce thin leaves, because with the same leaf dry weight there is much more leaf area, and thus much more light is intercepted. This light would otherwise not be utilised, it would simply fall to the floor.

Plant growth analysis is also used in the following example to understand the effects of light level on the growth of young cucumber plants.

RGR is higher at a higher light intensity, as a result of a dramatic increase in NAR when the plants are grown at high light intensities (Figure 2.5A). On the other hand, LAR is lower under high light intensity. Nonetheless, this decrease in LAR is less than the increase in NAR, hence RGR (= NAR · LAR) is higher. The decrease in LAR at higher light intensity is caused by a decrease in SLA (thicker leaves), whereas LWR is barely influenced by light intensity (Figure 2.5B). This reflects a more general trend: SLA is much more sensitive to environmental factors compared to LWR, which remains more constant.

2.2.2 Yield component analysis

To obtain a sufficient understanding of differences in crop yield, knowledge of the underlying processes is of the utmost importance. Yield component analysis provides this information by dissecting yield into its underlying components (Box 2.3). An example of applying this technique is given by Higashide and Heuvelink (2009). These authors studied the effect of plant breeding on greenhouse tomato yield improvement in the Netherlands over the past 50

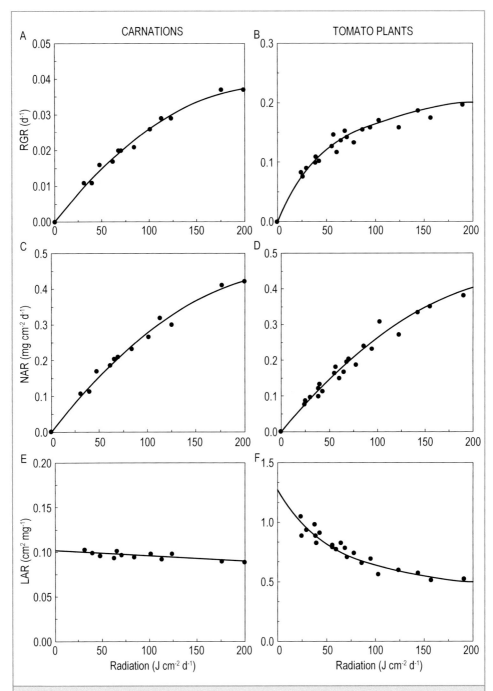

Figure 2.4. Relative growth rate (RGR), net assimilation rate and leaf area ratio (LAR) for young carnations (A,C,E) and tomato plants (B,D,F) as a function of the mean daily light integral R (PAR). Note that the y-axes are very different between species for RGR and LAR! (reprinted with permission from Bruggink, 1992).

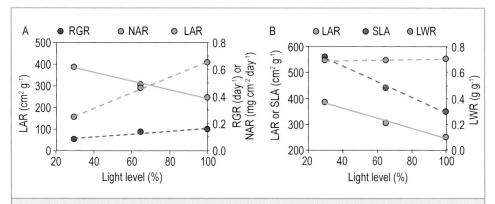

Figure 2.5. Growth analysis for young cucumber plants grown at three different light levels created with shading nets in a greenhouse.

Box 2.3. Think like a detective.

In yield component analysis it is important to think like a detective: what could possibly explain what you see? For example, this year the harvest fresh weight is 10% higher than last year; do the fruits contain more dry matter or is the dry matter simply diluted with more water? Has total dry matter production changed or only the fraction that is directed to the fruits? Did the crop intercept more light or did it use the same amount of light better? What explains these phenomena?

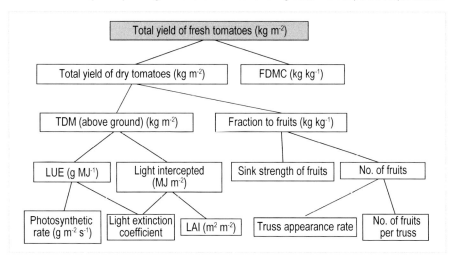

Tomato fruit yield and its underlying components. For example, tomato fruit fresh yield can be higher because the total fruit dry yield is higher and/or because of more dilution with water (lower fruit dry matter content).
FDMC = fruit dry matter content, LUE = light use efficiency, LAI = leaf area index.

years. A higher yield for modern cultivars did not result from a different partitioning compared to old cultivars, but rather was caused by an increase in light-use efficiency resulting from a decrease in the light extinction coefficient (a morphological change) and an increase in leaf photosynthetic rate (a physiological change). Note that a decrease in extinction coefficient results in a deeper light penetration (better vertical light distribution in the canopy) and at high LAI this effect is more important that the slightly lower light interception at decreased extinction coefficient (see also Section 7.1.2 on diffuse light).

2.2.3 Crop growth and yield: processes step by step

Plant growth analysis is especially suitable for explaining growth differences in young plants. RGR is not useful for adult plants as it is not constant once the exponential growth phase has ended; as stated previously, the absolute growth rate is a much more useful measure for older plants. During the linear growth phase, this absolute growth rate remains constant. An extensive literature review for horticultural crops on the processes underlying crop growth and yield was written by Marcelis *et al.*, (1998). Important processes, from light interception to crop yield, are shown in Figure 2.6.

2.2.4 Leaf area index

Leaf area index (LAI) is calculated as the square metres of green leaf area per square metre of ground area ($m^2\ m^{-2}$). When LAI is low, some of the sunlight is not intercepted by the crop and therefore not used for photosynthesis.

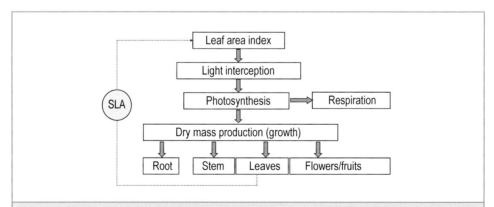

Figure 2.6. The steps in crop growth and yield formation. Only the light intercepted by the green leaf area leads to photosynthesis (production of assimilates). The photosynthetic products are used to make biomass (dry mass) and part of the photosynthetic products are used in respiration. Finally, biomass is distributed over the different plant organs. An increase in leaf dry weight results in an increase in leaf area (calculated by applying SLA = specific leaf area).

The relation between LAI and the fraction of intercepted light is curvilinear and expressed by the Lambert-Beer equation (Figure 2.7):

$$F_{int} = 1 - I/I_o = 1 - e^{-k \cdot LAI} \tag{2.17}$$

where F_{int} indicates the fraction of light intercepted, I represents the light level below the canopy (µmol m^{-2} s^{-1}), I_0 represents the light level above the canopy (µmol m^{-2} s^{-1}), k is an 'effective' extinction coefficient over the period considered and LAI is leaf area index. For many greenhouse crops, k has a value between 0.6 and 0.8. The extinction coefficient is species-specific and depends on leaf angle, leaf size and thickness and the pattern in which the leaves are placed around the stem. This also means that k is not strictly constant during the development of a crop, however, applying a constant k value provides an accurate enough estimation of light interception. A crop like gladiolus has a low extinction coefficient, especially because of its almost vertical leaf position. In that case, at the same LAI, much less light is intercepted compared to a more horizontal leaf position.

For an extinction coefficient of 0.8 and when LAI equals 1, e.g. in a young crop, light interception is only about 55%. Therefore, roughly half of the light is not intercepted and thus not used for photosynthesis. An LAI of 2 means about 80% interception and at an LAI of 3 about 90% is intercepted. With a low leaf area, the fraction of intercepted light is almost proportional to the LAI. In this case, the LAI is very important for the interception. If a crop already has a high LAI, a bit more or less hardly influences the light intercepted and therefore hardly influences crop growth.

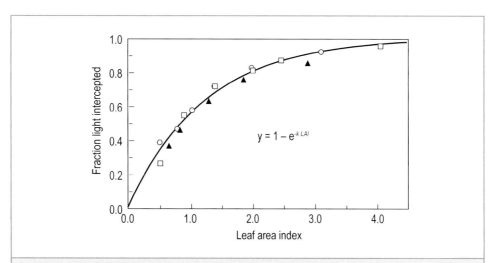

Figure 2.7. Influence of leaf area index (LAI) on the fraction of light intercepted by a tomato crop (k = extinction coefficient; 0.8 in this example); the symbols reflect measured data over time in different experiments (represented by the different symbols); the solid line is the Lambert-Beer equation.

In commercial farms in the Netherlands, LAI during the crop season has changed a lot (Figure 2.8). In 1993, the LAI fell after 120 days. Nowadays, growers are able to maintain an LAI higher than three throughout the season, obtained by maintaining extra stems on a fraction of the plants. In December, a planting density of 2.5 plants per m^2 is typically used, whereas in the spring an additional stem is kept on every second plant, such that stem density increases to 3.75 stems per m^2. A higher stem density means more leaves per m^2. Additionally, the fact that all tomatoes nowadays are grafted on vigorous rootstocks improves growth and therefore LAI, especially during the second part of the growing season (elevated temperatures, long stems, older plants). A detailed analysis of optimal leaf area for a tomato crop is provided by Heuvelink *et al.* (2005).

2.2.5 Light intercepted by a greenhouse crop

A detailed description of solar radiation and photosynthetic active radiation (PAR) is given in Jones *et al.* (2003), as well as in Chapter 3. Remember that not all radiation is useful for the plant. Global radiation (all short-wave radiation from the sun) has a wavelength between 300-2,500 nm, but only the range between 400-700 nm is used for photosynthesis. This is called PAR, which is also more or less the range of visible light. In this chapter, light is used as a synonym for PAR.

Despite the rather small range of wavelengths, PAR represents about 50% of the energy of global radiation. In other words, shorter wavelengths contain more energy. Due to the properties of a greenhouse, only about 70% of the PAR outside actually reaches the crop. The greenhouse cover material (plastic or glass) and construction parts intercept the light which

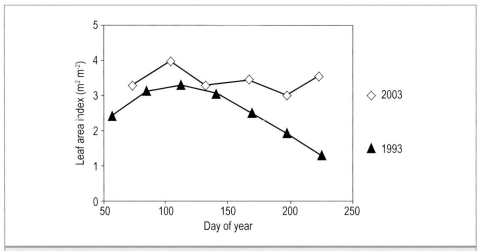

Figure 2.8. Leaf area index for tomato throughout the cultivation, measured at commercial farms in 1993 and 2003 (data from De Koning (1993) and Heuvelink *et al.* (2005)).

does not reach the crop (Chapter 3). For the most advanced greenhouses, transmissivity can be 80%, however the world-wide average for greenhouse transmissivity does not exceed 55-60%. Therefore, a field crop always receives much more light than a greenhouse crop at the same location.

On a nice summer day in the Netherlands, global radiation reaches 20 MJ m^{-2} d^{-1}. In the greenhouse, however, only 7 MJ m^{-2} d^{-1} PAR is available ($= 20 \cdot 0.5 \cdot 0.7$) above the crop. Note that intercepted light is not exactly equal to absorbed light. Intercepted light is the light above the crop I_0 minus the light that falls to the floor I_{floor}: $I_{intercepted}$ ($= I_0 - I_{floor}$). To calculate the absorbed light, we must also take into account that part of the incoming light is reflected by the crop ($I_{reflect}$). Furthermore, part of the light that falls on the floor is reflected back to the crop ($I_{reflect\ floor}$), especially when the floor of the greenhouse is covered with clean white plastic.

Only the absorbed light I_{abs} is relevant for photosynthesis.

$$I_{abs} = I_0 + I_{reflect\ floor} - I_{floor} - I_{reflect} \qquad (2.18)$$

Normally I_{abs} is about 95% of $I_{intercepted}$ and, therefore, in calculations, intercepted light is usually taken as if it were absorbed light.

2.2.6 Leaf photosynthesis

Photosynthesis is the basic process for plant growth: the plant produces sugars out of carbon dioxide and water and light provides the energy for this process.

$$6\ CO_2 + 6\ H_2O \rightarrow C_6H_{12}O_6 + 6\ O_2 \qquad (2.19)$$

Several equations can be used to describe the saturating response of photosynthesis rate as a function of light (Box 2.4). A detailed photosynthesis model is the Farquhar-von Caemmerer-Berry (FvCB) leaf-level photosynthetic rate model (Farquhar *et al.*, 2001). This model integrates various aspects of the biochemistry of photosynthetic carbon assimilation in C$_3$ plants into a mathematical model compatible with the studies of gas exchange in leaves. Almost all greenhouse crops are C$_3$ plants. All the photosynthetic pathways focus upon a single enzyme, namely ribulose-1,5-bisphosphate carboxylase/oxygenase, or Rubisco. This enzyme enables the primary catalytic step in the photosynthetic carbon reduction (or PCR cycle) in all green plants and algae (Ghannoum, 2018). Rubisco has a poor specificity for CO$_2$ as opposed to O$_2$. Binding O$_2$ results in so-called photorespiration. C$_4$ (e.g. maize) and Crassulacean acid metabolism (CAM; e.g. phalaenopsis and kalanchoe) variants of photosynthesis have metabolic concentrating mechanisms which enhance Rubisco performance by reducing photorespiration. Young kalanchoe plants behave in a C$_3$ way, whereas older plants show CAM characteristics. Some species can switch between CAM and C$_3$ e.g. *Guzmania monostachia* and *Vriesea* (*Bromeliaceae*): when the water supply is good they choose C$_3$. CAM plants typically have their stomata open at night and closed during the day. They mostly originate from hot, dry

---

Box 2.4. Formulae for the photosynthesis-light response.

Often, the negative exponential equation (Equation 2.20) provides a good description of the light response curve:

$$P_g = P_{g,max} \left(1 - e^{-\alpha \cdot I/P_{g,max}}\right) \tag{2.20}$$

P_g : actual gross photosynthesis rate (μmol m^{-2} s^{-1})
$P_{g,max}$: maximum gross photosynthesis rate (μmol m^{-2} s^{-1})
α : light-limited quantum efficiency (μmol μmol^{-1})
I : absorbed PAR (μmol m^{-2} s^{-1})

A more general formula is the 3-parameter non-rectangular hyperbola (Thornley, 1976):

$$P_g = \frac{\alpha \cdot I + P_{g,max} - \sqrt{(\alpha \cdot I + P_{g,max})^2 - 4 \cdot \theta \cdot \alpha \cdot I \cdot P_{g,max}}}{2 \cdot \theta} \tag{2.21}$$

where P_g, $P_{g,max}$, I and α have the same meaning as in Equation 2.20 and θ is a scaling constant for curvature; (when $\theta = 0.7$, the shape is very similar to Equation 2.20).

These equations are used for leaf photosynthesis but can also be applied for crop photosynthesis (the so-called 'big leaf approach'). Of course, parameter values differ between leaf and crop photosynthesis.

---

conditions where they would dry out quickly if the stomata were kept open during the day. Hence, for these plants, CO_2 enrichment during most of the day makes little sense, however, at the end of the day stomata may open and CO_2 enrichment could stimulate photosynthesis.

Photosynthesis can be viewed according to its electrical analogue (Ohm's law; Box 2.5). The rate of CO_2 uptake by a leaf is equal to the gradient in CO_2 divided by the resistance over that gradient. CO_2 must move from the greenhouse air into the leaf and then into the cells, to the site where the photosynthesis takes place: the chloroplast. There are several routes of resistance on this path:

› The boundary layer: a thin layer of non-moving air around the leaf. This resistance is much larger in a greenhouse than in an open field, as air movement is typically much lower in greenhouses (Chapter 6). Hairy leaves or large leaves (species-dependent) result in a thicker boundary layer and hence a higher boundary resistance.
› The stomatal resistance: the plant controls the opening of the stomata (small openings in the leaf epidermis) by shrinking or swelling the guard cells.
› The mesophyll resistance: this is the resistance between the air spaces in the leaf and the chloroplast and can greatly contribute to the total resistance. CO_2 must move from the air and dissolve into the water phase of a plant cell.

The history of a leaf determines its performance: a leaf that grew in sunny conditions is adapted to high light intensities, in contrast to a leaf that has grown in the shade (Figure 2.9). The latter reaches its maximum possible photosynthesis level at a much lower light intensity than

---

Box 2.5. Photosynthesis in analogy to Ohm's law.

Photosynthesis rate = $(CO_{2air} - CO_{2int}) / (r_s + r_b) = (CO_{2air} - CO_{2chloroplast}) / (r_s + r_b + r_m)$ (2.22)

CO_{2air} : concentration of CO_2 in greenhouse air (μmol mol^{-1})
CO_{2int} : concentration of CO_2 in the leaf (μmol mol^{-1})
$CO_{2chloroplast}$: concentration of CO_2 at the chloroplast (μmol mol^{-1})
r_s : stomatal resistance (m^2 s mmol^{-1}); stomatal conductance 1/r_s in mmol m^{-2} s^{-1}
r_b : boundary layer resistance (m^2 s mmol^{-1})
r_m : mesophyll resistance (m^2 s mmol^{-1})

The pathway resistances are explained and quantified in Section 6.1.

---

a sun leaf does. However, the initial slope of the curve (photosynthesis vs absorbed PAR) is not much different. This slope indicates the light-saturated quantum efficiency (α); and shade-adapted leaves have a lower light compensation point. Additionally, the differences between crops can be large: tomatoes can cope with a lot more light than potted plants, which are often shade plants in their original habitat.

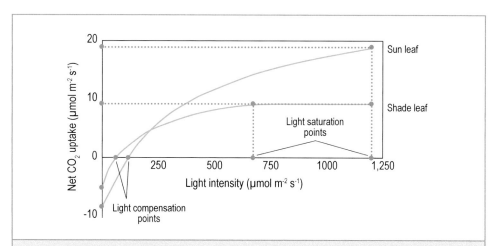

Figure 2.9. Light-photosynthesis response curves for sun and shade leaves. These can be, for example, winter and summer leaves in a tomato crop, but can also refer to shade-adapted plants (often pot plants originating from a tropical forest understore) and high light-adapted plant species (http://tinyurl.com/y9wb9sgz).

2.2.7 Crop photosynthesis

Most relations between environmental factors and photosynthesis have been investigated at leaf level, however for the grower's purposes we need to think at crop level. Crop photosynthesis can be calculated as a function of intercepted radiation (big leaf approach; similar functions as used for leaf photosynthesis, but different parameter values; e.g. the negative exponential equation) or more accurately by considering radiation absorption of different leaf layers in combination with a sub model for leaf photosynthesis. Patterns of leaf and crop photosynthesis as a function of light clearly differ (Figure 2.10). If leaf photosynthesis reaches its maximum value at 500 μmol m^{-2} s^{-1}, then crop photosynthesis reaches it only at 1000 μmol m^{-2} s^{-1}. The difference between leaf and crop photosynthesis also demonstrates that optimising leaf photosynthesis will not result in optimal crop growth and yield.

Crop photosynthesis shows a close relationship with the amount of intercepted light. Hence, the LAI is very important. At an LAI of one, only about half of the light is intercepted, therefore at every light level photosynthesis is rather low. An LAI of two results in about 80% interception (k = 0.8; see Figure 2.7). At high light intensities, crop photosynthesis will level off because the leaves at the top of the crop have reached light saturation; however, crop photosynthesis saturates at much higher light levels than leaf photosynthesis, as a crop's lower leaves are shaded and normally receive light intensities which are far below the saturating light intensity. This means that for these leaves, more light will still improve photosynthesis. In Dutch greenhouses, there is never too much light for crops such as tomato or cucumber, and more light would improve photosynthesis and yield. The only limitation is when the

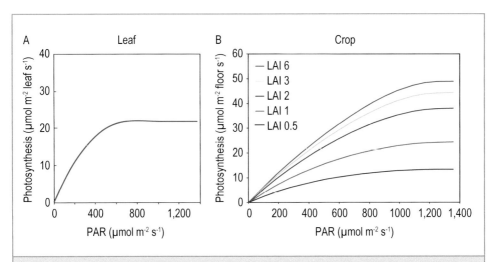

Figure 2.10. Leaf photosynthesis response to light (A) and simulated crop photosynthesis response to light (B) for 5 different values of the leaf area index (LAI). The crop is composed of leaves with a photosynthesis response as in (A).

temperature cannot be controlled properly and a lot of light (energy load on the crop) results in a temperature and/or vapour pressure deficit (VPD) that is too high.

Nowadays, the carbon dioxide level in the outside air is about 400 μmol mol^{-1}. In a greenhouse, this level can drop considerably when the amount used by the plant is not supplied (see Box 5.4): concentrations of 200 μmol mol^{-1} have been measured and such low levels reduce production drastically (Figure 2.11). That is why CO_2 dosing is common in modern greenhouses. A rise from, for example, 340 to 500 μmol mol^{-1} is very beneficial (Figure 2.11). Above a level of 800 μmol mol^{-1}, the effect of extra CO_2 is small, and it therefore does not make sense to exceed 1000 μmol mol^{-1}. Above this level, adverse effects have been observed: the stomata may partly close, which could result in excessively high leaf temperatures, leaf chlorosis and reduced photosynthesis.

The average increase in growth per 100 μmol mol^{-1} increase in CO_2 concentration, as shown in Figure 2.11, can be calculated by an empirical formula (Nederhoff, 1994):

Increase in growth per 100 μmol mol^{-1} = $1.5 \times 10^6 / [CO_2]^2$ (2.23)

in which $[CO_2]$ is the concentration taken as reference.

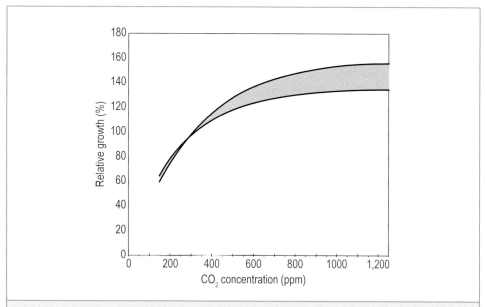

Figure 2.11. Relative crop production (%) at various levels of CO_2. The production at ambient level (340 ppm in 1985) is taken as 100%. The wide band is due to variation between crops and conditions. This graph is based on data from 60 publications of experiments worldwide in many greenhouse crops (Nederhoff, 1994).

From 250 to 350 µmol mol^{-1}: $1.5 \times 10^6 / 250^2 = 24\%$ extra growth. An increase from 350 to 450 µmol mol^{-1} gives 12% extra growth. From 1000 to 1,100 µmol mol^{-1} there is only 1.5% extra growth. The technology and options for CO_2 enrichment are analysed in Chapter 11. Nederhoff (2004) provides further information on CO_2 enrichment in greenhouses from a practical perspective.

Gross photosynthesis is hardly influenced by temperature in the range between 15 and 25 °C (Figure 2.12), at higher or lower temperatures, photosynthesis drops. Despite this, temperature strongly influences crop growth in the temperature range 15 to 25 °C. Other processes, such as respiration and development, are strongly influenced by temperature.

2.2.8 Dry mass production

Growth is usually expressed as an increase in dry mass. The sugars produced in the process of photosynthesis must be transformed into structural biomass, such as proteins, lignin, organic acids, fats, etc.; however, maintaining the already-existing biomass has highest priority. Therefore, part of the produced sugars is burned again in maintenance respiration. Processes such as the replacement of damaged enzymes and maintenance of ion gradients over membranes cost energy, but do not contribute to an increase in dry mass. This maintenance respiration doubles with every 10 °C-rise in temperature. As a rule of thumb, maintenance respiration is about 1.5%, which means the maintenance of 100 g of dry mass would use 1.5 g of sugars every day.

After the maintenance respiration has been taken care of, the remainder of the sugars from photosynthesis may be used for dry mass production. This conversion into structural materials costs energy; which is called growth respiration. The energy costs for ion uptake are often considered as part of the growth respiration.

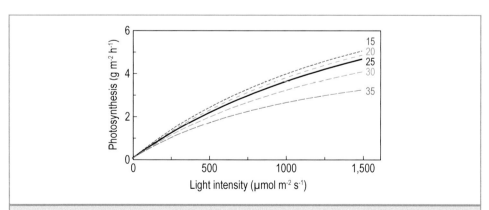

Figure 2.12. Influence of temperature (°C) on crop photosynthesis (in CH_2O units; simulated by TOMSIM at CO_2 concentration 340 µmol mol^{-1}, LAI = 3; Bertin and Heuvelink, 1993).

From photosynthesis and respiration to daily crop growth rate

Equation 2.24 shows how daily crop growth rate can be calculated from daily crop photosynthesis and maintenance respiration:

$$dW/dt = C_f(P_{gd} - R_m)$$

(2.24)

where dW/dt represents crop growth rate (g m^{-2} d^{-1}), C_f is a conversion efficiency (g dry matter per g CH$_2$O [g g^{-1}]), P_{gd} is crop gross assimilation rate in CH$_2$O units (g m^{-2} d^{-1}) and R_m is crop maintenance respiration in CH$_2$O units (g m^{-2} d^{-1}). Note that photosynthesis and respiration can be expressed in CO$_2$ or CH$_2$O units. 1 g CO$_2$ means 30/44 = 0.68 g CH$_2$O. These conversions are part of the steps from CO$_2$ assimilaton to fresh yield (Box 2.6).

C_f is about 0.7 for most greenhouse crops, which means that 1 g of sugars (remaining after use for maintenance respiration) results in 0.7 g of dry mass (the loss is called growth respiration).

Box 2.6. From CO$_2$ assimilation to fresh yield.

(photo tomato: Paolo Battistel; photo lettuce: Cecilia Stanghellini)

One kg of assimilated CO$_2$ does not result in one kg of product; first, growth respiration (conversion efficiency of 0.7; Equation 2.24) and unit conversion have to be taken into account. When CO$_2$ is converted to CH$_2$O in sugars, one oxygen atom is exchanged for two much lighter H-atoms: the weight is reduced to 68% (M_{CH_2O} / M_{CO_2} = 30/44). These two effects combined result in 1 kg of fixed CO$_2$ converting to approximately 0.5 kg of dry matter (70% × 68% = 48%). Part of this goes to organs that are not marketed, such as roots and often stems and leaves. A fraction, the harvest index (HI), is stored in the product (fruits, flower stems or leaves, such as lettuce). The HI of tomatoes is about 66%. Therefore, you get about 0.325 kg of fruit dry matter out of 1 kg of assimilated CO$_2$. Tomatoes have a dry matter content around 6%, meaning the fresh yield is about 5 kg (0.325/0.06): 1 kg of fixed CO$_2$ leads to roughly 5 kg of fresh tomatoes. The HI of lettuce is about 90% and the dry matter content of the leaves is around 7%, which means 1 kg of assimilated CO$_2$ results in more than 6 kg of fresh lettuce.

A linear relationship between dry mass production (g m^{-2}) and the cumulative intercepted light sum (MJ m^{-2}) has been observed for many crops. The slope of this relationship is called the crop light use efficiency (LUE; g MJ^{-1}). Light use efficiency in fact takes into account the processes of gross photosynthesis and respiration without detailing them (Box 2.7). The crop parameter 'light use efficiency' can be used to directly calculate crop growth rate from the amount of intercepted light (Equation 2.25). In literature, we often find the term 'radiation use efficiency' (RUE), which is the ratio between cumulative biomass production and cumulative solar radiation. Since energy in PAR is about half of that in solar radiation (see Chapter 3), a RUE of 1.5 g MJ^{-1} global radiation is the same as a LUE of 3 g MJ^{-1} PAR.

Box 2.7. Light use efficiency: how many g per mol light?

We can estimate the average LUE based on annual yield and received light during the annual crop cycles. The LUE of *Chrysanthemum* is between 3 and 6 g dry weight per MJ PAR (Figure 2.13). Considering that 1 MJ PAR natural light is about 4.6 mol (Box 10.3), this gives a LUE of about 1 g per mol PAR.

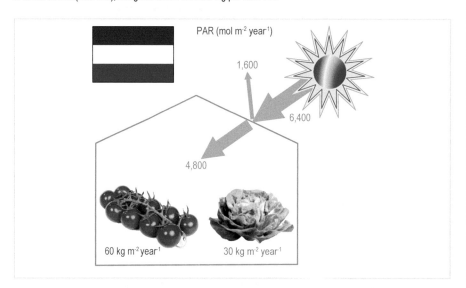

PAR (mol m^{-2} year^{-1})

1,600

6,400

4,800

60 kg m^{-2} year^{-1} 30 kg m^{-2} year^{-1}

Using the numbers above we can calculate the LUE of tomatoes and lettuce produced in Dutch glasshouses. From the previous box, we know that 1 kg total dry weight is about 10 kg tomato and about 12 kg lettuce. This gives a LUE of approximately 1.25 g mol^{-1} for tomatoes and 0.52 g mol^{-1} for lettuce.

As 2 kg CO_2 are fixated for each kg dry weight that is produced, the CO_2 fixation rate of tomatoes in a Dutch greenhouse is 2.5 g mol^{-1} PAR and of lettuce about half as much.

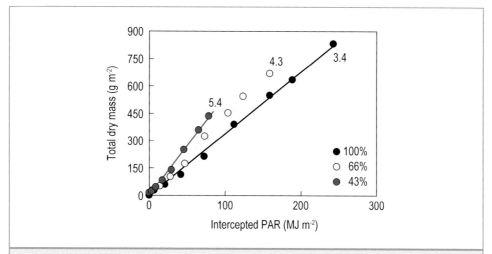

Figure 2.13. Light use efficiency (LUE; g MJ⁻¹), the slope of the linear relationship between total dry mass and intercepted PAR (numbers on top of each line) in summer at 3 light levels (shade screens) for cut chrysanthemum (recalculated from Heuvelink *et al.*, 2002).

From light use efficiency to daily crop growth rate

$$dW/dt = \text{LUE} \cdot (1 - e^{-k \cdot LAI}) \cdot I \tag{2.25}$$

where dW/dt is crop growth rate (g m⁻² d⁻¹), LUE is light use efficiency (g MJ⁻¹ PAR), k is light extinction coefficient, LAI is leaf area index and I is daily light sum (MJ m⁻² d⁻¹). Note that $(1 - e^{-k \cdot LAI})$ represents the fraction of intercepted light (Equation 2.17).

Equation 2.25 assumes a constant LUE, which is usually true for field crops. Field crops are sown, or planted, in (late) spring and harvested in late summer or early autumn, whereas in a greenhouse crops are also grown in winter. In a greenhouse, LUE is not necessarily constant, as CO_2 enrichment takes place (more biomass produced for the same amount of intercepted light) and light levels in a greenhouse (year-round cultivation) vary much more over a year compared to a field crop growing season. Figure 2.13 shows the LUE for three different light levels in a cut chrysanthemum crop in summer. At 100% light, the LUE is 3.4 g MJ⁻¹. Shading improves the light use efficiency: at a light level of 66%, the LUE rises to 4.3 g MJ⁻¹ and at 43% light, it is 5.4 g MJ⁻¹. Therefore, the efficiency depends on the light level, which is a logical response since at high light levels, closer to saturation, a reduction in LUE occurs (Figure 2.10B).

2.2.9 Leaf area increase

For crop growth and yield, LAI plays a key role. If the extinction coefficient is known LAI can be estimated from light measurements above and below a canopy applying the Lambert-Beer law (Equation 2.17). But, how do we simulate the LAI when we want to simulate crop growth? For field crops, LAI is often calculated as a function of the plant developmental stage and this is determined by the temperature sum. In the greenhouse, this simple relation does not hold true, because the grower can regulate temperature independently from the light level. LAI can be predicted based on the leaf dry matter (Equation 2.26), taking into account the SLA (Equation 2.14). This is the leaf area per leaf dry weight in $cm^2 g^{-1}$, and it varies between crop species (Table 2.1) as well as with plant age, season and environmental conditions (e.g. water stress decreases SLA), but it is nonetheless often taken as a constant (Box 2.8).

2.2.10 Assimilate partitioning

Assimilate partitioning is an important determinant of crop yield. Usually, only a part of plant growth results in economically important products (Table 2.2), such as the fruit of tomato plants or cucumber. In these crops, partitioning of the produced assimilates between the fruit and the vegetative parts (leaves, stem and roots) is mainly determined by the number of growing fruits.

Table 2.1. Typical values of SLA (cm^2 per g dry weight).

Carnation	130-170
Cucumber	130-180
Chrysanthemum	320-420
Roses	280-350
Lettuce	520-600
Tomato	150-350

Box 2.8. Simulation of LAI based on daily crop growth rate.

$$LAI_t = LAI_{t-1} + GR \cdot F_{leaves} \cdot SLA / 10,000 \qquad (2.26)$$

LAI_t, LAI_{t-1}	: leaf area index on day t and day t-1 ($m^2 m^{-2}$)
GR	: crop growth rate (g $m^{-2} d^{-1}$)
F_{leaves}	: fraction partitioned to the leaves
SLA	: specific leaf area ($cm^2 g^{-1}$)
$10,000$	: factor between cm^2 and m^2: 1 m^2 = 10,000 cm^2

Table 2.2. Harvest index (HI) and dry matter content (DMC; %) of harvestable product for some greenhouse crops (partly from Heuvelink and Challa, 1993).

Crop	HI	DMC	References
Roses	0.65-0.80	20-25	Kool (1996)
Chrysanthemum	0.85	11-13	Lee (2002)
Pot plants	0.9-1.0	10-12	
Lettuce	0.85	5	Seginer *et al.* (2004)
Cucumber	0.7	2.5-4.0	Marcelis (1994)
Sweet pepper	0.8	8.5	Wubs (2010)
Tomato	0.5-0.7	4-8	Heuvelink (1996, 2018)

All organs attract assimilates, but their relative sink strength determines which part of the assimilates they receive (Marcelis, 1996). Organ sink strength can be quantified by its potential growth rate, that is the growth rate under non-limiting assimilate supply. For cucumber, we obtain this with only one fruit on the plant, and for tomato with one fruit per truss. A fruit that has been growing in competition with other fruits quickly reaches the potential growth rate after removing the competing fruits (Figure 2.14). This clearly shows that the potential growth rate (and not the potential relative growth rate), is a measure for sink strength.

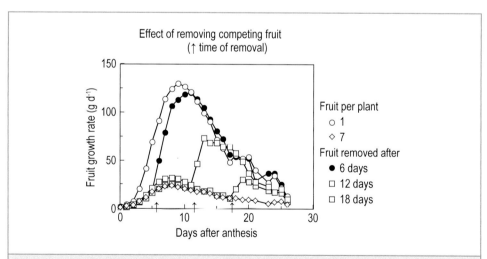

Figure 2.14. The effect of the removal of competing fruits on the growth rate of the cucumber fruit at axil 9. Either one or seven fruits were retained on the plant, or six fruits were removed from plants with seven fruits at 6, 12 or 18 days after anthesis at axil 9; dates of removal are indicated by arrows (reprinted, with permission, from Marcelis, 1993).

Relative sink strength for the calculation of assimilate partitioning (Box 2.9):

$$\text{Fraction to fruits} = \frac{SS_{fruits}}{SS_{fruits} + SS_{veg}} \qquad (2.27)$$

where SS_{fruits} is the sink strength of all fruits together and SS_{veg} is the sink strength of all vegetative parts (leaves, stems, roots) together. The number of fruits growing on a plant at the same time strongly determines partitioning. For example, in tomato plants, if all trusses have just one fruit, only 20% of the total biomass is allocated to the fruit (after a growth period of about 100 days). If all trusses have seven fruits, about 64% is allocated to the fruits (Heuvelink, 1997).

The partitioning is not influenced by the total availability of assimilates. Whether there is a lot or only a little to divide, the same part always goes to the fruits, according to their relative sink strength.

Box 2.9. Example of calculating partitioning.

Assume that for a section of a tomato plant, leaves (+ internode) require 1 g of assimilates and the fruit truss requires 7 g of assimilates. Based on Equation 2.27, the fruits will obtain 70% of the available assimilates when three leaves are present for each truss. If however, we remove one out of three young leaves, this will result in a higher partitioning to the fruits (77%). Note that in the calculation we assume that removing 1/3 of the leaves at an early stage reduces vegetative sink strength by 1/3; although this would actually only be true for a new cultivar with two instead of three leaves between trusses (thus also two instead of three stem internodes).

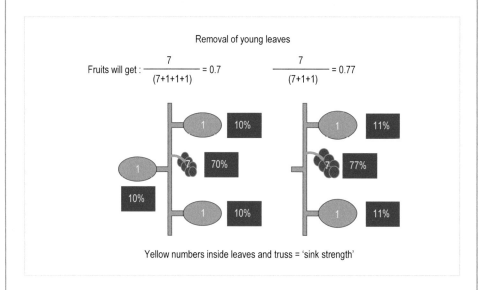

Yellow numbers inside leaves and truss = 'sink strength'

2.2.11 Fruit dry matter content

Until now, this chapter has focused on dry matter production. However, horticultural products are not sold as dry matter, but rather as fresh products. These products contain a lot of water and only a small proportion of dry matter (Table 2.2), yet, this amount of dry matter determines the potential fresh yield. To transfer dry weight into fresh weight, dry weight is divided by the dry matter content. When 4 kg of tomato fruit dry matter has been produced, with a dry matter content of 5%, this represents $4/0.05 = 80$ kg fresh tomatoes (Box 2.6). A complicating factor is that dry matter content is not a constant. For example, for cucumber in a 2-month experiment it varied over time between 2.5 and 4% (Marcelis *et al.*, 1998). Furthermore, cultivars may greatly differ in fruit dry matter content, e.g. for cherry tomatoes 8% would be normal, whereas for some modern standard tomato cultivars only 4% is common. This has significant consequences for the outcome of the calculations and is one of the reasons why cherry tomatoes have a relatively low fresh yield compared to standard tomato cultivars.

2.3 Concluding remarks

For a clear understanding of how climatic factors influence crop growth and yield, it is important to distinguish between development (mainly influenced by temperature) and growth (mainly influenced by light and CO_2). Plant growth analysis is a useful tool for distinguishing between photosynthetic and morphogenetic reasons for the differences in plant growth due to the environment or genotype. Yield analysis identifies underlying processes such as light interception, light use efficiency, biomass partitioning and leaf area development. This chapter provides important concepts for understanding crop growth and yield, but before we can use this knowledge in managing the greenhouse climate, it is imperative to obtain a good understanding of the energy and mass balances in a greenhouse. This will be provided in the following chapters.

Chapter 3
Radiation, greenhouse cover and temperature

Photos: Cecilia Stanghellini (left) and Frank Kempkes, Greenhouse Horticulture, Wageningen University & Research (right).

All over the world greenhouses are built to protect crops, increase productivity and make more efficient use of water, fertilisers and crop protection. But when you put a shelter over a crop, there is a lot more going on than simply sheltering. The whole internal climate is changed. These changes are predictable when physical laws are taken into account, so greenhouse production depends on a basic knowledge of physical principles. We will start with the most important property of a crop shelter: what it does to solar radiation, and how this is related to the interior temperature.

We will begin with the simplest possible shelter – an empty un-ventilated 'greenhouse' – and then make it gradually more realistic (and complex) in subsequent sections. To be able to appreciate what happens inside an unventilated greenhouse, some understanding is required of the properties of light/radiation, the properties of the greenhouse cover, the principles that determine heat transfer, the Law of Stefan-Boltzmann and the energy balance.

3.1 Solar radiation

Solar radiation reaching the Earth's surface has wavelengths ranging from approximately 300 to 2,500 nm (Figure 3.1). This is also called 'light' although, strictly speaking, light is what we can see with our eyes, between 380 and 750 nm, approximately. However, in everyday speech we usually mix up the terminology and incorrectly use terms like ultraviolet light and infrared light, when what we actually mean is radiation.

For crop production it is necessary to split the solar spectrum into three wavelength ranges (wavebands): ultraviolet (UV, the range between 300 and 400 nm); photosynthetic active radiation (PAR, 400 to 700 nm), and near infrared (NIR, 700 to 2,500 nm). PAR is the part of the radiation spectrum that plants use for photosynthesis (and is also in the range we can see). UV and NIR do not contribute to photosynthesis but some crop development processes may depend on them. About half of the energy in the solar radiation is located in the NIR wavelength range. As the fraction of energy contained in the UV range is small (a few percent), it is fair to say that half of the energy from solar radiation is in the PAR range and half in the NIR range. Recent research has shown that there are plant development processes that respond to specific wavelengths or narrow wavebands (for a review: Van Ieperen, 2016). However, the three ranges PAR, NIR and UV are good enough for most purposes.

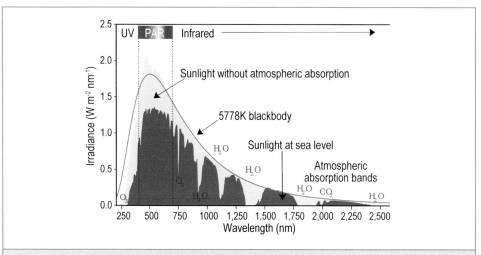

Figure 3.1. The solar radiation spectrum both at the top of the Earth's atmosphere (represented by the yellow area) and at sea level (red area). Regions for ultraviolet, PAR and infrared radiation are indicated. The sun produces radiation with a distribution similar to that expected from a 5,778 K (5,505 °C) blackbody (see Law of Stefan-Boltzmann), which is approximately the sun's surface temperature. As radiation passes through the Earth's atmosphere, some is absorbed by gases with specific absorption bands. These curves are based on the American Society for Testing and Materials (ASTM) Terrestrial Reference Spectra (https://commons.wikimedia. org/wiki/File:Solar_spectrum_en.svg).

Radiation consists of elementary particles/wavelets, or photons. The energy content of a photon is inversely proportional to its wavelength (Box 3.1). A blue photon, for instance, contains almost twice the energy of a red photon. However, photons in all wavelengths within the PAR range contribute (almost) equally to photosynthesis (Box 3.2). Therefore, photosynthesis is

---

Box 3.1. The energy of radiation: the link between Joules and micromoles.

The energy E of a photon of wavelength λ is $E = h\,c\,\lambda^{-1}$. If we use the values for Planck's constant, h, and the speed of light, c, we get $E \cong 2 \times 10^{-25}$ J λ^{-1}, with E in J and λ in m. The energy of a micromole ($\cong 6 \times 10^{17}$) photons of wavelength λ is thus $\cong 1.2 \times 10^{-7}$ J λ^{-1}.

If we quantify the wavelength λ in nm, this is approximately $120/\lambda$.

The average wavelength of PAR radiation in the solar spectrum is about 500 nm (green), so that 1 µmol in the sun's PAR range carries about 0.24 J. But then we must account for the fact that only about 47% of the energy of sun radiation is in the PAR range, so that

1 W m^{-2} of sun radiation $\approx$ 2 µmol m^{-2} s^{-1} in the PAR range. (1 J $\approx$ 2 µmol $\rightarrow$ 1 MJ $\approx$ 2 mol)

This is only (almost) true for solar radiation. For lamps, it is necessary to know the wavelength distribution before the link between energy and photosynthetic activity can be calculated (see Box 10.3).

Box 3.2. The human eye versus the plant.

Artificial light is everywhere and measurement of light is therefore dominated by the requirements for illuminating houses, offices, factories, streets, etc. The human eye is roughly sensitive to the same waveband as the PAR range. However, we can see some colours a lot better than others. Yellow looks very bright to our eyes, blue looks dark. The 'lux' unit of measurement takes into account all the peculiarities of our eyes and brains.

This means that lux is not at all suitable for characterising the benefits of light/radiation for plants. The photosynthetic response curve of leaves for radiation in the PAR range has a very different pattern than the sensitivity of the human eye for the different wavelengths. So we need a very different measuring unit for plants: $\mu mol\ m^{-2}\ s^{-1}$. However, the output of assimilation lamps is sometimes indicated in lux.

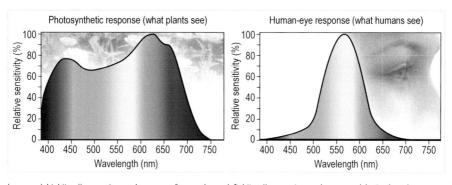

(source: right: https://sunmastergrowlamps.com/lamp-science; left: https://sunmastergrowlamps.com/plant-science)

only dependent on the number of photons in the PAR range, and not on the energy they carry. That is why when we measure radiation in terms of its effect on photosynthesis we use a number (moles), but whenever we are concerned about its effect on temperature we need to measure the energy content (J). Obviously we use flux densities ($\mu mol\ m^{-2}\ s^{-1}$ and $W\ m^{-2}$) when the speed and intensity of a process are relevant.

The relationship between what is measured by an ideal PAR meter (quantum meter or photometer, $\mu mol\ m^{-2}\ s^{-1}$; Box 3.3) and an ideal pyranometer (solarimeter, $W\ m^{-2}$) depends on the spectral distribution of the radiation source being measured, the link being the energy of a photon. Since we know the spectral distribution of solar radiation (Figure 3.1), we also know that $1\ W\ m^{-2}$ is approximately two $2\ \mu mol\ m^{-2}\ s^{-1}$ (Box 3.1). In addition to some approximation in the calculation itself, it should be noted that the spectral distribution of sunlight is affected by atmospheric conditions (clouds, smog), so this is indeed an approximate value. Moreover, in the real world the spectral response of the sensor also plays a role (Box 3.3).

As natural light is partially scattered in the atmosphere, solar radiation at the Earth's surface also has a spatial distribution: in daytime we get light from all spots in the sky. From the spot where the sun is, we get direct radiation, the rest is diffuse radiation, i.e. solar radiation

Box 3.3. How do we measure radiation?

The right sensor for measuring radiation depends on the application. The energy carried by solar radiation is measured by a pyranometer (or solarimeter, left), whose output is W m^{-2}, whereas the radiation available for photosynthesis can be measured using the quantum meter or photometer (right), whose output is µmol m^{-2} s^{-1}.

The measuring principle is the same: the difference in temperature between a black and a highly reflecting surface is related to the amount of incoming radiation energy. A waveband pass filter in the screen placed above the sensor (for instance, double and transparent above left and whitish above right) determines which waveband is measured. Since no filter is perfect (three commercial quantum sensors are shown below left), one should be aware of the response of the filter, particularly when measuring (virtually) monochromatic light sources.

Both sensors give 'spot' measurements, so many measurements may be necessary whenever illumination is not uniform (for instance, under a greenhouse roof), or within/below a crop. Particularly for the latter, a line sensor can be useful (below right, about 1 m long). As the sensor area is much larger, the readings are less dependent on the exact place of measurement. This is ideal for measuring light within a canopy.

Lambert's Cosine Law states that radiant intensity under a perfectly uniform (Lambertian) surface is proportional to the cosine of the angle between the incoming light and the surface. Light sensors usually include a correction for light sources at any angle within the hemisphere of measurement. Such a correction is not very accurate in line sensors, which usually underestimate light coming from angles close to the horizon.

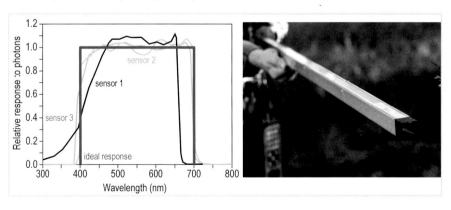

scattered by the atmosphere. The lower the sun's elevation on the horizon, the longer the path of its light in the atmosphere and the larger the fraction diffuse radiation. This is a consequence of latitude (high latitude → low sun elevation), season and time of day. Diffuse radiation is measured by shading a spot sensor with a band that mirrors the course of the sun in the sky.

In addition to this geometrical effect, the scattering effect of the atmosphere can be increased by clouds, water vapour and smog (Figure 3.2). For instance, on a cloudy day (fully overcast sky), we only have diffuse radiation, and that is at most 30% of the radiation at the top of the atmosphere. Conversely, on a very clear day (global radiation 70% or more of extra-atmospheric radiation) about 80% of the radiation at the Earth's surface is direct radiation.

Altogether, for instance, in the Netherlands only 20% of global radiation is direct in winter; in summer it is 35%. In Arizona in summer 80% of the radiation is direct and in winter this is still 70%, while in southern Italy it is 60% in summer and 40% in winter.

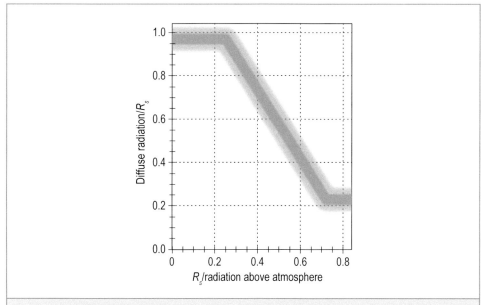

Figure 3.2. Fraction of the diffuse component of global radiation, as a function of the ratio of global (R_s) to extra-atmospheric radiation (i.e. a measure of atmospheric turbidity/cloudiness). Redrawn from Roderick (1999) who compiled (published and unpublished) data spanning 50 degrees latitude on both sides of the equator and more than 500 m altitude.

3.1.1 Reflection, absorption and transmission of solar radiation

Whatever the nature of the greenhouse cover is, glass or plastic, it always does three things: reflects, absorbs and transmits the solar radiation (Figure 3.3), which is like saying that the sum of the absorption, reflection and transmission coefficients has to be one. The transmission should be as high as possible, so the reflection and absorption have to be small.

Reflection can be modified by coatings or structures on the surface of hard materials. The reflection coefficient ρ (ratio of reflected to incoming radiation) is used to characterise reflection. The absorption depends on the 'turbidity' of the material and its thickness. In particular, the radiation (I_d) under a turbid medium of thickness d is given by:

$$I_d = I_{sun}\, e^{-kd} \qquad\qquad\qquad\qquad \text{W m}^{-2} \text{ or } \mu\text{mol m}^{-2} \text{ s}^{-1} \quad (3.1)$$

where I_{sun} is the incoming radiation and k, the extinction coefficient, is a property of the material. So, the suitability of a material for greenhouse covers depends on the combined effect of its strength (which determines thickness) and its transparency. For instance, glass, a very transparent material (low extinction coefficient) has to be used in 4 mm panes and acts in a similar way to polyethylene (PE), which is much less transparent but can be used in films of 0.1 mm. Please observe that the radiation absorbed in the cover material is bound to warm it up.

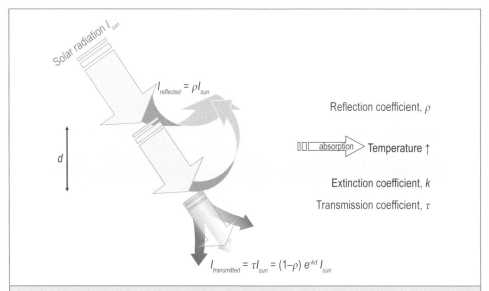

Figure 3.3. Schematic representation of the relevant properties of greenhouse cover materials. Reflection happens at both surfaces of the pane (or film) and transmitted light may have different spectral and geometrical (scattering) properties from incident light.

The total transmissivity of a greenhouse cover is indicated by the symbol τ, where:

$$I_{transmitted} \equiv \tau I_{sun} = (1 - \rho)\, e^{-kd}\, I_{sun} \qquad \text{W m}^{-2} \text{ or } \mu\text{mol m}^{-2}\, \text{s}^{-1} \quad (3.2)$$

The symbol $\equiv$ indicates that this is a definition (of τ).

Perpendicular radiation gives the least reflection and absorption. When the angle of the incoming radiation is not 90°, the path of the radiation in the material d is longer, so there is more absorption. Nevertheless, as absorption is small, what decreases transmission at high angles of incidence is the higher reflection on the outer and inner surfaces of the material. In general, only a minimal (if any) fraction of solar radiation on a greenhouse with any shape on any position on earth is perpendicular to (a fraction of) the greenhouse cover, so the perpendicular transmittance is a serious over-estimate (at least 10%, but there is a large variation among materials, Table 3.1) of the fraction of solar radiation below a greenhouse cover. Therefore, a better indication of the average amount of light on crop level is given by the hemispherical transmittance, which is measured either under a uniformly lit (Lambertian) dome or by averaging transmittance measured under many angles of incidence of light.

Once inside the greenhouse, the radiation has two effects: it is used in photosynthesis (the PAR component); and the energy it carries (both the PAR and the NIR component) heats up everything, the crop and all parts of the greenhouse and equipment included. The first is always good, the second only to a certain extent (depending on where the greenhouse is situated in the world and on the season).

Table 3.1. Typical PAR transmission (perpendicular and hemispherical) of greenhouse cover materials (after Hemming et al., 2005).[1]

	Light transmission (%)	
	Perpendicular	Hemispherical
Traditional glass	89-90	82
Low-iron glass	90-91	83-84
AR glass	95-97	89-91
Diffuse glass	90-91	76-82
AR diffuse glass	95-97	85-91
PE/EVA films	85-90	78-82
PE/EVA films diffuse	85-90	73-80
ETFE (F-clean)	93	86
ETFE (F-clean) diffuse	93	81
PC sheet	76-80	60-70
PMMA sheet	89-92	76

[1] AR = anti-reflection; ETFE = ethyl tetra fluor ethylene; EVA = ethylene-vinyl-acetate; PC = polycarbonate; PE = polyethylene; PMMA = poly methyl methacrylate.

There is almost never too much PAR for light loving crops. A single leaf can reach the limit of its photosynthetic capacity, but the whole canopy is seldom light-saturated (Figure 2.10). Next to that, in most climate zones there is a relatively dark season when light becomes the limiting factor for crop growth. That is why growers are especially keen on the transmission of the cover when they invest in a new greenhouse. Indeed, there is enough evidence to suggest that for all common greenhouse crops, every fractional increase in available light results in a comparable increase in production (Marcelis *et al.*, 2006).

In addition, modern techniques (diffusing greenhouse covers; Box 3.4) to increase scattering without decreasing (much) transmissivity (such as surface structures in glass, coupled to anti-reflection coatings) have been proven to increase the productivity of Dutch greenhouse crops (Li *et al.*, 2014). Usually light scattering from transparent panes (such as car glasses) is quantified by 'haze', whose definition reflects the fact that in most cases haze is undesired. Research (Hemming *et al.*, 2016) has shown that haze does not give enough information on the geometrical distribution of light. Therefore, the new norm on characterisation of optical properties of greenhouse coverings and screens (NEN2675) introduces 'Hortiscatter' as the most indicated measure for greenhouse covers.

Since diffusive covers affect only the direct component of solar radiation, one would expect a larger effect wherever the direct component of solar radiation is larger than in the Netherlands, i.e. at lower latitudes. However, whereas glass (the reference material in the Dutch studies) has no scattering, the 'clear' plastic mostly used as the standard greenhouse cover in warmer regions has a scattering around 20%, so the improvement to be obtained by diffusing covers should be assessed against this standard. Next to that, not much research has been done on diffuse light effects under high light intensity conditions.

3.1.2 Shading

What can happen is that the energy carried into the greenhouse by the radiation that is transmitted by the cover results in too high a temperature (Chapter 4). Whenever ventilation cannot cool the greenhouse sufficiently (Chapter 5), it may be necessary to reduce the input of radiation, i.e. provide shade (Box 5.3). It should be noted that this method would also reduce photosynthesis, so shading is really 'the lesser evil', and when designing greenhouses, maximum consideration should be given to a good ventilation capacity.

The most widely applied method of shading is whitewash, a white paint that reflects a fraction of sun radiation. Usually it is a water-soluble, calcium carbonate (chalk)-based paint, that is applied in late spring until early autumn, and is then washed away. The white colour ensures high reflectivity, without the cover getting warm. The amount of reduction in transmissivity (τ) is very roughly determined by the thickness of the paint. Typically, the reduction is 30 to 50%. The white paint has a scattering effect, so the light in the greenhouse is diffuse, which may somewhat mitigate the negative effect of the shade on production.

Box 3.4. Light-diffusing covers: what scattering does and what material technology can do.

A diffusing cover scatters an incoming beam of light into a wider angle. A high scatter means that a large fraction of incoming light is deflected by a relatively large angle. So, a (small) fraction will be scattered back. This means that increasing diffusivity must decrease the overall transmissivity of the cover, as we see here below.

Each point shows the transmittance (hemispherical) vs the diffusivity of different materials: plastics (left, redrawn from Hemming, 2015) and glasses (right, redrawn from Hemming *et al.*, 2014). Since Hortiscatter has been introduced after most of the data were collected, haze is in the x-axis here. The decreasing trend in transmittance with increasing haze was expected.

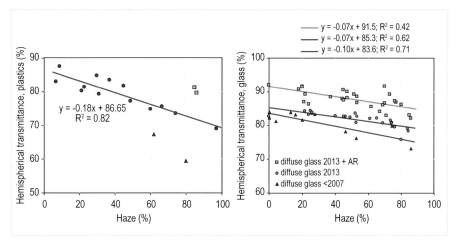

However, both plots show what material technology can do: the plastics indicated by red triangles and green squares in the left plot deviate significantly from the general trend shown by the blue line (1.8% less transmittance for 10% increase in haze). The two 'green' materials perform better, the 'red' ones much worse.

The plot about glass shows the improvement that has been attained in commercial glass covers within the first six years following the introduction of diffuse glass in horticulture. Whereas transmittance would decrease by 1% for each 10% increase in haze in 2007, the loss was only 0.7% in 2013, thanks to the application of newly developed structures on glass (non-flat glass surfaces). In addition, transmittance of the basic material was also improved by some 2%, by the introduction of low-iron ('white') glass (the intercept shows the transmission of clear glass). The green points represent commercial materials available in 2013 that were treated with anti-reflection layer(s). The best-fit line shows that such treatment increases transmittance in average by more than 6% (in structured glass as well as clear glass) and has no effect (neither good nor bad) on the trend with haze (0.7% less transmission per 10% increase in haze).

Various improvements in the paint formula have been proposed. Some spectral selective temporary coatings are more reflective in the NIR wavelength range than in the PAR range, which in theory ensures a higher PAR transmission ($\rightarrow$photosynthesis) at a given energy cut-off (τ) (see also Box 3.5). Some temporary coatings become more transparent when wet, which should ensure higher transmissivity when there is too little (rather than too much) radiation, such as in rain or mist. However, in both cases the spectral performance of the

---

Box 3.5. NIR filters: absorption or reflection?

In principle, a material filtering the NIR component of solar radiation, yet fully transparent to PAR, would result in a much cooler greenhouse ambient temperature, about half of the crop transpiration, without affecting photosynthesis or production in any way. No wonder then that there is a lot of interest in NIR-selective filters, particularly for greenhouses in warm regions.

A filter can work either by absorbing or reflecting radiation in the desired wavelengths.

Radiation that is absorbed warms up the filter material (the cover), and quite a large fraction of the energy will end up in the greenhouse anyhow, as heat or thermal radiation. Reflection seems a better choice. However, leaves already have by nature quite a high reflectivity for NIR radiation (about 50%), so that all NIR radiation transmitted by an imperfect 'mirror' will end up being trapped between two mirrors. The outcome will be less than what could be expected from the properties of the filter (Stanghellini *et al.*, 2011).

Both reflective and absorption NIR filters are available on the market. The ideal material, however, has not yet been produced: absorption filters have significant absorption also in the PAR range, and the reflection filters achieve 50% reflectivity in the NIR at best, although better materials are presently being investigated.

In addition, for unheated greenhouses in regions with low temperature seasons, the warming-up caused by the energy carried by NIR radiation is necessary for part of the growing season. That was shown by research in Almeria (Spain), by Garcia Victoria *et al.* (2012).

---

formulas presently marketed is not optimal. Recently also coloured temporary coatings have been introduced in order to influence specific morphological effects (e.g. stem length of roses).

Shading by means of movable screens has the advantage that they can be folded whenever the temperature allows for it (such as in the morning and evening, or on cloudy days). However, screens in front of the ventilation openings will limit air exchange and thus work against the cooling effect of ventilation. This could partly be solved by external screens, which also prevent energy coming into the greenhouse in the first place. These, however, are quite sensitive to wind damage and dust accumulation, and thus rarely installed. An additional issue in both cases is the shadow cast by the screen, when folded. Care should be taken to ensure a screen package as small as possible.

Finally, one relevant issue is the material used. The most commonly used multi-purpose screens (used for shading as well as for night-time insulation) have alternated aluminium strips and open spaces, to attain the desired level of cut-off with minimal reduction of ventilation. However, aluminium is right for night-time insulation, since its low emissivity (see Section 3.2) prevents radiative cooling (see also Figure 3.4), a mechanism particularly effective in clear sky conditions, certainly very useful in the conditions a shading screen may be needed. Therefore, aluminium should never be used in shading screens, but rather highly reflective (and diffusive) white strips of a high-emissivity and high IR transmissivity (thus low IR reflectivity) material.

3.2 Thermal radiation: the laws of Stefan-Boltzmann and Wien

Anything that is warmer than absolute zero (0 K = -273.15 °C) releases energy in the form of radiation. The Stefan Boltzmann law quantifies the amount of radiation emitted by a so-called 'black body' (an idealised physical body that absorbs all incident electromagnetic radiation, regardless of frequency or angle of incidence). In particular, the Law of Stefan-Boltzmann states that the total amount of emitted radiation is proportional to the fourth power of the absolute temperature (K), through a universal constant of proportion $\sigma = 5.67 \times 10^{-8}$ [W m^{-2} K^{-4}]

$$R = \sigma T^4 \hspace{8cm} \text{W m}^{-2} \quad (3.3)$$

The wavelength of the emitted radiation depends on the temperature of the body. In particular, the Law of Wien states that the wavelength of the maximum intensity of radiation is inversely proportional to temperature:

$$\lambda_{max} \approx \frac{2.9 \times 10^6}{T} \hspace{7cm} \text{nm} \quad (3.4)$$

with λ in nm and the temperature in K. The sun is about 6,000 K and emits radiation between 300 and 2,500 nm (Figure 3.1). Bodies at common temperatures on the Earth's surface (~300 K) emit radiation which we cannot see (unless we wear very special glasses) and lie in the range of 6,000 to 20,000 nm (= 6-20 μm). This wavelength range is called far infrared or Thermal InfraRed (TIR). Sometimes it is called longwave radiation, as opposed to shortwave radiation, i.e. solar radiation (300-2,500 nm).

3.2.1 Emissivity

Something that is not strictly 'black' (sometimes called a 'grey body'), emits less radiation. Emissivity (ε) is defined as the ratio between the total radiation emitted by a grey body and the radiation emitted by a black body at the same temperature, and is obviously contained between 0 and 1. Emissivity is a property of the surface of the material, so something which is wet, for instance, will have an emissivity equal to 1 (like water), whatever its emissivity is when dry. Most natural materials have an emissivity above 0.7, most construction materials even above 0.9. The only affordable material with low emissivity is aluminium, which has an emissivity between 0.03 and 0.2, depending on how it is polished.

3.2.2 Net radiation

Since all things emit radiation, everything receives radiation from its surroundings as well. There is a huge energy exchange going on all the time, so what is important is the net amount of energy a body is receiving or losing. A leaf in the top of a greenhouse canopy, for example, receives and loses energy from/to the greenhouse cover, other leaves/stems, the soil and any heating system that may be there. What is important, in this case, is how much the leaf 'sees' of each one of the elements with which it is exchanging radiation. This is expressed as the view factor ($F_{1,2}$). For instance, $F_{leaf,cover}$ indicates how big a fraction of the 'leaf's horizon' is taken up by the greenhouse cover. Obviously, the sum of all view factors for one point must be one. The net thermal radiation of the leaf can then be calculated as:

$$Net_{TIR} = \sigma \sum_{i=1}^{n} F_{leaf,i} (T_{leaf}^4 - T_i^4) \qquad\qquad \text{W m}^{-2} \quad (3.5)$$

where i indicates all possible things present in the 'leaf's world'. For relatively small temperature differences, linearization (i.e. the first term of the Taylor expansion; Box 3.6) can suffice, in particular:

$$Net_{TIR} \approx 4\,\sigma\,T_{mean}^3 (T_{leaf} - \sum_{i=1}^{n} F_{leaf,i}\,T_i) \qquad\qquad \text{W m}^{-2} \quad (3.6)$$

which means that an 'apparent' radiation temperature of the leaf's world can be estimated as the mean of the temperatures of each element, weighted by the relative size of each element in that world. The term $4\,\sigma\,T_{mean}^3$ has the dimensions of a heat transfer coefficient (α_R, W m^{-2} K^{-1}) and varies between 4.13 (at -10 °C) and 7.0 (at 40 °C), for instance. So, a very first rule of thumb is that the net radiation energy exchanged between two things at 'terrestrial' temperatures is about 6 W m^{-2} for each degree difference in temperature.

With this in mind it is possible to estimate the radiative losses of a greenhouse to the atmosphere. The drier the air, the more transparent the atmosphere for thermal radiation, and the colder the apparent temperature of the sky (the temperature decreases a lot with height). Some meteorological stations do measure apparent sky temperature (Box 3.7), otherwise one has to rely on empirical models that estimate the apparent temperature of the sky, accounting for air temperature, humidity and cloud cover (for instance: Cheng and Nnadi, 2014)

Box 3.6. Linearization.

The net radiation exchanged between two objects is the difference between the fourth power (y-axis) of their temperatures (x-axis). This is given by the temperature difference multiplied by the slope of the fourth power line on the temperature range. In the temperature interval, there is always a point (T_m), where the tangent line has the same slope as the line going through the extremes. As we know, the slope of the tangent line at one point is given by the derivative calculated at that point:

$$\frac{\Delta y}{\Delta x} = \left[\frac{\partial y}{\partial x}\right]_{x=T_m} \quad \Delta y = 4\sigma T_m^3 (T_2 - T_1)$$

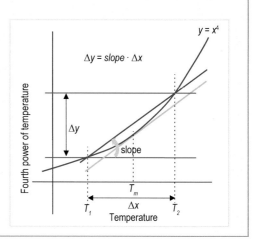

On a clear, cold night there can easily be a difference of 50 °C between a greenhouse and the upper layers of the atmosphere, so that radiative losses alone can be some 150 W m⁻² (the sky is only half of the horizon of a greenhouse). This is the reason a good energy screen must contain aluminium, much more than 50%. It is good to bear in mind, however, that the efficacy of such a screen is much reduced when the aluminium becomes wet or dirty.

3.3 Heat transfer – conduction, convection and thermal storage

In addition to radiative exchange, there is also heat exchange. Whenever there is a temperature difference between two bodies, there will be heat transfer from the warmest to the coldest. We say there is 'conduction' of heat when the two bodies are in contact, and we call it convection

---

Box 3.7. How do we measure net radiation?

A pyrgeometer is based on the same principle as a pyranometer, the only difference being the waveband that the filter lets through (>2,500 nm). As the sensor itself emits radiation in this waveband, the measurement is the net radiation exchanged by the sensor and the sky (if it is facing upwards, for instance).

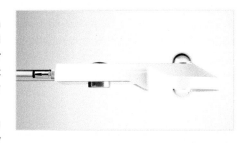

An absolute measurement is possible only by measuring (or controlling) the temperature of the sensor. This is how sky temperature is measured in meteorological stations. If two sensors facing opposite sides are coupled, subtracting the two signals gives the net radiation exchanged between what is above and what is below the sensors. An even more complete instrument also includes two pyranometers facing opposite sides.

---

when heat is released to a fluid (such as air) that then may warm up other, colder bodies. In all cases, the rate of heat transfer is proportional to the difference in temperature, the constant of proportionality being called the heat transfer coefficient, usually indicated by α, with the dimension of W m⁻² K⁻¹, or W m⁻² °C⁻¹.

A greenhouse is usually warmer than outside, so additional energy is lost through the cover:

$$H_{in,out} = \alpha_{cover} \frac{A_{cover}}{A_{soil}} (T_{in} - T_{out}) \qquad\qquad \text{W m}^{-2} \quad (3.7)$$

where:

$H_{in,out}$ W m^{-2} is the heat flux density from the inside to the outside of the greenhouse, per
                 m^2 soil
A_{cover} m^2 is the surface of the cover
A_{soil} m^2 is the surface of the soil
T °C or K   the temperature with*in* and *out*side, respectively
α_{cover} W m^{-2} K^{-1} the heat transfer coefficient of the cover

The heat transfer coefficient of the cover depends on the material – multiple layers have a
much lower heat transfer coefficient than single layers – but also on wind that removes the
warm air from the cover. In a first approximation one can assume that, for a single layer cover,
heat transfer increases linearly with wind speed.

In greenhouse design the U coefficient is used (see also Section 8.1.1), which has the units
of a heat transfer coefficient (W m^{-2} K^{-1}) and accounts for both heat and thermal radiation
losses. The U coefficient is used for dimensioning heating installations, and depends not
only on the greenhouse construction and cover material, but also on where the greenhouse is
situated (cloudiness of the place, average wind speed, etc.). The U coefficient is an average
overall heat transfer coefficient, and should not be used for calculating the temperature in the
greenhouse at a given time (or under given external conditions).

Similarly, there can be (and usually is) heat transfer between the greenhouse air and the soil
(which is usually colder than the greenhouse air during the day and warmer at night).

$$H_{in,soil} = \alpha_{soil} (T_{in} - T_{soil}) \hspace{4cm} \text{W m}^{-2} \quad (3.8)$$

where:
$H_{in,soil}$ W m^{-2} is the heat flux density from the inside air to the soil, per m^2 soil
T °C or K   is the temperature with*in* the greenhouse and of the *soil* surface, respectively
α_{soil} W m^{-2} K^{-1} is the heat transfer coefficient from the air to the soil

The latter depends on how insulated the soil surface is (for instance, concrete, plastic or non-
woven mulch) and how much air movement there is near the soil surface.

The energy going into the soil warms up the top layer of soil which, in turn, warms up the
underlying material, and so on, so that some heat is stored in each layer and the heat that is
transferred has a delay and a temperature smaller in amplitude with respect to the course of
temperature at the soil surface. At night the energy flow is reversed, if the air in the greenhouse
is colder than the soil surface (usually not in a heated greenhouse), so that the energy released to
the air comes from the cooling of the top layer, and the heat released by the layers underneath.
So one can imagine that at a certain depth the diurnal temperature excursion will no longer be
felt, and there is a depth at which not even the annual temperature excursion will be apparent.
In short, a relatively shallow layer of soil exchanges energy daily with the greenhouse air,
whereas a much thicker one stores energy in summer and releases it in winter. The thickness

Chapter 3   Radiation, greenhouse cover and temperature

of each layer is determined by the thermal capacity of the soil, which depends on the type of soil (sand, loam, organic content, rock) and on its water content. A soil whose pores are filled with water can store much more energy than when dry. Of course, both the soil composition and its water content change with depth, so estimating the temperature of each layer on the assumption of soil properties constant with depth gives rather approximate results.

3.4 Energy balance and greenhouse temperature

One of the fundamental laws of physics is the conservation of energy. That is, in a steady state, the energy entering a greenhouse must equal the energy leaving it. When conditions change, in the time that it takes to reach a new equilibrium, energy is either stored or released in/by the greenhouse, in other words the greenhouse either warms up or cools down. Since the energy fluxes leaving the greenhouse (and, to a lesser extent, the fluxes entering it) depend on the temperature of the greenhouse, it is clear that the temperature is the variable that changes in order to maintain the energy balance. We will call the greenhouse temperature T_{tunnel} to remind us that there is no ventilation and we assume cover temperature equal to air temperature.

Figure 3.4 gives a summary of the energy fluxes in an unheated, unventilated greenhouse. If we combine all the information that we have seen in this chapter, we can calculate the temperature in an unventilated greenhouse (like a closed tunnel), without a crop. In an equilibrium situation the radiation from the sun, coming through the cover, must equal the losses to the sky, to the outside air and to the soil.

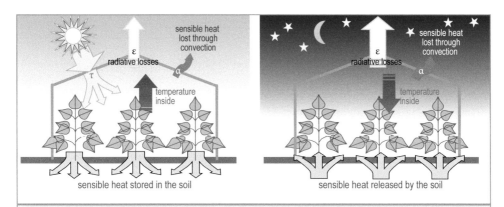

Figure 3.4. The energy fluxes in and out of an unventilated and unheated greenhouse, during the day (left) and night (right). The symbols τ, ε, and α represent the three very important properties of the greenhouse cover, respectively: the transmissivity for solar radiation; the emissivity for thermal radiation; and the heat transfer coefficient.

Greenhouse horticulture   69

$$\tau I_{sun} = F_{tunnel,sky}\, \varepsilon_{cover}\, \alpha_R\, (T_{tunnel} - T_{sky}) + \frac{A_{cover}}{A_{soil}}\, \alpha_{cover}\, (T_{tunnel} - T_{out})$$

$$+ F_{tunnel,soil}\, \varepsilon_{soil}\, \alpha_R\, (T_{tunnel} - T_{soil}) + \alpha_{soil}\, (T_{tunnel} - T_{soil}) \qquad \text{W m}^{-2}{}_{soil}\quad (3.9)$$

where:

τ	–	Transmissivity [hemispherical] of the cover for sun radiation	Equation 3.2
I_{sun}	W m^{-2}	Solar radiation	
$F_{a,b}$	–	View factor of b from a	Equation 3.5
ε_{cover}	–	Emissivity of the cover, including effect of screen(s)	Equation 3.3
α_R	W m^{-2} K^{-1}	Apparent transfer coefficient for thermal radiation	Equation 3.6
T_{tunnel}	K or °C	Temperature of the unventilated greenhouse	
T_{sky}	K or °C	Apparent temperature of the sky (depends, among other things, on cloudiness)	
T_{out}	K or °C	Temperature of the air outside the greenhouse	
T_{soil}	K or °C	Temperature of the soil surface	
$A_{cover,soil}$	m^2	Surface of cover, soil	Equation 3.7
α_{cover}	W m^{-2} K^{-1}	Heat transfer coefficient of the cover material (including possible screens)	Equation 3.7
α_{soil}	W m^{-2} K^{-1}	Heat transfer coefficient of the soil (including possible mulches)	Equation 3.8

For a greenhouse in a field, unobstructed by high buildings, we can assume $F_{tunnel,sky} = F_{tunnel,soil} = 0.5$. Of course, when the external conditions and the properties of the cover (and of the soil mulch) are known, Equation 3.9 can be solved to give the temperature within the unventilated greenhouse, with no crop inside.

$$T_{tunnel} - T_{out} = \frac{\tau I_{sun} - F_{tunnel,sky}\, \varepsilon_{cover}\, \alpha_R\, (T_{tunnel} - T_{sky})}{\dfrac{A_{cover}}{A_{soil}}\, \alpha_{cover}}$$

$$- \frac{F_{tunnel,soil}\, \varepsilon_{soil}\, \alpha_R + \alpha_{soil}}{\dfrac{A_{cover}}{A_{soil}}\, \alpha_{cover}}\, (T_{tunnel} - T_{soil}) \qquad \text{°C}\quad (3.10)$$

$$= \frac{\tau I_{sun} - radiation\ to\ sky \pm soil\ flux}{\dfrac{A_{cover}}{A_{soil}}\, \alpha_{cover}}$$

where soil flux is considered positive when directed into the soil ($T_{tunnel} > T_{soil}$).

Since the sky is always much colder than the air at the Earth's surface (e.g. Monteith and Unsworth, 2014: p72), Equation 3.10 makes clear that in an unheated greenhouse at night it can be warmer than outside only to the extent that the soil is warm and radiative losses to the sky are (made) small. Piscia *et al.* (2013) combined this energy balance with computational fluid dynamic (CFD) simulations for a more in-depth study of the effects of cover properties on the night-time temperature of a unheated greenhouse. Under a clear sky the predicted temperature gain by using a high reflectance material (that is: low transmissivity and low emissivity for thermal radiation) compared to the standard polyethylene used in unheated greenhouses, was approximately 6.5 °C, with a soil heat flux gain of 25 W m⁻².

Management of the radiative and insulation properties of the cover to get the best possible greenhouse temperature, requires usually compromises between transparency, insulation and clever use of thermal storage, both in passive greenhouses (Box 3.8) and in heated greenhouses (Box 3.9). A very educative tool to examine the effect of various screens and greenhouse covers on the energy required to maintain a desired temperature within a heated greenhouse and the vertical gradient of the temperature of a crop inside, is available on line (De Zwart, 2016).

Box 3.8. A traditional Chinese solar greenhouse.

(photos: Cecilia Stanghellini)

A traditional Chinese greenhouse has one south-facing slope, and a thick earthen wall on the north side. Here the excess of solar energy is stored during the day. At night the wall releases the energy to the air, whereas the thermal conductivity of the cover is reduced by rolling out straw mats over the plastic. This clever use of natural thermal storage coupled to management of cover properties significantly lengthens the growing season, into the very cold continental winter of inner China, without relying on heating. Growing welfare and care for quality are, however, bringing about much needed improvements to this simple design.

Box 3.9. Energy saving: balancing insulation and light transmission.

There is both an economic and a social pressure on decreasing the (fossil) energy requirements of heated greenhouses. That means maximal utilisation of solar energy and any additional energy that may be needed must be renewable. Electricity sustainably generated will most likely power additional light and heat pumps (for heating, cooling and de-humidification) in the greenhouses of the future. In the shorter term it is necessary to reduce heating requirements through smarter climate management and improved insulation properties of the cover (De Gelder *et al.*, 2012). Two prototype covers whereby very high insulation was achieved without jeopardising the PAR transmission have recently been developed and tested (Hemming *et al.*, 2017). The VenloW-e® greenhouse is equipped with new double glazing, in very large panes, with Argon gas sealed inside, three surfaces coated with an anti-reflection product and one of the two inner surfaces with a low emissivity one. The 2SaveEnergy® greenhouse has lower investment costs. The cover is single, low-iron glass, both sides coated with an anti-reflection product, and a 50 μm diffuse ethyl tetra fluor ethylene (ETFE) film is fitted by means of original profiles on the inner surface of the glass panes.

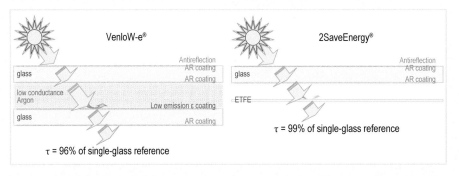

Both prototypes were fitted with additional energy saving measures, such as energy screens and active de-humidification with sensible heat regain. The table shows their performance, with respect to a single-gas reference, similarly fitted (Kempkes *et al.*, 2017a).

	Reference single-glass	VenlowEnergy double-glass	2SaveEnergy glass-film
Gas consumption ($m^3 \, m^{-2} \, y^{-1}$)	24.8	14.8	19
Electricity ($kWh \, m^{-2} \, y^{-1}$)	6.5	6.7	6.9
Water condensation on roof ($kg \, m^{-2} \, y^{-1}$)	83	3	23
Light sum ($mol \, m^{-2} \, y^{-1}$)	5,737	5,706	5,888
Light sum in winter ($mol \, m^{-2}$)	667	712	738
Dry matter production in winter (% i.r.t. reference)	100	108	115

3.5 Concluding remarks

In summary, the materials for the cover and additional screen(s) must be chosen for their spectral characteristics: reflection, transmissivity and emissivity. Desirable properties depend on the spectral range (PAR, NIR and TIR). In the PAR range transmissivity τ must be as high as possible. What is desirable in the infrared range depends on the region in which the greenhouse is located. If keeping cool is an issue, NIR reflection must be high, otherwise it must be low. Also for the TIR range, the best solution depends on the region. In warm parts of the world high emissivity may be welcome to enable cooling down at night (and to limit warming-up during the day). Some crops really do need colder nights, or a substantial difference between day and night temperatures. Wherever greenhouses are heated at night, emissivity should be as small as possible (aluminium screens). The same is true for unheated greenhouses where low night-time temperatures may be a factor limiting productivity, such as winter crops in the Mediterranean area.

Chapter 4
Properties of humid air and physics of air treatment

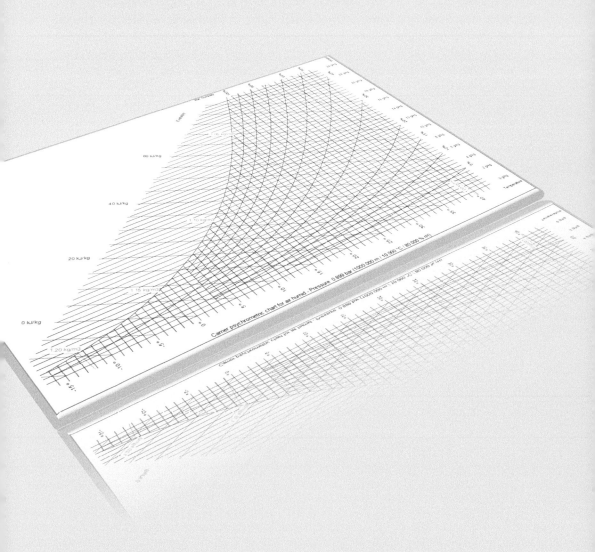

Photo: Bert van 't Ooster

Psychrometrics, psychrometry, and hygrometry are names for the field of engineering concerned with the physical and thermodynamic properties of gas-vapour mixtures (Gatley, 2004). The term comes from the Greek *psuchron* (ψυχρόν) meaning cold and *metron* (μέτρον) meaning means of measurement. The most common use of psychrometrics, and the most relevant here, is an air-vapour mixture. Psychrometrics is used here to describe the states and processes of humid air. In order to be able to regulate the greenhouse climate, it is essential to understand the properties of humid air and air-conditioning processes. The Mollier (and/or Carrier) chart is an important tool applied in heating, ventilating, air-conditioning and meteorology, but understanding and solving its equations gives a more precise picture.

4.1 Properties of humid air

Heating, cooling and evaporating water all change the state of greenhouse air. These are called psychrometric processes. The psychrometer shown in Figure 4.1 is the traditional measuring instrument, and is the basis for the concept of 'wet' and 'dry' bulb temperature.

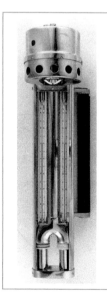

Figure 4.1. A traditional Assmann psychrometer (thermo-schneider.de). Of the two thermometer bulbs one is dry, and the other is wrapped in a cloth, moist with distilled water. The fan on top forces air along the thermometers to evaporate water from the cloth. Because evaporation takes away energy, the wet bulb is colder than the dry bulb. Modern instruments (Box 4.1) feature PT100 instead of thermometers that are usually connected to data loggers. Temperature and humidity are measured very accurately (Gieling, 1998), but one drawback is the high maintenance required to keep the cloth moist. This prevents its application in modern practice.

Box 4.1. Measuring temperature and humidity in greenhouses.

Temperature and humidity sensors give information on the thermal climatic state in a greenhouse. These sensors are usually mounted in a ventilated box. The most important electronic temperature sensing principles are resistance change with temperature, as in thermistors and resistance temperature detectors (RTDs), or the Seebeck effect which is applied in thermocouples. For humidity the electronic measuring principles are capacitive, resistive, thermal and gravimetric.

Principles for electronic temperature sensing:

- RTD elements (wikipedia and Kiyriacou (2010)) normally consist of a length of fine wire wrapped around a ceramic or glass core. The fine wire is a pure material, either platinum, nickel or copper, that has an accurate resistance/temperature relationship which is used to provide an indication of temperature. RTD elements are normally housed in protective probes. Examples are Pt100, Pt500, and Pt1000.

- Thermocouples consist of two dissimilar electrical conductors forming electrical junctions at different temperatures. A thermocouple produces a voltage because of the thermoelectric effect resulting from the temperature difference between the junctions. This voltage can be interpreted to measure temperature. Thermocouples are a widely used type of temperature sensor, but increasingly replaced by RTDs.

Principles for humidity sensing – hygrometers

- Capacitive hygrometers measure the effect of humidity on the dielectric constant of a thin film polymer or metal oxide material. Calibrated sensors have an accuracy of ±2% relative humidity (RH) in the range 5-95% RH. Capacitive sensors are robust against condensation and temporary high temperatures. They are suitable for many applications.

- Resistive hygrometers measure the change in electrical resistance of a material such as salts and conductive polymers due to humidity. Resistive sensors are less sensitive than capacitive sensors and require more complex circuitry. The sensor must be combined with a temperature sensor because material properties also tend to depend on temperature. Robust, condensation-resistant sensors exist with an accuracy of up to ±3% RH, but accuracy and robustness vary depending on the resistive material.

- Thermal hygrometers measure the change in thermal conductivity of air due to humidity. These sensors measure absolute humidity rather than RH.

- A gravimetric hygrometer measures the mass of an air sample compared to an equal volume of dry air. This is considered the most accurate primary method for determining the moisture content of the air. National standards based on this type of measurement have been developed in the US, UK, EU and Japan. The inconvenience of using this device means that it is usually only used to calibrate less accurate instruments.

- Dew point temperature is a measure for humidity when combined with a temperature sensor. Traditionally dew point temperature is determined by chilled mirror hygrometers. By shining a light beam at a chilled mirror and measuring the reflected light, the reflection changes instantly at dew point temperature due to condensation of water vapour. Stable thin film capacitive sensors developed since the 1980s allow measurement of dew points at a fraction of the chilled mirror cost (Roveti, 2001). Calibration data for each specific sensor are incorporated in the sensor for improved accuracy.

The temperature of the air (the sensible heat content) can be measured, but an important aspect remains concealed: the latent heat. The water vapour in the air contains a lot of energy; the energy needed for a phase change from water (liquid) to vapour is approx. 2.5 MJ kg^{-1}.

The amount of vapour that air can contain increases exponentially with its temperature (Figure 4.2). Therefore, if warm, humid air is cooled down, there will be a certain temperature at which the air cannot hold all the water in vapour form and part of it will condense. This point is called the dew point.

Important measurements defining the properties of the air are as follows:
  ‣ Dry bulb temperature, T_{db} (°C), is the real air temperature measured with a (dry) thermometer.
  ‣ Wet bulb temperature, T_{wb} (°C), results from the wet thermometer. This helps define air humidity (e.g. specific humidity and relative humidity (RH)).
  ‣ Dew point temperature, T_{dp} (°C), is the temperature at which condensation takes place.

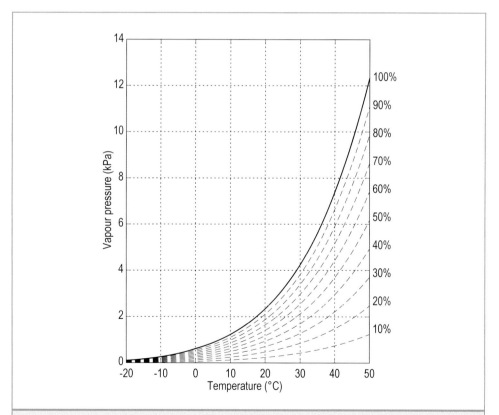

Figure 4.2. Vapour pressure as a function of temperature for different relative humidities (curves). Curve for 100% relative humidity (RH) represents saturated vapour pressure.

Other relevant properties of humid air are (Albright, 1990; ASHRAE, 2017):

- ➤ Density ρ in kg m^{-3}. Hot air has a lower density than cold air. Humid air has a lower density than dry air at the same temperature. Higher in the atmosphere air density is lower.
- ➤ Partial vapour pressure: p_v in Pa (Pascal). This is the part of the air pressure exerted by the vapour component. This is a small fraction of atmospheric pressure: usually less than 5 kPa.
- ➤ Humidity ratio: x or w in kg kg^{-1}. This is a mass-based concentration representing the mass of water vapour in a unit mass of dry air, so it is dimensionless, but is sometimes represented in kg kg^{-1} (or, more commonly, in g kg^{-1}) to emphasise concentration.
- ➤ Specific humidity: $q = w/(1 + w)$ is the mass of water vapour in a unit mass of moist air expressed in terms of humidity ratio.
- ➤ Absolute humidity: χ in kg m^{-3} or, more commonly, g m^{-3}, the volume-based vapour concentration.
- ➤ Saturation deficit Δp_v in Pa or Δx in kg kg^{-1} representing the amount of vapour that can be added to the air at unchanging T_{db} to reach saturation with water vapour.
- ➤ RH in % is the ratio of the actual vapour pressure to the saturated vapour pressure at the actual temperature. Only when the temperature is known can RH be transformed into vapour pressure, specific or absolute humidity.
- ➤ Degree of saturation DoS in % is comparable with RH but it gives the ratio of concentrations instead of vapour pressures.

Table 4.1 provides a summary of commonly used symbols and units for properties of moist air. The thermodynamic psychrometric constant γ will be explained later in this chapter (Section 4.1.7).

Knowledge of air vapour mixtures is all basic physics, but there is a clear link to greenhouse technology. The crop transpires a lot of the irrigation water: 70-90% of the total uptake (Section 13.1.1). Therefore the greenhouse air is usually very humid. The vapour contains L kJ kg^{-1} of latent energy. In fact the crop is a converter of solar energy into latent heat. A greenhouse with a crop is therefore cooler than one without a crop. Also a fogging system, (Chapter 9) which brings in water that evaporates immediately, converts sensible into latent heat. Opening the vents in a greenhouse means exchanging air, but also exchanging heat, both sensible and latent heat.

Modern heat exchange systems bring the warm outgoing greenhouse air into contact with cold incoming outside air. So the sensible energy of the outgoing air is transferred to the incoming air that heats up. If the incoming air is under the dew point of the outgoing air, vapour will condense, and latent heat will be released and transformed into sensible heat. Heat recovery systems can, as their name indicates, recover latent heat by condensing humid air against a cold surface.

Vapour is lighter than air (Box 4.2). As a consequence it rises; this is called buoyancy. In the greenhouse vapour travels from the crop to the upper parts of the greenhouse. The more vapour it contains, the lower the density of air and the more buoyancy. But remember, the air can only contain a limited amount of vapour (Figure 4.2).

Table 4.1 Common symbols of psychrometric properties used in literature (Albright, 1990; ASHRAE, 2017; Gatley, 2004; Singh and Heldman, 2014).[1]

Symbol	Description	Unit
P, P_{tot}	atmospheric pressure	Pa
e^*, p_{vs}	saturation vapour pressure	Pa
e, p_v	vapour pressure	Pa
R	general gas constant	J mol^{-1} K^{-1}
DB, T_{db}	dry bulb temperature	°C
WB, T_{wb}	wet bulb temperature	°C
DP, T_{dp}	dew-point temperature	°C
H, h	total energy, enthalpy	kJ kg^{-1}
x, w	humidity ratio	kg kg^{-1}
q	specific humidity	kg kg^{-1}
χ	absolute humidity, vapour concentration	kg m^{-3}
$\rho, \rho_a, \rho_{da}, \rho_{ha}$	air density (dry or humid)	kg m^{-3}
L	evaporation energy of water (latent heat)	kJ kg^{-1}
RH	relative humidity	%
DoS, DS	degree of saturation	%
γ	thermodynamic psychrometric constant	Pa K^{-1}

[1] With respect to the latent heat in water vapour, L, there is a slight effect of temperature equals 2,501 kJ kg^{-1} at 0 °C and 2,257 kJ kg^{-1} at 100 °C. In determining the total energy content it is assumed that water evaporates at 0 °C.

Box 4.2. Molar mass of dry air and vapour.

Standard air contains 78.08% nitrogen, 20.95% oxygen, 0.04% carbon dioxide and also very small volumes of argon, neon, helium, methane, sulphur dioxide and hydrogen. The weighted average of all the molecules gives an apparent molar mass of dry air M_{da} of 28.9645 g mol^{-1}.

The molar mass of water (and vapour) M_v is 18.01534 g mol^{-1}.

As a mole of any gas occupies a given volume at a given temperature, the density of vapour is lower than the density of dry air, and humid air has a lower density than dry air at the same temperature.

Another psychrometric effect results from the atmospheric pressure. This pressure depends on the elevation above sea level. At sea level standard pressure and temperature is 101.325 kPa and 15 °C, at an elevation of 500 m 95.461 kPa and 11.8 °C, at 1000 m 89.874 kPa and 8.5 °C, and at 3,000 m standard pressure is only 70% of atmospheric pressure at sea level, 70.108 kPa, and standard temperature is -4.5 °C. This follows the standard atmosphere definition (Cavcar, 2000; US Standard Atmosphere, 1976). So, it matters where a greenhouse is situated; the elevation above sea level affects the air properties. At a higher level the amount of molecules of air decrease and mass-based concentrations change at constant vapour pressure. The maximum vapour pressure is not influenced by greenhouse altitude but mass-based concentration is. The explanation for this phenomenon is that vapour saturation does not depend on the amount of air but on interaction of the water molecules.

4.1.1 Vapour pressure

In the past scientists have measured the vapour pressure at a whole range of temperatures (Haynes, 2014) in order to get a clear idea of vapour pressure characteristics. It was found that vapour pressure at saturation is a function of the temperature only (Figure 4.2). The vapour pressure is limited by molecular characteristics. Water sprayed in a vacuum will evaporate very quickly, but the space cannot contain more vapour than the saturation pressure allows at a certain temperature. This pressure is low at -20 °C, namely 0.2 kPa and increases exponentially to above 12 kPa at 50 °C (Figure 4.2). To give an idea of the order of magnitude of these pressures, 1 atmosphere is 101.3 kPa. At a higher temperature the water molecules move faster and stay longer in the vapour phase. In this book vapour pressure is indicated with the symbol e, and saturated vapour pressure by e^*.

The Magnus equation (Equation 4.1) is an empirical equation often used for calculating saturation pressure because it fits the data for a wide range of temperatures (-30-60 °C). There are two forms of this equation: for temperatures above 0 °C and below 0 °C:

$$e^* = \begin{cases} 10^{2.7857+\frac{9.5 \cdot T}{265.5+T}} & T<0 \ (above \ ice) \\ 10^{2.7857+\frac{7.5 \cdot T}{237.3+T}} & T \geq 0 \ (above \ water) \end{cases} \qquad \text{Pa} \quad (4.1)$$

More accurate equations for limited temperature ranges are given in the literature, e.g. Buck (1981) and Alduchov and Eskridge (1996). Figure 4.2 shows that actual vapour pressure e at 30 °C and 60% RH is 2.5 kPa. The saturation pressure e^* at this temperature is found by moving vertically upwards until the saturation line (RH 100%) is reached. At this point the pressure e^* is 4.2 kPa. The ratio is these two pressures 2.5/4.2 is of course 0.6 (RH = 60%).

The vapour pressure deficit (VPD) is the amount of vapour that can still be stored in the air until the saturation point is reached (at the same temperature):

$$VPD = e^* - e \qquad \text{Pa} \quad (4.2)$$

In the example this $4.2 - 2.5 = 1.7$ kPa. *VPD* is favourite amongst growers discussing greenhouse climate. It gives an indication of the possibilities of the crop to keep on transpiring at a given air temperature T_{db} without the danger of condensation, especially on the crop itself, because free water significantly increases the risk of fungal attack.

When cooling air and maintaining vapour pressure at the same level, at some point the gas mixture will be saturated and condensation will take place. This point is called the dew point.

It is thus equivalent to any absolute measure of vapour content, such as vapour concentration, vapour pressure, specific humidity. We will see that it is useful to quantify humidity in terms of dew point when dealing with humidity management in greenhouses, as there is condensation on any spot in the greenhouse (the cover, the crop, a cooling device, etc.) that is colder than the dew point. To find it, move horizontally in Figure 4.2 until the 100% RH curve is reached. The dew point depression is the difference between the actual temperature and the temperature at dew point.

4.1.2 The ideal gas law

The usability of Figure 4.2 is limited as it presents temperature and pressure only. It doesn't give any information on energy content and other properties of humid air. Therefore we move in this paragraph step by step towards a more complicated but far more useful graph: the Psychrometric chart, also called Mollier chart or Carrier (Simha, 2012) chart.

First, we assume that air and water vapour behave like perfect gases in which case the ideal gas law applies. The ideal gas law is as follows: $PV = nRT$, in which P is the pressure in Pa, V the volume m³, n the amount of moles in the gas, R the general gas constant in J mol⁻¹ K⁻¹ (Table 4.1) and T the absolute temperature in K. Please note that the temperature must be in Kelvin! The other units only need to be consistent.

For dry air the equation is: $P_a V = n_a RT$. For the vapour the equation is: $eV = n_v RT$, in which e is (as before) the vapour pressure. We combine these two for the mixture of dry air and vapour: $PV = nRT$. The atmospheric pressure is the sum of partial pressures of dry air and vapour: $P = P_a + e$. The total amount of moles: $n = n_a + n_v$, the mole fractions are: $n / n = n_a / n + n_v / n$.

The humidity ratio is the mass of water vapour in a unit of mass of dry air. To get the mass in kg the mole fractions are multiplied by the molar mass of dry air $M_{da} = 28.9645$ g mol⁻¹ and vapour $M_v = 18.01534$ g mol⁻¹ to get the mass per mole dry air. Now with substitution the humidity ratio x is found:

$$x = \frac{M_v}{M_{da}} \cdot \frac{n_v}{n} \cdot \frac{n}{n_a} = \frac{M_v}{M_{da}} \cdot \frac{e}{P - e} \qquad \text{kg kg}^{-1} \quad (4.3)$$

Instead of a mass-based concentration, a volume-based concentration is often used as the measure of vapour content of air. This is χ, the absolute humidity in kg m⁻³ or g m⁻³. The link between [vapour] pressure e and concentration χ is given by the ideal gas law, which we will re-write as:

$$eV = \frac{m_v}{M_v} RT \text{ and } \chi = \frac{m_v}{V} \rightarrow \chi = \frac{M_v}{RT} \cdot e \cong 7.4 \, e \qquad \text{g m}^{-3} \quad (4.4)$$

Remember in Equation 4.4, temperature T must be expressed in Kelvin, R is the general gas constant (8.314 J mol⁻¹ K⁻¹), e is in Pa except for $\cong 7.4 \, e$, where e is in kPa, m_v is the mass of water vapour in volume V. The link between vapour concentration and humidity ratio (g kg^{-1}) is obviously given by the density of dry air.

Now calculate for yourself the humidity ratio of humid air at sea level with T_{db} = 18 °C and RH = 80%. The standard atmospheric pressure is 101.325 kPa (Answer: 10.3 g kg^{-1}).

4.1.3 Enthalpy

The energy content of the air, i.e. the sum of sensible and latent heat, is called enthalpy (kJ kg^{-1}). Obviously only dry air at 0 K does not contain enthalpy. However, to have the zero scale of enthalpy as such is rather impractical, so the zero point of enthalpy is taken for dry air at 0 °C. This may be misleading, but since we deal with enthalpy in terms of differences (enthalpy added or taken out of air) it does not matter. Figure 4.3 shows the energy content as a function of air temperature when saturated with water vapour.

The specific heat of dry air is close to 1 kJ kg^{-1}. For example at 20 °C the sensible heat in kJ kg^{-1} is almost equal to the temperature in °C. However, when it is saturated, the amount of energy in the vapour is at saturation with 37.4 kJ kg^{-1} almost twice that of dry air. The enthalpy is the specific heat of dry air multiplied by the temperature, plus the latent and sensible energy in the vapour:

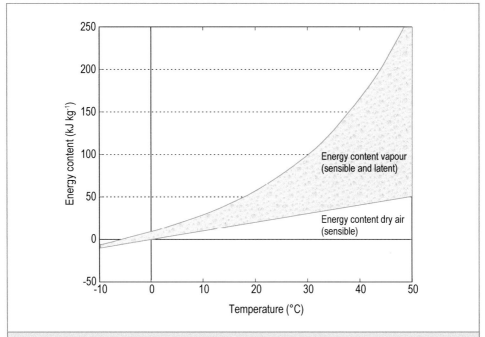

Figure 4.3. Energy content (enthalpy) of humid air, which is the sum of energy in dry air (sensible heat, the lower line) and energy in vapour, the shaded area; the upper line is for saturated air.

$$h = c_{da} \cdot T_{db} + x \, (L + c_v \cdot T_{db}) \qquad\qquad\qquad\qquad\qquad \text{kJ kg}^{-1} \quad (4.5)$$

In the example in Box 4.3, latent heat is higher than sensible heat, as it usually is in a greenhouse.

The vapour in the greenhouse air can be a result of solar irradiation causing the crop to transpire (Chapter 6). So the solar energy 'hides' in the vapour. When, for example, solar radiation is 700 W m^{-2} and the transmissivity of the greenhouse is 70%, 490 W m^{-2} is available in the greenhouse. Half of it, or more, can be converted into vapour by the transpiring crop so only 245 W m^{-2} remains for raising the temperature. The transpiration of the crop functions as a cooling mechanism.

4.1.4 Setup of a psychrometric chart

As observed on Figure 4.2, giving vapour pressure as a function of temperature has its limitations. Physicists therefore chose a different approach in which the focus is not on temperature but on the relationship between enthalpy and humidity ratio, the h-x diagram. Equation 4.5 can be used to draw linear isotherms $h = a + b \cdot x$ with intercept $(a = c_{da} \cdot T_{db})$ and slope $(L + c_v \cdot T_{db})$ increasing with temperature.

For $T_{db} = 0$ °C the equation becomes very simple. The isotherm passes through the origin with intercept 0 and slope L. For $T_{db} = 40$ °C, the intercept is $c_{da} \cdot 40 = 40.2$ and the slope is $L + c_v \cdot 40 = 2572.2$. So the slope slightly increases, which means that the isotherms are not exactly parallel (Figure 4.4A). The next step towards a useful psychrometric chart is rotating the horizontal gridlines by an angle L. These gridlines are the isenthalpic lines. We create a new enthalpy axis and convert the former enthalpy axis to a new temperature axis as shown in Figure 4.4B. The horizontal axis is the vapour (humidity ratio), a new diagonal axis is the enthalpy-axis. The line for 0 °C is horizontal, but the rest of the temperature lines diverge from the grid.

Box 4.3. Example – calculation of enthalpy.

In the greenhouse the temperature is 18 °C and RH is 80%. What is the enthalpy h?

The only unknown parameter is humidity ratio x. Using Equation 4.1 and 4.3:

$$e^* = 10^{2.7857 + \frac{7.5\,T_{db}}{237.3 + T_{db}}} = 2{,}063 \text{ Pa};$$

$$e = e^* \cdot RH = 2{,}063 \cdot 0.8 = 1{,}650 \text{ Pa};$$

$$x = \frac{0.622\,e}{P - e} = 0.0103$$

Enthalpy:	$h = 1.006 \cdot 18 + 0.0103 \cdot (2{,}500 + 1.805 \cdot 18) = 44.2$ kJ kg^{-1} with
sensible heat:	$h = 1.006 \cdot 18 + 0.0103 \cdot 1.805 \cdot 18 = 18.4$ kJ kg^{-1}
latent heat:	$x \cdot L = 0.0103 \cdot 2{,}500 = 25.8$ kJ kg^{-1}

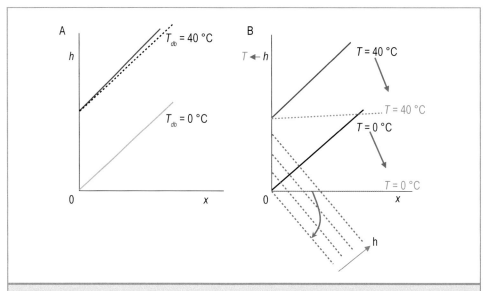

Figure 4.4. Set-up of the psychrometric chart. The h-x diagram basis (A) and the h-x diagram with rotated horizontal axis (B).

What the new graph (Figure 4.4B) lacks is RH. The Magnus equation Equation 4.1 indicates that vapour pressure at saturation only depends on temperature and the temperature information is in the graph. For each temperature the vapour pressure at saturation is calculated by Equation 4.1. But there is no vapour pressure in the graph, only the concentration, which is found using Equation 4.3. This combination is shown below (Box 4.4). When x_s is substituted in Figure 4.4B and the range between $x = 0$ and $x = x_s$ is split in 10 equal parts the RH lines result as shown in Figure 4.5, the Mollier diagram.

The *RH*-lines in the Mollier diagram actually indicate the *DoS*. For a broad range of air conditions there is hardly any difference between *RH* and *DoS*, but at high temperatures or

---

Box 4.4. Finding humidity ratio at saturation.

$$e^* = 10^{2.7857 + \frac{7.5\,T}{237.3 + T}}$$

$$x_s = 0.622 \frac{e^*}{P - e^*}$$

In Equation 4.3 *P* is the atmospheric or total air pressure. The higher above sea level, the lower this pressure is. So the x_s lines move to the right.

(In the Carrier chart in Figure 4.6, the x_s lines move upwards with decreasing *P*).

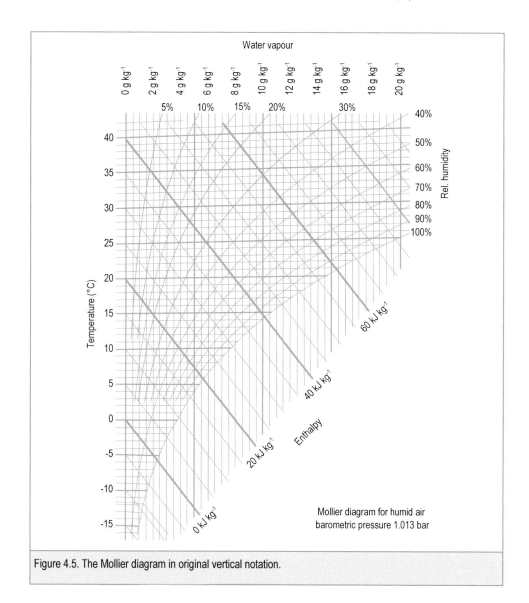

Figure 4.5. The Mollier diagram in original vertical notation.

very low vapour pressures the deviation is significant (though never more than 2%). The correction can be found by solving the equation:

$$DoS = \frac{P - e^*}{P - e} \cdot RH \qquad \{P \gg e^* \text{ and } P \gg e, \text{ then } DoS \approx RH \qquad \% \quad (4.6)$$

The correction factor in front of RH is close to 1, because P is far greater than e and e^*. In most cases RH can readily be substituted by DoS.

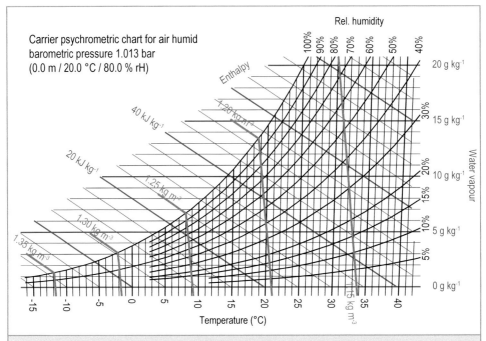

Figure 4.6. Carrier psychrometric chart for humid air. The diagram depends on the atmospheric pressure (here 1.013 bar), which has to be mentioned with the chart.

4.1.5 The Mollier diagram and the Carrier chart

The Mollier diagram is a thermodynamic diagram, named after Richard Mollier (1863-1935). The vertical axis is temperature, the horizontal axis is humidity ratio and the diagonal is enthalpy. Nowadays the Carrier chart, named after Willis H. Carrier (1876-1950), is most frequently used (Figure 4.6). It is the same diagram, but in contrast to the Mollier diagram, temperature is on the horizontal axis. The vertical axis is humidity ratio and the diagonal is enthalpy. The green lines indicate the density of the air (colder air is heavier). The Carrier chart is a mirrored and rotated Mollier diagram.

Looking up data in a graph is often easier than calculating them, especially when there is no need for great accuracy. The psychrometric charts contain a lot of information and it takes some time to find your way. An example below clarifies the utility and the possibilities of the chart.

Dry bulb temperature at 20 °C and RH at 70% is indicated as a black circle in Figure 4.7. The dew point is found by moving horizontally to the left until the saturation line is hit. Then go down to the temperature axis (light blue arrow) to read T_{dp} = 14.2 °C. The wet bulb temperature is approximately found by following the isenthalpic line until the saturation line is hit. Vertically down from this point you will find the wet bulb temperature on the temperature

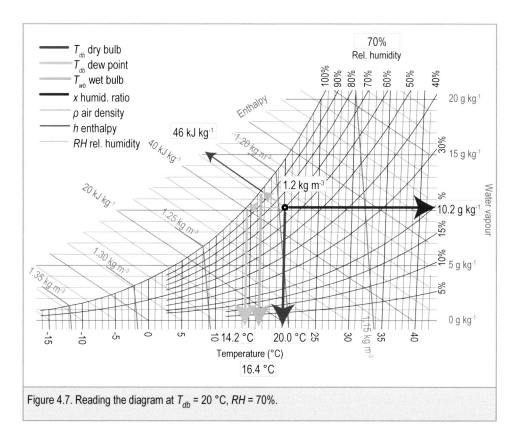

Figure 4.7. Reading the diagram at T_{db} = 20 °C, RH = 70%.

axis (green arrow), T_{wb} = 16.4 °C. To find the vapour concentration, go horizontally to the right (dark blue arrow), x = 10.2 g kg^{-1}. The energy content – the enthalpy – can be read from the isenthalpic gridline (red). The point is just above the enthalpy line of 45 kJ kg^{-1}, so h is about 46 kJ kg^{-1}.

The air density is indicated by the thin green lines. The point is very close to the density line of 1.20 kg m^{-3}. If the concentration in g m^{-3} is needed, you will have to multiply air density by x, resulting in χ, the absolute humidity. To find vapour pressure from the chart a calculation is needed: $e = P \cdot x / (0.622 + x)$ Pa. Reading the diagram is a quick way to find the order of magnitude for all properties of humid air.

4.1.6 Calculating psychometric variables in models

If the outcome of psychrometric properties of humid air has to be accurate, e.g. in modelling of the climate, calculations are necessary. This section presents the equations.

The density of dry air, i.e. the mass of dry air in one m^3 humid air, is derived from the ideal gas law. This law is rewritten as:

$$\rho_{da} = \frac{PM_v}{RT_{db}(0.6222+x)} \qquad \text{kg m}^{-3} \quad (4.7)$$

R is the general gas constant (8.314 J mol^{-1} K^{-1}) and T_{db} is in Kelvin (!). P is atmospheric pressure in kPa. M_v is molar mass of vapour, x is given in kg kg^{-1} (not in g kg^{-1}). If we add the vapour, then for the humid air the equation of air density looks like this:

$$\rho_{ha} = \rho_{da} \cdot (1+x) = \frac{PM_v}{RT} \cdot \frac{1+x}{0.6222+x} \qquad \text{kg m}^{-3} \quad (4.8)$$

This is the density of dry air multiplied by $(1+x)$, which normally results in only a small increase. For really dry air, substitute $x = 0$ in Equation 4.8.

Conversions of vapour pressure e (Pa) to absolute humidity χ (g m^{-3}) and to humidity ratio x (g kg^{-1}) follow from Equation 4.4. Table 4.2 gives an impression of the magnitudes of different variables. See also Equation 4.3 for a different definition of x.

It is useful to remember that:
> The saturation pressure (when expressed in hPa) is the same order of magnitude (but higher) than the temperature T (°C).
> M_v / RT is almost a constant, about 0.74 at 20 °C.
> The vapour concentration (g m^{-3}) is usually higher than humidity ratio (g kg^{-1}).

The dew point temperature only depends on vapour pressure e. The empirical equation for this value is:

$$T_{dp} = -35.96 - 1.8726 \cdot \ln(e) + 1.1689 \cdot \ln^2(e) \qquad 0 \le T_{db} \le 70\,°C \quad °C \quad (4.9)$$

With e in Pa. Table 4.3 gives an impression of the magnitudes of e and T_{dp}.

Table 4.2. Conversion of different absolute humidity parameters. Vapour pressure e is given in hPa here, to see the decreasing order of magnitude in the numerical values of e^*, χ^* and x^*.

T	e^*	$e^* \to \chi^*$	χ^*	$\chi^* \to x^*$	x^*
Conversion:		M_v/RT		$1/\rho_{da}$	
°C	hPa	g m^{-3} hPa^{-1}	g m^{-3}	m^3 kg^{-1}	g kg^{-1}
8	10.72	0.771	8.27	0.80	6.58
10	12.27	0.765	9.39	0.80	7.53
15	17.05	0.752	12.82	0.82	10.46
20	23.37	0.739	17.28	0.83	14.35
25	31.66	0.727	23.02	0.84	19.44

Table 4.3. Order magnitude of vapour pressure and dew point temperatures in the temperature and humidity range of the greenhouse climate.

T_{db} (°C)	RH (%)	e (kPa)	T_{dp} (°C)
30	50	2.12	18.3
	80	3.93	26.1
20	50	1.17	9.1
	80	1.87	16.3
10	50	0.61	0.2
	80	0.98	6.6

4.1.7 The psychrometric equation

The psychrometric equation (Equation 4.10) translates the dry and wet bulb temperature into an estimate of vapour pressure or vapour concentration (Allen *et al.*, 1998):

$$e = e^* (T_{wb}) - \gamma \cdot (T_{db} - T_{wb}) \qquad\qquad \text{kPa} \quad (4.10)$$

The psychrometric constant, γ, is the ratio of specific heat of moist air c_p in MJ kg⁻¹ °C⁻¹ to latent heat of vaporisation L in MJ kg⁻¹ (Brunt, 2011):

$$\gamma = c_p \cdot P / (\varepsilon \cdot L) \qquad\qquad \text{kPa °C}^{-1} \quad (4.11)$$

Where P is atmospheric pressure kPa, $\varepsilon = M_v / M_{da} = 0.622$ (the ratio of molar masses) and c_p = 1.005 kJ kg⁻¹ °C⁻¹ for dry air, but actually c_p varies with air composition. A more accurate energy balance based approach is given in Box 4.5.

4.2 Physics of air treatment

In the greenhouse several typical psychrometric processes can take place: heating, cooling, adiabatic cooling, ventilation, mixing of two air flows or two volumes of air, (de)humidification and (de)humidification combined with heating or cooling at the same time.

These processes can be understood by psychrometrics if only we look at the air. Remember psychrometrics is about air and not the thermal behaviour of solids enclosing the air. This means that energy exchanges between the air and greenhouse walls, cover, crop or equipment for example, cannot be calculated by psychrometrics (this energy and mass exchange is solved by energy and mass balances discussed in other chapters). The above processes are discussed one by one, including examples.

Chapter 4 *Properties of humid air and physics of air treatment*

Box 4.5. Physical basis of the energy balance based psychrometric equation.

The psychrometric equation is basically an energy balance in the boundary layer of the air around the wet bulb of the psychrometer (ASHRAE, 1989). Air flows along the wet thermometer (of the psychrometer). The air picks up vapour and carries it away. The vapour and the incoming air both contain energy. When the air flow has passed beyond the wet bulb, it has a different energy than when it came in. It is assumed that the air just leaving the boundary layer of the wet bulb fabric is saturated at wet bulb temperature. So one has to deal with three enthalpies: the enthalpy of the incoming air, the enthalpy of the water in the boundary layer and the enthalpy of the outgoing air. The derivation of the psychrometric equation then looks as follows: enthalpy of incoming air h plus enthalpy of water added to the air $(x_s^* - x) \cdot h_w^*$ equals enthalpy of outgoing air h_s^*:

$$h + (x_s^* - x) \cdot h_w^* = h_s^*$$

Condition: x_s^*, h_w^* and h_s^* are all valid at one T^* that satisfies the equation. This T^* is equal to T_{wb}. Substituting $h = c_{da} \cdot T_{db} + x(L + c_v \cdot T_{db})$ and $h_s^* = h(T_{wb}, x_s(T_{wb})) = c_{da} \cdot T_{wb} + x_s(T_{wb}) \cdot (L + c_v \cdot T_{wb})$ in the energy balance gives:

$$\underbrace{c_{da} \cdot T_{db} + x(L + c_v \cdot T_{db})}_{h} + (x_s(T_{wb}) - x) \cdot \underbrace{c_w \cdot T_{wb}}_{h_w^*} = \underbrace{c_{da} \cdot T_{wb} + x_s(T_{wb}) \cdot (L + c_v \cdot T_{wb})}_{h_s^*} \Rightarrow$$

$$x \cdot (L + c_v \cdot T_{db} - c_w \cdot T_{wb}) = c_{da} \cdot (T_{wb} - T_{db}) + x_s(T_{wb}) \cdot (L + c_v \cdot T_{wb} - c_w \cdot T_{wb}) \Rightarrow$$

$$x = \frac{(L + (c_v - c_w) \cdot T_{wb}) \cdot x_s(T_{wb}) - c_{da} \cdot (T_{db} - T_{wb})}{L + c_v \cdot T_{db} - c_w \cdot T_{wb}} \qquad \text{kg kg}^{-1} \quad (4.12)$$

$$x = \frac{(2501 - 2.381 \cdot T_{wb}) \cdot x_s(T_{wb}) - 1.005 \cdot (T_{db} - T_{wb})}{2501 + 1.805 \cdot T_{db} - 4.186 \cdot T_{wb}}$$

Equation 4.12 is exact since it defines the thermodynamic wet bulb temperature T^*.

4.2.1 Heating

Whatever the energy source, heating has to be distributed into a greenhouse. The heat can be transported by fans, hot pipes or hot wires. Figure 4.8 gives an example of a direct convection heater. The rise of the air temperature depends on the added energy and the airflow.

The added energy at a known change of enthalpy for a given mass flow of ventilation air can be calculated with:

$$E = f_m \cdot \Delta h \cdot \frac{1000}{3,600} \text{ (kg h}^{-1}\text{ m}^{-2} \cdot \text{kJ kg}^{-1} \cdot \text{h ks}^{-1}) \qquad \text{W m}^{-2} \quad (4.13)$$

Where E is the energy input flux in W m^{-2}, f_m is the air mass flux in kg h^{-1} m^{-2}. The volume flow of ventilation air f_v is obviously the ratio between mass flow and air density:

$$f_v = f_m / \rho \qquad \text{m}^3\text{ h}^{-1}\text{ m}^{-2}$$

90

Greenhouse horticulture

Figure 4.8. A direct convection heater with air recirculation.

For a given energy input the enthalpy change can be calculated:

$$\Delta h = \frac{E}{f_m} \cdot 3.6 \qquad \text{kJ kg}^{-1} \quad (4.14)$$

In the Carrier chart (Figure 4.6) the enthalpy can be found by moving horizontally to the right since the humidity ratio is not affected. Equation 4.14 delivers enthalpy change Δh, which is all sensible heat. The increase in temperature generated by heating is found by projecting the vector connecting inlet enthalpy h_{in} with outlet enthalpy $h_{in} + \Delta h$ on the horizontal temperature axis. So only the intersection of the horizontal line (constant humidity ratio x) with the new enthalpy line has to be found to know the temperature T_{db} and other air properties of the air leaving the heater.

4.2.2 Mixing of air volumes

In the greenhouse colder air from outside is often mixed with warmer inside air, for instance by air mixing units connected with an air distribution system. The enthalpy of the mixed air is:

$$h_m = \frac{f_A \cdot h_A + f_B \cdot h_B}{f_A + f_B} \qquad \text{kJ kg}^{-1} \quad (4.15)$$

Equation 4.15 states that the enthalpy of the air mix h_m is the weighted average of the two fluxes (f_A and f_B). The unit of the mass flow is not important as long as they are the same. The mass flows may be substituted by air masses, which occurs for instance when air below and above a thermal screen mix after the screen is opened. The air mass above a closed energy screen is cold, and below it is warm. Opening the screen will mix the two air masses. The equilibrium enthalpy can be determined, if the volume of the two air masses is known.

To know the equilibrium air condition fully, we need at least two properties of the mixed air. The same equation as Equation 4.15 applies for the humidity ratio of the air mix:

$$x_m = \frac{f_A \cdot x_A + f_B \cdot x_B}{f_A + f_B} \qquad \text{kg kg}^{-1} \quad (4.16)$$

Enthalpy and humidity ratio of the mixed air define one point in the Psychometric chart. Figure 4.7 allows you to read the other air properties. If we apply the same equation for the temperature of the mixed air, we will have a small error because of the increasing slope of the isotherms (Figure 4.4). Though the deviation is small it is not advised to use the mixing equation for temperature T_{db}. Instead, derive T_{db} from Equation 4.5 or read it from the chart.

A mixing problem can also be solved using the Psychrometric chart only. This is illustrated in Figure 4.9, where the condition of the mixing air is found based on vector length.

Please note: in Figure 4.10, $f_A / (f_A + f_B)$ is the fraction of air A fr_A, just as $f_B / (f_A + f_B)$ is the fraction of air B fr_B in the total flow, and that $fr_A + fr_B = 1$.

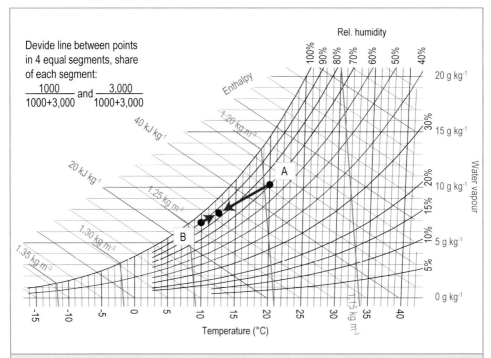

Figure 4.9. Mixing of two airflows A: 1000 kg s⁻¹ warm air (T_{db} = 20 °C, RH = 70%) and B: 3,000 kg s⁻¹ cold air (T_{db} = 10 °C, RH = 90%). The mixing point will be on the line connecting the points A and B. Because of the ratio of the two mass flows this point will be on one quarter of the total length of the line, closest to the condition of the cold air, i.e. the largest airflow of 3,000 kg s⁻¹. This concept is supported by the calculation shown in Figure 4.10.

$$\left|x_A - x_m\right| = x_A - \frac{f_A \cdot x_A + f_B \cdot x_B}{f_A + f_B}$$

$$= \frac{x_A \cdot (f_A + f_B) - f_A \cdot x_A + f_B \cdot x_B}{f_A + f_B}$$

$$= \frac{f_B \cdot (x_A - x_B)}{f_A + f_B}$$

$$\frac{\left|x_A - x_m\right|}{x_A - x_B} = \frac{f_B}{f_A + f_B}$$

$$\left|x_B - x_m\right| = x_B - \frac{f_A \cdot x_A + f_B \cdot x_B}{f_A + f_B}$$

$$= \frac{x_B \cdot (f_A + f_B) - f_A \cdot x_A - f_B \cdot x_B}{f_A + f_B}$$

$$= \frac{f_A \cdot (x_B - x_A)}{f_A + f_B}$$

$$\frac{\left|x_B - x_m\right|}{x_A - x_B} = \frac{f_A}{f_A + f_B}$$

$$\frac{\left|x_A - x_m\right|}{\left|x_B - x_m\right|} = \frac{f_B}{f_A}$$

Figure 4.10. Maths behind graphic solution of mixing two air flows of different conditions. The vectors are projected on the humidity ratio axis x. The x-component vector length represents $|x_A - x_m|$ or $|x_B - x_m|$.

4.2.3 Fogging

The nozzles of a fogging system spray water in very fine droplets (as shown in Figure 4.11), causing it to evaporate. This process humidifies and cools the greenhouse air at the same time. It is called adiabatic cooling: no energy enters or leaves the system concerned. To find the temperature and humidity we move up along the isenthalpic line. The amount of water added, mass flow f_w in g s^{-1} or mass m_w in g, and the dry air mass flow f_m in kg s^{-1} or mass of air m_{da} in kg determine the increase in air humidity in g kg^{-1} dry air and how far the vector follows the isenthalpic line according to $\Delta x = f_w / f_m$ or $\Delta x = m_w / m_{da}$ in g kg^{-1}.

The maximum end point of vector A-B in Figure 4.12 is of course the intersection with the 100% RH line. Then the greenhouse air is cooled down to the wet bulb temperature where no more water can evaporate. If more water is supplied, droplets will remain in the air. When fogging is active, ventilation rate is minimised.

Figure 4.11. Example of a fog system actively spraying fine droplets in a greenhouse.

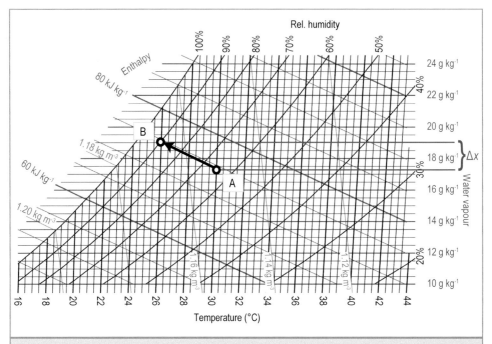

Figure 4.12. Process of adiabatic cooling as the result of a fogging system. A warm airflow of 1 kg s⁻¹ with condition A (T_{db} = 30 °C, RH = 65%, h = 74.5 kJ kg⁻¹, x = 17.3 g kg⁻¹) receives a fogging water flow f_w equal to 1.7 g s⁻¹ which results in Δx = 1.7/1 g kg⁻¹ and Δh ≈ 0. So, after fogging, the new condition B is (T_{db} = 26 °C, RH = 90%, h =74.5 kJ kg⁻¹, x = 19.0 g kg⁻¹).

4.2.4 Forced cooling and dehumidification with a cold surface

When a surface in the greenhouse is below dew point, condensation will take place: the air will get drier. The decrease in temperature depends on how much energy is taken away from the air, which in turn depends on the air condition, T_{db} and RH, and temperature of the cold surface (Figure 4.13).

4.2.5 Pad and fan cooling

This system is used a lot in regions with a warm and dry climate. A fan forces the outside air through a wet pad (often cellulose). The air picks up moisture and is cooled. There are horizontal and vertical pads kept wet by a recirculating water flow. If no energy is added or withdrawn from the water, it will assume the wet bulb temperature. As the air leaving the pad will be (at best) as cold as the wet bulb of external air, a large wet bulb depression is needed for cooling by pad and fan.

So how much can warm and dry air be cooled down? Hot desert air of 45 °C and 10% RH, can theoretically be cooled down to the wet bulb temperature of close to 21 °C and reach

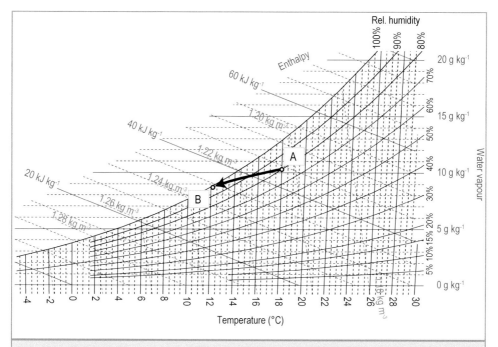

Figure 4.13. Dehumidification on a cold surface. In this example air A has condition (T_{db} = 18 °C, *RH* = 80%, *h*= 44 kJ kg⁻¹), airflow is 10 kg s⁻¹ and the capacity of heat removal is 100 kW. Enthalpy removal is 100/10= 10 kJ kg⁻¹. This enthalpy removal leads to condensation. So, we have saturated air. The resulting air condition B is found at the intersection of *RH*=100% and the isenthalpic line of 34 kJ kg⁻¹.

saturation $x_s(T_{wb})$. This air condition can be found by following the isenthalpic line until the 100% RH is reached. In reality the temperature drop will be less because contact between air and water is relatively short. An example is given in Figure 4.14. Chapter 9 gives details on how to design a pad and fan system. Figure 4.15 gives an indication of pad and fan cooling and water use over time in a simulation for Almeria, Spain.

4.2.6 Sensible and latent heat load

The last psychrometric process dealt with in this book is that of heat load. In greenhouses the solar energy causes a sensible heat load but also a latent heat load because the crop transpires. The net energy added to the greenhouse is the transmitted solar radiation in W m⁻². Greenhouse air will transport a substantial part of this energy resulting in Δh_{sens} kJ kg⁻¹ sensible energy increase and Δh_{latent} kJ kg⁻¹ latent energy increase as a result of humidity increase. Equation 4.14 converts an energy flow and a ventilation flow to a change in enthalpy. The ratio between change in sensible and latent energy increase determines the direction of the line in the psychrometric chart that shows the direction of change of the air condition.

$$\frac{\Delta h_{sens}}{\Delta h_{latent}} = \frac{c_{da} \cdot \Delta T_{db} + c_v\,(x_2 T_{db,2} - x_1 T_{db,1})}{\Delta x \cdot L} \qquad \text{kJ kg⁻¹} \quad (4.17)$$

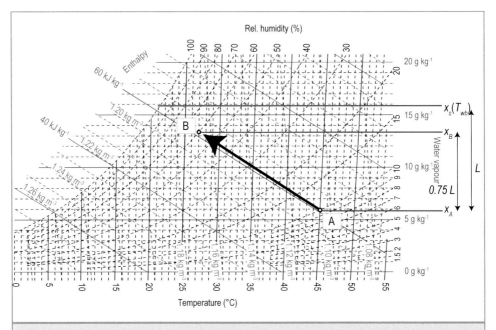

Figure 4.14. Carrier chart showing adiabatic cooling with a pad and fan cooling system. The example shows cooling of desert air A with condition (T_{db} = 45 °C, RH= 10%) with a saturation efficiency of 75%. This means outlet humidity ratio $x_B = x_A + 0.75 \cdot (x_s(T_{wb}) - x_A)$ or the actual vector length is 0.75 times maximum vector length L.

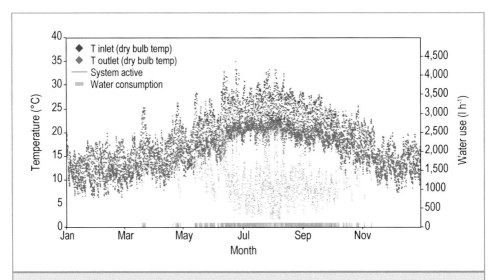

Figure 4.15. The effect of pad and fan cooling on the greenhouse inlet air temperature when activated at outside temperature T_{db} = 23 °C in Almeria (Spain). Left y-axis T_{db} (°C), Right y-axis, water consumption in l h^{-1} ha^{-1}.

Δh_{sens} is the net sensible energy taken up by the greenhouse air per kg dry air, represented by the horizontal red vector in Figure 4.16, and Δh_{latent} is the net latent energy taken up by the greenhouse air, which is represented by the vertical red vector in Figure 4.16. The green vector indicates the resulting change in air condition as the result of a heat load.

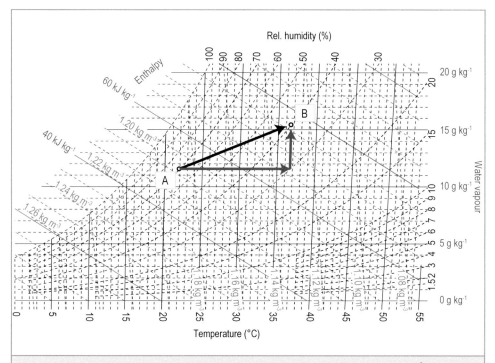

Figure 4.16. Effects of net sensible and net latent heat loads on greenhouse air. An airflow of 1 kg s⁻¹ with condition A (T_{db} = 22 °C, RH = 70%, x = 11.55 g kg⁻¹) changes to condition B (T_{db} = 35.6 °C, RH= 40%, x = 15.44 g kg⁻¹). It receives a sensible heat load of 25 kW (15 kW sensible heat and 10 kW latent heat).

4.2.7 Heat recovery

Heat recovery by heat exchangers is a combination of forced cooling and heating. With equal airflow rates, the enthalpy decrease in the cooled air is equal to the enthalpy increase in the heated air. If the counterflow airflow rates are not equal, the conservation of energy indicates that $f_{m1} \cdot \Delta h_1 = f_{m2} \cdot \Delta h_2$. An additional criterion to the temperature efficiency of a heat recovery system (see Chapter 8) defines the condition of the used outlet air of the heat exchanger.

4.3 Concluding remarks

For certain scientific and experimental work and in practice, particularly in the field of air conditioning, heat transfer and climate technology, some air properties that are not measured are required. Psychrometrics provides both calculation and reading of all relevant air properties of a measured air condition. If two properties are known, other properties can be calculated or read from the psychrometric chart. When air treatment processes are classified properly, air treatment processes can be designed with the help of psychrometrics. Both the (change in) air condition after a treatment and the required thermal and (de)humidification actions for a desired air condition can be quantified. A major limitation is that psychrometrics focusses exclusively on air. The effects of energy and mass transport in other components of the greenhouse are not considered. The energy and mass balance of a greenhouse do take into account these other components and are dealt with in the following chapters.

Chapter 5
Ventilation and mass balance

Photo: Paolo Battistel, Ceres s.r.l.

We have seen that a greenhouse is a solar collector: the energy from solar radiation warms everything up, very often too much. Ventilation is the process by which temperature is controlled. Excess energy is discharged by exchanging hot and humid air from inside the greenhouse with the air outside, which has a lower enthalpy. Ventilation is the cheapest, and often the most effective, means of controlling the temperature.

5.1 Introduction

Ventilation usually replaces warm, humid air with colder, drier air. Besides the obvious removal of heat, the fact that vapour is also removed from the greenhouse greatly increases the effects of ventilation. We need to remember that a large amount of energy is used for the phase change from water to vapour (the latent heat, usually indicated by $L \cong 2,500$ J g^{-1}) and as long as the vapour is in the greenhouse, it may condense and release that energy. When it is removed, the corresponding amount of energy is also taken out of the greenhouse. Therefore, with the greenhouse temperature T_{in} and vapour concentration χ_{in} (g m^{-3}), and the outside air temperature T_{out} and vapour concentration χ_{out}, the energy removed through ventilation can be calculated using the enthalpy difference:

$$energy = \rho c_p (T_{in} - T_{out}) + L (\chi_{in} - \chi_{out}) \qquad \text{J m}^{-3} \quad (5.1)$$

where ρ is the density of air $\cong 1.2$ kg m^{-3} and c_p is the specific heat of air $\cong 1000$ J kg^{-1} K^{-1}. The volumetric heat capacity of air ρc_p is thus approximately 1,200 J m^{-3} K^{-1}. In order to know the amount of energy that can be carried out of the greenhouse in a given amount of time, E (energy) must be multiplied by the specific ventilation rate g_V in m s^{-1} (m^3$_{air}$ m^{-2}$_{soil}$ s^{-1}) (Box 5.1).

When air exchange is operated by fans (usually placed on the side walls or gables of a greenhouse), we speak of forced/mechanical ventilation. On the other hand, natural ventilation (which is by far more common) relies on two natural processes creating the pressure differences required for the air to be exchanged: buoyancy and wind pressure.

Box 5.1. Definitions of the ventilation rate.

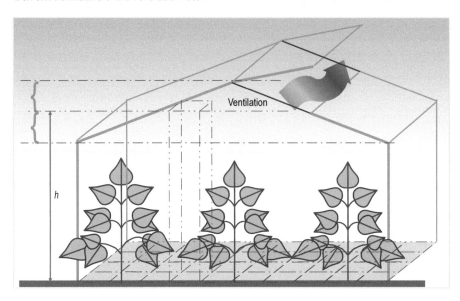

The ventilation rate of a greenhouse can be expressed in several ways: total ventilation in cubic metres per hour ($m^3 h^{-1}$); air changes per hour, expressing how many times per hour the volume of the greenhouse is refreshed (h^{-1}); and the specific ventilation rate, g_V which is given in $m^3_{air} m^{-2}_{soil} s^{-1}$. If we know the volume of the greenhouse (soil surface times the mean height), all definitions are interchangeable. Therefore, to convert from air changes per hour to the specific ventilation rate, we need to multiply by the volume of the greenhouse, divide by the soil surface (this is equal to multiplying by the mean height of the greenhouse) and divide by 3,600, the number of seconds in an hour.

5.2 Ventilation performance

The actual ventilation at a given time (the ventilation rate) depends on the ventilation capacity and the prevailing conditions. The capacity depends on the amount, size, form and position of the vents. The prevailing conditions that need to be considered are the actual opening of the vents, the wind speed and wind direction and the difference in temperature and humidity between inside and outside.

Even if there is no wind, there will still be an air exchange due to the buoyancy, also called the 'chimney effect'. The greenhouse air is usually lighter than the outside air, since it is both warmer and more humid. As we know, air is a mixture of various gases and behaves similarly to a perfect gas with the molecular weight M_{air} = 29. Water vapour (M_{H_2O} = 18) is lighter. In places where there is no (or little) wind, greenhouses need to be designed to take advantage of this, i.e. like chimneys, with openings on top (Figure 5.1). This is, however, not enough, since for the air to escape there must also be side openings (as low as possible) to let air in.

Figure 5.1. Greenhouses in tropical low land (near Kuala Lumpur, Malesia), designed (Campen, 2005) to take advantage of buoyancy in the absence of wind (photo: Anne Elings, Greenhouse Horticulture, Wageningen University & Research).

Given the importance of the side inlet, such greenhouses cannot be wider than 50-60 m at most (Baeza *et al*., 2008).

Whenever wind speed exceeds about 2 m s^{-1}, the effect of wind pushing air through (or sucking air from) the vents becomes dominant (Baeza *et al*., 2014). The ventilation rate increases linearly with wind speed (Figure 5.2). In this case as well, it is the size and the position of the openings that predominantly determines the ventilation capacity (e.g. Box 5.2). Vents facing the wind are more efficient (there is more air flow per unit of vent surface) than vents on the leeward side of the roof. However, vents facing the wind are also more likely to be damaged by gusts of wind and the penetration of air may be non-homogeneous, meaning that leeward flaps are most commonly preferred and are the first to be used in greenhouses that have computer-controlled flaps on both sides of the roof.

As the left panel of Figure 5.2 shows, side wall openings do contribute to wind-driven ventilation. It goes without saying, however, that their importance decreases with the size of the greenhouse, mainly because the openings are on each span, and the surface available for roof openings therefore increases more than for side wall openings. Figure 5.3 shows the effect of increasing the number of spans in a greenhouse when the ventilation is purely wind-driven (Kacira *et al*., 2004) or purely buoyancy driven (Baeza *et al*., 2009).

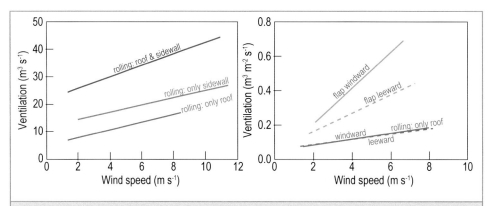

Figure 5.2. Ventilation rates measured in a Spanish parral greenhouse (soil surface 882 m^2; roof slope 12°) with various configurations of the vents, as indicated. Left: total ventilation, right: ventilation per unit of surface vents. Windward and leeward ventilation of roof rolling ventilators were not significantly different (redrawn from Pérez Parra *et al.*, 2004).

Box 5.2. Local climate and ventilation capacity.

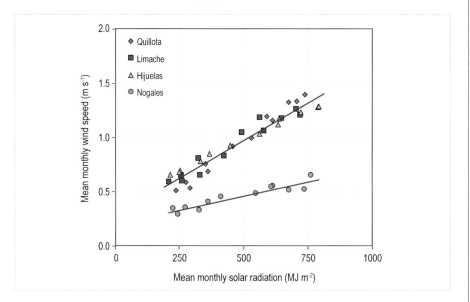

The graph above shows the monthly mean wind speed vs the monthly mean sun radiation for four villages within 30 km of each other in Central Chile. Vegetables and flowers for nearby Santiago are produced in greenhouses there. All greenhouses there look very much the same, see for instance Figure 6.2, right. Is this wise?

To answer this question it is helpful to observe that each point of the plot represents one month in the indicated village. The x-axis indicates the 'energy load' of the greenhouse and the y-axis the 'ability' of air to carry that load out of a vent.

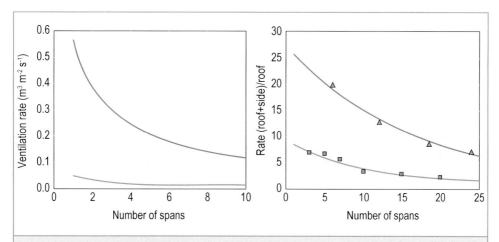

Figure 5.3. Compilation of computational fluid dynamics (CFD) calculations on buoyancy-driven (red, from Baeza *et al.*, 2009) and wind-driven (blue, from Kacira *et al.*, 2004) ventilation. Left, the effect of the number of spans on the ventilation rate when both roof and side vents are present. Right, the ratio of the ventilation rate with both roof and side vents to the rate with only roof vents. The points are calculated, the lines are best-fit. The assumed shape of the greenhouse (width of spans, size of vents) differed in the two papers.

Less obviously, the slope of the roof also has an effect up to about 30°, as shown by Baeza, 2007, Figure 5.4. The conclusion was that increasing the roof slope could greatly improve ventilation performance of the parral greenhouses typical in the region of Almeria, whose slope is low, about 12°.

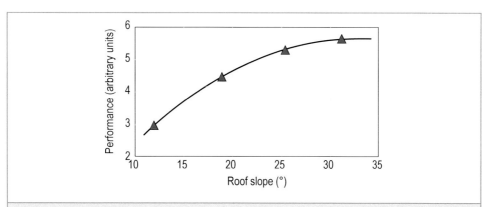

Figure 5.4. Effect of roof slope on the ventilation performance of a greenhouse, determined through CFD (computational fluid dynamics). The y-axis shows the linear coefficient of the ventilation flow vs wind speed function. Drawn from Baeza, 2007.

A very important determinant of ventilation capacity is whether or not the openings are fitted with nets against insects. Perez Parra *et al.* (2004) combined their own measurements of the reduction of the ventilation rate caused by a screen with the published results of three authors to create an appealingly simple equation of the porosity of the screen, ε:

$$\frac{g_{V,screen}}{g_V} = \varepsilon\,(2 - \varepsilon) \tag{5.2}$$

where $g_{V,screen}$ and g_V are the ventilation rate with and without a screen, respectively, and ε is the area of holes per unit area of screen, which is dimensionless. The reciprocal of the right-hand side of Equation 5.2 is the factor by which the ventilation area should be increased when insect screens of porosity ε are installed, in order not to reduce the ventilation rate. Typically, the porosity of an anti-thrip screen is 25% and that of an anti-aphid screen 40%, which translates into a reduction of the ventilation rate by 56 and 36% respectively.

Finally, the ventilation rate at a given wind speed depends, rather obviously, on the measure of the opening of the vents. For flap vents, this is quantified by the opening angle, sometimes as a percentage of the maximal opening. In fact, the ventilation rate is usually less than linear with the opening angle, as Figure 5.5 shows. A tool for calculating the ventilation rate of a multi-tunnel, depending on wind speed and characteristics of the vents is available on line (Muñoz *et al.*, 2011).

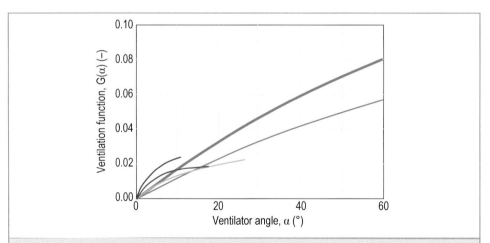

Figure 5.5. The ventilation function, usually denoted by G, describing the relationship between the ventilation flux per unit window area of the cover surface and the wind speed vs the opening angle (α) of the roof vents. The blue lines refer to parral greenhouses in Almeria; thick is windward and thin is leeward (redrawn from Perez Parra *et al.*, 2004). The red, violet and green lines refer to Venlo glasshouses and various types of roof vents and much more limited opening angles (redrawn from Stanghellini and De Jong, 1995).

5.3 The temperature in a ventilated greenhouse

We are now able to write the energy balance of a ventilated greenhouse by adding the energy carried out of the greenhouse by ventilation to Equation 3.10:

$$\tau I_{sun} = F_{tunnel,sky}\, \varepsilon_{cover}\, \alpha_R\, (T_{in} - T_{sky}) + \frac{A_{cover}}{A_{soil}}\, \alpha_{cover}\, (T_{in} - T_{out})$$

$$+ F_{tunnel,soil}\, \varepsilon_{soil}\, \alpha_R\, (T_{in} - T_{soil}) + \alpha_{soil}\, (T_{in} - T_{soil}) \qquad \text{W m}^{-2} \quad (5.3)$$

$$+ g_V\, [\rho c_p\, (T_{in} - T_{out}) + L\, (\chi_{in} - \chi_{out})]$$

where most symbols have been defined above and:

T_{in}	K or °C	Temperature inside the ventilated greenhouse	
ρc_p	J m^{-3} K^{-1}	Volumetric heat capacity of air	Equation 4.7 and 4.11
g_V	m s^{-1}	Specific ventilation rate	after Equation 5.1
L	J g^{-1}	Latent heat of evaporation of water	Equation 4.5
χ	g m^{-3}	Vapour concentration of air, respectively in- and out-side	Table 4.1

Adding and subtracting T_{tunnel} to the terms containing T_{in} then grouping the terms in order to extract Equation 3.10, then adding and subtracting T_{out} and rearranging, we get:

$$\frac{T_{in} - T_{out}}{T_{tunnel} - T_{out}} = \qquad\qquad\qquad\qquad\qquad\qquad\qquad\qquad\qquad (5.4)$$

$$\frac{F_{tunnel,sky}\, \varepsilon_{cover}\, \alpha_R + \dfrac{A_{cover}}{A_{soil}}\, \alpha_{cover} + F_{tunnel,soil}\, \varepsilon_{soil}\, \alpha_R + \alpha_{soil}}{F_{tunnel,sky}\, \varepsilon_{cover}\, \alpha_R + \dfrac{A_{cover}}{A_{soil}}\, \alpha_{cover} + F_{tunnel,soil}\, \varepsilon_{soil}\, \alpha_R + \alpha_{soil} + g_V \left[\rho c_p + L\, \dfrac{(\chi_{in} - \chi_{out})}{(T_{in} - T_{out})}\right]}$$

Equation 5.4 represents a hyperbola which obviously starts at one when the ventilation rate g_V is zero (since this is our definition of *tunnel*), and asymptotically approaches 0 ($T_{in} = T_{out}$) when the ventilation rate becomes very large (Box 5.3). The trend between these two extremes depends on the properties of the cover and on the ratio of latent to sensible heat difference between inside and outside air (see Section 4.1.3). The slope of the hyperbola is largest when g_V is near 0: for example, an increase of g_V from 0.01 to 0.02 m s^{-1} lowers the temperature in the greenhouse by much more than an increase from 10.01 to 10.02 m s^{-1}.

In summary, the difference in temperature between the greenhouse and outside increases with input of solar radiation and decreases with thermal radiation loss, the ventilation rate, the heat transfer coefficient of the cover and the difference in humidity between inside and outside.

Box 5.3. Shading and ventilation.

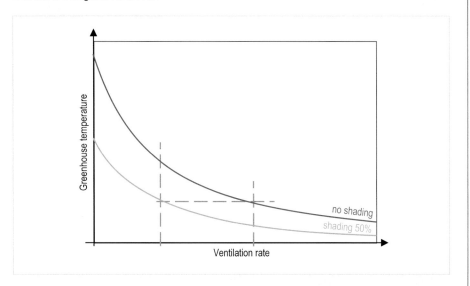

Whenever there is sufficient ventilation, there should be no need for shading. In the example shown here, the red line represents the trend of temperature with ventilation in a greenhouse, and the blue line shows when shading is applied to reduce input of solar radiation by 50% (all else being equal). The temperature drop that is obtained by shading at a given ventilation rate could have been attained even without shading by adequately increasing the ventilation rate.

5.4 Mass balance

It has already been said implicitly that ventilation not only removes energy from the greenhouse, it also removes vapour. In fact, whenever there is a difference in the concentration of any gas (not only water vapour, but also, for instance, carbon dioxide) between the air inside and outside the greenhouse, ventilation will also move the gas from the high to the low concentration. In other words, there is a mass flow of the gas, M_{gas}, with the ventilation flow.

$$M_{gas} = g_V (C_{in} - C_{out}) \qquad\qquad \text{kg m}^{-2} \text{ s}^{-1} \quad (5.5)$$

where g_V [m s^{-1}] is the specific ventilation rate defined above and C indicates a volumetric concentration, *in*side and *out*side the greenhouse respectively. If the concentration is given in kg m^{-3}, the mass flow will be in kg m$^{-2}_{\text{soil}}$ s^{-1}. However, very often kg is not the most obvious unit to use. For instance, above we used g m^{-3} for water vapour in the air, since in natural conditions air will contain grams (and not milligrams or kilograms) of water vapour. Another frequently used unit is the molar concentration (mol m^{-3}), which is equivalent. The two are interchangeable once the molecular weight of the gas is known. It is also possible to indicate concentration as the fraction of volume occupied by the gas (m^3_{gas} m^{-3}_{air}), which is again equivalent, thanks to the ideal gas law. Finally, since the concentration of most gases in the air (except nitrogen) is small, $\mu mol\ mol^{-1}$ or $cm^3\ m^{-3}$ (which are equivalent) are often used, sometimes indicated as vpm: volume parts per million.

As with energy, there is a general law for the conservation of mass: in a steady state, the mass of a substance entering a greenhouse must equal the mass leaving it. When conditions change, in the time that it takes to reach a new equilibrium mass is either stored or released in/by the greenhouse air, that is, the concentration of the substance in the air will change until a new equilibrium is reached (Figure 5.6). As Equation 5.5 shows, the mass fluxes leaving (or entering) the greenhouse depend on the difference in concentration between the air inside and outside the greenhouse, whereas the transpiration and assimilation depend on the difference between the concentration within the crop and the air of the greenhouse. Therefore, it can be understood that the concentration in the greenhouse is the variable that changes in order to maintain the mass balance (Box 5.4).

The primary function of ventilation in a greenhouse is to get rid of excess energy. However, water vapour is usually also released in the process, and carbon dioxide is let into or out of greenhouses without and with a CO_2 supply, respectively (Box 5.5).

Indeed, in the absence of carbon fertilisation, all three of the above functions are very good reasons for ventilation, even if the energy balance (i.e. air temperature in the greenhouse) is usually the only one that most climate control systems would consider. Most of the time there would be no conflict, except on sunny, cold winter days, when ventilation would not be wise in unheated greenhouses. However, significant CO_2 depletion may then occur, as discussed by

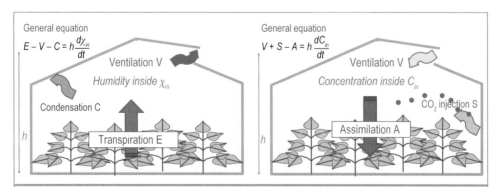

Figure 5.6. Schematic representation of the water vapour (left) and carbon dioxide (right) balance in a greenhouse. 'h' represents the mean height of the greenhouse. The crop is a source of water vapour and a sink of carbon dioxide. Besides ventilation, condensation can remove vapour from the air. In the absence of carbon dioxide enrichment, all the carbon dioxide that is assimilated by the crop must come from outside, through the vents.

Box 5.4. Ventilation rate and carbon dioxide balance.

The ventilation rate affects the photosynthesis rate. Of course this happens indirectly, through the concentration of carbon dioxide within the greenhouse. This results from the balance of all mass flows: the assimilation by the crop, the carbon dioxide inflow or outflow through the vents and, possibly, the supply rate of carbon dioxide.

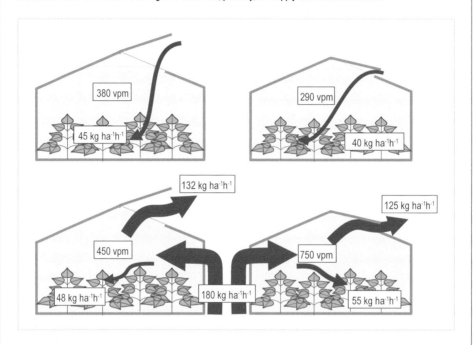

In the example above the crop assimilation is shown, as it follows from the carbon dioxide balance for a greenhouse without (top) and with carbon supply rate of 180 kg ha^{-1} h^{-1} (bottom), in the case that air is refreshed 20 h^{-1} (left) and 3 h^{-1} (right). A mean height of the greenhouse of 6 m and an external concentration of 400 vpm have been assumed. For the assimilation rate, Equation 12.3 has been used, with A_{MAX} = 72 kg ha^{-1} h^{-1}, non-limiting light, and the response to carbon dioxide concentration [CO_2] from Stanghellini *et al.* (2008):

$$m([CO_2]) = \frac{[CO_2]}{[CO_2] + 230}$$

Stanghellini *et al.* (2008). They quantified the yield loss for a winter crop in the Mediterranean region as a consequence of either limiting ventilation and allowing CO_2 depletion or ventilating (and cooling) the greenhouse in order to prevent depletion and estimated yield loss in both cases around 10% of the yield that could be attained by keeping CO_2 at ambient level and temperature at the level one could get without ventilation.

Box 5.5. How to 'measure' ventilation.

The ventilation rate under given (steady state) conditions can be measured through the balance of a tracer gas, Equation 5.5. There are two techniques: the 'continuous injection method' measures the flow that is required to maintain a pre-set constant concentration inside the greenhouse, whereas the 'decay method' measures the time that it takes for the concentration to drop to a given fraction of an initially high concentration (Baptista *et al.*, 1999). Most literature reporting ventilation models (for instance, Figure 5.2 and 5.5) is based on these kinds of measurements repeated under varying conditions (wind speed, vent types, vent openings).

Another approach is computational fluid dynamics (CFD), estimated by means of computer packages that account for the relevant physical processes (also on a very small scale) and defining the geometry of the greenhouses, vents and boundary conditions such as wind, temperature and radiation load. Such methods have been used in the literature referred to in Figure 5.3 and 5.4.

However, models based on either of these methods are imperfect predictors of the ventilation rate of a given greenhouse at a given moment, both because the geometry of the greenhouse and the vents may not be exactly the same as in the model applied and because of the 'noisy' nature of wind and its direction.

In principle, the energy balance Equation 5.3 and/or the mass balance of either water vapour or CO_2 (Figure 5.6) could be solved for the ventilation rate g_V. In fact, there are more unknowns (such as condensation, assimilation, energy exchange with the soil, etc.) that make this approach impractical. Bontsema *et al.* (2008) proposed reducing the problem of 'unknowns' by simultaneously solving the three equations online, using data available in the climate computer. Such a 'ventilation monitor' has been applied in the routine for the optimal supply of CO_2 (Stanghellini *et al.*, 2012b) and other process controllers that have recently been introduced in a commercial climate computer (Tsafaras and De Koning, 2017).

5.5 Concluding remarks: [semi]-closed greenhouses

In conclusion, ventilation is necessary but has its drawbacks. The most obvious one is, of course, whenever it prevents carbon fertilisation or limits the efficacy thereof. In addition, it consumes water that may be scarce, as Katsoulas *et al.* (2015) have shown, quantifying the inverse relationship between water-use efficiency and ventilation requirements of greenhouses. Open vents also allow a way in for unwelcome pests and a way out for predators that may have been introduced for biological control.

However, the largest drawback of ventilation is that it throws away excess energy that might be useful later. This refers to the day/night cycle: in most unheated greenhouses in the 'mild winter region', winter nights are often too cool for satisfactory production, whereas ventilation is often required during the day. The drawback of heated greenhouses is that heating is provided by burning fossil fuels. Yet, even at the latitude of the Netherlands (52°N), greenhouses have a yearly surplus of energy i.e. the energy that is carried out by ventilation (Figure 5.7) is more than the energy that is supplied by heating. The problem is the phase shift: there is an excess of energy when there is sunshine and heating is usually required in the dark and on cold days.

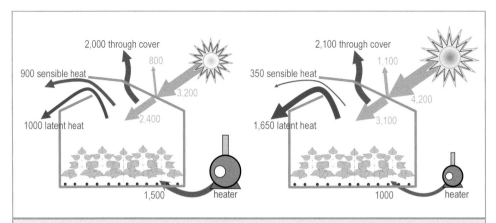

Figure 5.7. Typical energy budget (MJ m^{-2} y^{-1}) of a reasonably insulated glasshouse in the Netherlands (left) and a plastic greenhouse in Central Italy (right).

Of course, in order to limit ventilation, there must be a means of getting rid of the excess energy that would otherwise escape through the vents. This can be achieved through cooling and dehumidification (Chapter 9). Moreover, in order to close the loop, at least a fraction of what is 'cooled down' must be stored and retrieved when needed. This is the reasoning behind the 'closed greenhouse'. Instead of using ventilation, heat is removed from the greenhouse air through cold water circulated in heat exchangers and stored in an underground aquifer (the 'warm well'). Then, in the winter, water is drawn from there, circulated through the heat exchangers to warm up the greenhouse air and thus cooled, and pumped into a 'cold well', which is then the source of cold water in the summer. As the temperature differences in the heat exchanger(s) would otherwise be small (one does not want the greenhouse to get warmer than, say, 28 °C or cooler than 15 °C), this is best done through heat pumps (see also Section 9.2). Getting rid of heat pumps is impractical, since it would require a lot of very efficient heat exchangers (Bakker *et al.*, 2006).

It is true that the cooling load may be reduced in order not to collect more energy than would be needed for heating (Box 9.4). In this case, ventilation is not fully prevented, but must be applied on top of cooling in the warmest hours. This is the energy-related reasoning behind semi-closed greenhouses, of which there are quite a few in commercial operation in the Netherlands.

In a review of experiments and the practical implementation of semi-closed greenhouses, De Zwart (2012) concluded that the largest benefits by far (with current prices of fossil fuels) are the increased yield that follows from the ability to maintain higher-than-ambient CO_2 concentrations (see also Figure 12.5), particularly in bright sunshine, and the fact that the highest cost is the electricity that the system requires.

Chapter 6
Crop transpiration and humidity in greenhouses

More than 90% of the water taken up by plants is transpired. The remaining fraction (less than 10%) is fresh weight growth. Transpiration, together with root zone conditions, is an important process influencing fresh weight. In addition, transpiration is the process whereby the crop has a very big effect on the greenhouse climate: the crop absorbs energy and uses part of it to transform water into vapour, thereby greatly affecting the energy budget of the greenhouse.

6.1 Transpiration from a leaf surface

The physics of the process of transpiration (energy balance; vapour and heat transfer) allow us to write an explicit equation for crop transpiration rate E, as firstly done by Penman (1948) and modified by Monteith (1965).

We will first analyse transpiration from a simple leaf surface, and then we will use it to determine crop transpiration by analogy.

The steady state energy balance (Figure 6.1, left) reads:

$$R_n = H + LE \qquad\qquad\qquad\qquad\qquad \text{W m}^{-2} \quad (6.1)$$

where:

R_n	W m^{-2}	is the net radiation of the leaf, that is the balance of intercepted and reflected sun radiation plus the balance of incoming and outgoing long-wave radiation
L	J g^{-1}	is the latent heat of vaporization of water, almost constant and about 2,450 for air temperature $T_a = 20\ °C$
E	g m^{-2} s^{-1}	is the evaporation flux density
H	W m^{-2}	is the density of sensible heat flux to or from the leaf surface

Transfer of sensible and latent heat (Figure 6.1, right):

$$H = \rho c_p \frac{T_s - T_a}{r_b} \qquad\qquad\qquad\qquad \text{W m}^{-2} \quad (6.2)$$

$$E = \frac{\chi_s - \chi_a}{r_s + r_b} \qquad\qquad \text{g m}^{-2}\text{ s}^{-1} \quad (6.3)$$

where:

ρc_p	J m^{-3} K^{-1}	is the volumetric heat capacity of air $\cong$ 1,200 J m^{-3} K^{-1}, being ρ the density of air ($\approx$ 1.2 kg m^{-3}) and c_p the specific heat ($\approx$ 1000 J kg^{-1} K^{-1});
T	K or °C	is temperature, the subscript a indicates air; s is for leaf surface;
χ	g m^{-3}	is the vapour concentration, the subscript a indicates air; s is for leaf surface;
r_b	s m^{-1}	is the resistance of the leaf boundary layer (Box 6.1);
r_s	s m^{-1}	is the stomatal resistance of the leaf (Box 6.2);

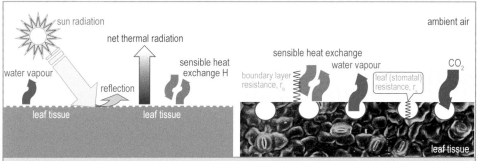

Figure 6.1 Schematic representation of the energy balance of a leaf surface (left) and of the mass and energy transfer to and from it (right). Net radiation is the balance of all short- and longwave incoming and outgoing fluxes; heat and vapour have to cross the layer of air affected by the presence of the leaf (the boundary layer). In addition, water vapour has to leave the stomatal cavities, which is represented by an additional resistance (the stomatal resistance). The photograph shows the stomatal openings in a real leaf (photo: Sjaak Bakker, Wageningen University & Research).

Box 6.1. Boundary layer resistance.

The boundary layer resistance decreases with wind speed and increases with leaf dimensions. At very low wind speeds air movement is created by the difference in temperature and humidity content between the leaf and the air (the chimney effect), the larger the difference, the smaller the resistance. Stanghellini (1987) determined the parameters of a model for the boundary layer resistance of a tomato crop in a greenhouse.

She also showed that the relatively small variations in air movement within a greenhouse ensure that the boundary layer resistance is fairly constant. In addition, she showed that in such greenhouse conditions, the transpiration rate is fairly un-sensitive to the boundary layer resistance, so that there is not much loss of accuracy in using a constant value for it.

Box 6.2. Stomatal resistance.

Stomatal resistance is known to change much during a day, particularly in response to light. Stomatal resistance is very high at night (stomata are closed), typically 1000 s m^{-1} or higher, and decreases with light to a crop-specific minimum value. Factors such as water shortage or high/low temperatures may cause [partial] stomatal closure, so that the minimum value may be never reached. Jarvis (1976) proposed a model of the form:

$r_s = r_{s,min} f_1 (I_{sun}) f_2 (T) f_3 \text{ (vapour deficit) } f_4 (CO_2)$

where $r_{s,min}$ is a crop-specific minimum value of the stomatal resistance and the functions f_{1-4} (always larger than 1) describe the increase of it in dependence of light, temperature, humidity and carbon dioxide concentration. A review of $r_{s,min}$ for many crops can be found in: Körner, *et al.* (1979).

It has been shown that often stomatal behaviour may be accurately reproduced by simpler models, not accounting for all four variables: the effect of CO_2 is usually small. In addition, given the high correlation between air temperature and vapour deficit in natural conditions, usually accounting for one of the two would suffice. Even worse, accounting for both may result in large overestimates of the resistance (and underestimates of transpiration) under high vapour pressure deficit (VPD) (e.g. Montero *et al.*, 2001). In addition, Fuchs and Stanghellini (2018) have advanced doubts about the functional dependence of stomatal resistance and vapour deficit.

What is sure is that the effect of sun radiation must be taken into account, and an additional dependence on either temperature or vapour deficit may add accuracy.

There is plenty of literature about empirical determination of the parameters of the functions f_{1-4} for all major greenhouse crops. When using the numbers, one should be careful not to extrapolate much from the conditions of the experiments in which the values were determined.

Even assuming that the properties of the leaf are known, there are more unknowns (temperature and vapour concentration of the leaf surface, and sensible and latent heat fluxes) than equations. A fourth equation is obtained by assuming that water is freely available in the sub-stomatal cavities, that is: that the evaporating surface is saturated at its temperature:

$$\chi_s \equiv \chi^* (T_s).$$

Since saturated vapour concentration has an exponential trend with temperature, the system formed by the 4 equations cannot be solved analytically. It can be done by numerical iteration, of course. Nevertheless, Monteith applied the linearization method described in Box 3.6, which is equivalent to using the first terms of the Taylor expansion with $T_s \approx T_a$

$$\chi_s \equiv \chi_s^* \cong \chi_a^* + \chi' (T_s - T_a) \qquad \qquad \text{g m}^{-3} \quad (6.4)$$

where $\chi' \equiv d\chi/dT_a$, [g m^{-3} K^{-1}], is the slope of the saturated vapour concentration curve at air temperature. $\chi' L$ [J m^{-3} K^{-1}] is the variation of latent heat of saturated air for a change of one unit in temperature. The corresponding variation in sensible heat is, obviously, the volumetric heat capacity of air ρc_p [J m^{-3} K^{-1}] and the dimensionless ratio of the two is indicated by the symbol ε (Box 6.3).

Box 6.3. The slope of saturated vapour functions.

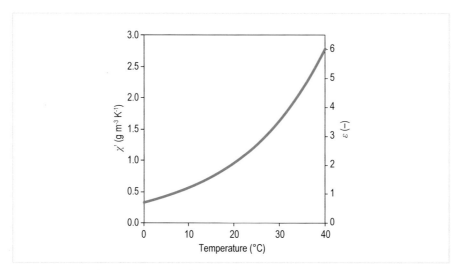

The slope of the saturated vapour concentration curve (χ', left axis) and the ratio of the slope of latent heat of saturated air to the slope of sensible heat (ε, right axis), plotted against the air temperature.

Useful approximate functions, for the range shown here are:

$\chi' \cong 0.36 \exp(0.05 T_a)$

$\varepsilon \cong 0.7156 \exp(0.0533 T_a)$

Substituting Equation 6.4 into Equation 6.3; then Equation 6.2 into Equation 6.1 and the new version of Equation 6.1 into the new version of Equation 6.3:

$$E = \frac{1}{r_b + r_s}\left[\left(\chi_a^* - \chi_a\right)+\chi'\frac{r_b(R_n - LE)}{\rho c_p}\right] = \frac{\chi_a^* - \chi_a}{r_b + r_s}+\varepsilon\frac{r_b}{r_b + r_s}\left(\frac{R_n}{L}-E\right) \Rightarrow$$

$$\Rightarrow \frac{r_b + r_s + \varepsilon r_b}{r_b + r_s}E = \frac{\chi_a^* - \chi_a}{r_b + r_s}+\frac{\varepsilon r_b}{r_b + r_s}\frac{R_n}{L} \Rightarrow$$

$$E = \frac{\chi_a^* - \chi_a + \varepsilon\, r_b \dfrac{R_n}{L}}{\left(1+\varepsilon\right)r_b + r_s} \qquad\qquad \text{g m}^{-2}\text{ s}^{-1} \quad (6.5)$$

Equation 6.5 shows that transpiration from a simple leaf surface increases both with the saturation deficit of the air (the ability of air to absorb the vapour that is produced) and with the radiation energy that is available. One could expect that transpiration would decrease with increasing resistances, and it certainly does with increasing stomatal resistance. The effect of boundary layer resistance is more complex since it appears also on the numerator. In fact, an increase in boundary layer resistance means that the leaf has to get warmer, in order to discharge a given amount of sensible heat, Equation 6.2. If it gets warmer (and water is still freely available in the sub-stomatal cavities), the vapour concentration difference between the leaf and the air increases, which should increase transpiration, at least in the measure that the increase of boundary layer resistance does not prevent it, Equation 6.3. This process is called the 'thermal feed-back', and ensures that transpiration is less than proportional to a change in either resistance. Indeed, if the stomatal resistance increases, transpiration would decrease, then sensible heat loss has to increase, Equation 6.1, which means that temperature needs to increase, Equation 6.2, and then the vapour concentration difference between the leaf and the air increases, which should increase transpiration, as described above.

6.2 Transpiration of a crop

Obviously the physics for transpiration of a surface are the same for a complex structure as a crop. The main difference is that there is some amount of leaf surface (with varying orientations) for heat, vapour and radiation exchange, for any square meter of soil surface. Leaf area index (LAI), is defined as the leaf area, m^2, per square meter soil, so it is dimensionless. Then we have 2LAI area per square meter soil for heat and vapour exchange. Some authors use only LAI for vapour exchange, since most crops have stomata only on one side of the leaves. However, since there is also some water loss from the leaf cuticula, not through stomata, and using different surfaces for vapour and heat exchange adds some complication, the most widely accepted representation is to use 2LAI as the surface exchanging both heat and vapour, and thus:

$$H = 2LAI\,\rho c_p \frac{T_{crop} - T_a}{r_b} \qquad \text{W m}^{-2} \quad (6.6)$$

$$E = 2LAI \frac{\chi_{crop} - \chi_a}{r_s + r_b} \qquad \text{g m}^{-2}\,\text{s}^{-1} \quad (6.7)$$

whereby an ideally homogeneous crop has been defined, with one temperature, T_{crop} and one vapour concentration χ_{crop}, and apparent stomatal and boundary layer resistances equivalent to the resistances of a single leaf, divided by 2LAI. This representation is often called 'big-leaf model'. It has some limits, but it has been shown to be adequate for estimating crop transpiration (see, for instance: Raupach and Finnigan, 1988).

The balance of energy must hold for a crop as well as for a leaf. However, the net radiation of a crop is not so easily determined as for a leaf surface. An approach suitable for field crops is to place a net radiometer (measures the difference between downward and upward radiation of all wavelengths; Box 3.7) far above the crop, and one (or more) meter(s) for heat flux into the soil. Then one can state that the amount of radiation absorbed by the crop is the difference between what is measured by the net radiometer and the soil flux meter. This would not work in a greenhouse: there may be an energy source (the heating system) between the two meters, besides the obvious limitation to having a net radiometer 'far enough' from the crop.

So the approach required for greenhouse crops is to determine the net amount of radiation (both short- and longwave) that is really absorbed by the crop. One way to measure it is to measure net radiation on a plane above and one below the crop. In view of the large variations (shadows, presence of heating elements, etc.) a large number of measurements are required to get a representative value. The other is to estimate it through a model that accounts for the relevant properties of the canopy (Box 6.4).

Once it is clear what is meant by it, one may retain the symbol R_n and the energy balance equation of a crop is formally the same as Equation 6.1, whereas the valid transfer equations

---

Box 6.4. Net radiation of a crop.

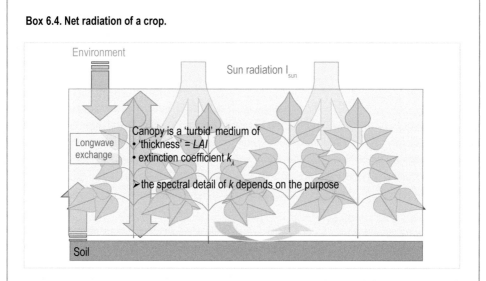

For the purpose of estimating radiation exchange, the canopy is represented as a 'turbid medium' (Ross, 1975). Transmission of sun radiation, I_{sun} is described similarly to what happens in a semi-transparent (turbid) medium, Equation 3.1, the *LAI* having the function of the thickness d of the medium. The coefficient of extinction for radiation of wavelength λ, k_λ, depends on the inclination of the leaves, and their reflectance.

$$\frac{I_{sun,below}}{I_{sun,top}} = e^{-k_\lambda LAI} \tag{6.8}$$

For instance, most crops have a high (≈50%) reflectance in the near infrared (NIR) and low (≈5%) in the photosynthetic active radiation (PAR), which means that NIR radiation penetrates deeper into a canopy (has a lower extinction coefficient) than PAR radiation.

As they are almost all water, leaves are 'black' for thermal radiation. In addition, as we have assumed one temperature, T_{crop}, for all the leaves, there is no net exchange of thermal radiation between leaves, and the exchange with 'the rest of the world' will be like that of a 'big leaf', at temperature T_{crop} with the shape of the horizontal projection of all leaves, that is called the 'soil cover'.

---

are Equation 6.6 and 6.7. The procedure used for getting Equation 6.5, gives the equation for the transpiration of a crop:

$$E = \frac{2LAI}{(1+\varepsilon)r_b + r_s}\left[\chi_a^* - \chi_a + \frac{\varepsilon\, r_b}{2LAI}\frac{R_n}{L}\right] \qquad\qquad \text{g m}^{-2}\,\text{s}^{-1} \quad (6.9)$$

where:

LAI		is the leaf area index of the crop;
ε		is the ratio of the latent to sensible heat content of saturated air for a change of 1 °C in temperature;
χ_a	g m^{-3}	is the vapour concentration of air. χ_a^* is saturated vapour concentration, a function of air temperature. The difference $\chi_a^* - \chi_a$ is thus the vapour concentration deficit of air;
L	J g^{-1}	is the latent heat of vaporization of water, very weakly dependent on temperature;
r_b	s m^{-1}	is the resistance to heat transfer of the leaf boundary layer;
r_s	s m^{-1}	is the stomatal resistance;
R_n	W m^{-2}	is the net radiation of the crop, that is the balance of intercepted and reflected solar radiation plus the balance of incoming and outgoing long-wave radiation.

Transpiration from a crop, increases with the vapour deficit of the air and with available radiation, as we have observed for transpiration from a simple leaf surface. However, the relative importance of the two factors is changed: soil cover is never 100%, so the exchange of radiation will be less than that of a uniform surface, whereas the relevance of air vapour deficit becomes large, in the measure that LAI exceeds 0.5, as the area 'pushing' vapour into air is [much] larger than the area exchanging radiation with the surroundings.

6.3 Calculating transpiration of a greenhouse crop

Stanghellini extended Ross's analogy to account for reflection both of the canopy itself and of the soil surface, which is relevant for a crop cultivated in rows, such as in greenhouses. She determined experimentally the relevant coefficients for a tomato crop on a white-mulched soil. Her final model can be simplified with little loss of accuracy to:

$$R_n = 0.86\,(1 - e^{-0.7\,LAI})\,I_{sun} \qquad\qquad \text{W m}^{-2} \quad (6.10)$$

where I_{sun} is global radiation at the top of the canopy. Since usually global radiation is measured outside the greenhouse, one has to multiply it by the transmissivity of the greenhouse, typically between 50 and 70%. Similar formulas with specific parameters for other crops can be found in the literature, such as: Baille *et al.* (1994) for roses, or Medrano *et al.* (2005) for cucumbers.

Soil cover (the projection of all leaves on the horizontal soil surface) is, in fact, the complement of the 'sun flecks' on the soil surface under a crop of fully black (non-reflecting, non-transmitting) leaves. Indeed, under such a crop, only the light that is not intercepted by any leaf would reach the soil. After Equation 6.8 we can write:

Fraction of intercepted radiation ≡ soil cover = $1 - e^{-k_{black}LAI}$ (6.11)

Ross (1975) has shown that the extinction coefficient of black leaves (k_{black}) is only dependent on their average angle of inclination, and values for it are reported in Table 6.1. One useful observation is that leaves are 'black' in the thermal infrared range, so the extinction coefficients in Table 6.1 apply for longwave radiation. The extinction coefficient in other wavebands is smaller, in the measure that leaves reflect and transmit radiation.

For the stomatal resistance, Van Beveren *et al.* (2015) proposed a slight modification of the model parametrised by Stanghellini (1987), in order to make it more general. The equation they proposed is:

$$r_s = 82 \, (1 + 6.95 \, e^{-\alpha \, I_{global}/LAI}) \, (1 + 0.023 \, (T_a - 20)^2) \qquad \text{s m}^{-1} \quad (6.12)$$

where I_{global} is radiation at the top of the canopy, possibly including supplemental light. The parameter a accounts for the different behaviour of crops when light increases, so it is crop-specific. Reportedly, for tomato $a = 0.4$ [m^2 W^{-1}], whereas roses have a less steep response $a = 0.008$. Typical values of $r_{s,min}$ for some greenhouse crops are given in Table 6.2.

In field conditions the boundary layer resistance is determined by wind speed, which is forcing the removal of vapour from the surface of leaves. However, air velocity in greenhouses is usually very small (5 to 50 cm s^{-1}) and not much variable. The fact that it is small ensures

Table 6.1. Theoretical extinction coefficient in a crop of black leaves, oriented with the angles shown (with respect to horizontal), subject to perfectly diffuse (Lambertian, Box 3.3) radiation. Spherical distribution means that all angles are equally appearing (such as the surface elements of a sphere) and conical means that leaves have a preferential inclination (compiled from Goudriaan, 1977; Monteith and Unsworth, 2014; Ross, 1975; soil cover is calculated through Equation 6.11).

Leaf angle distribution		Extinction coefficient	Soil cover with LAI=3
Horizontal		1	0.95
Conical	30°	0.87	0.93
	45°	0.83	0.92
	60°	0.54	0.80
Vertical		0.4	0.70
Spherical		0.68	0.87

Table 6.2. Typical values of the minimal stomatal resistance, the boundary layer resistance and leaf area index (LAI) of major greenhouse crops (Bakker, 1991; Stanghellini, 1987).

Crop	$r_{s,min}$ [s m^{-1}]	r_b [s m^{-1}]	LAI
Tomato	200	200	3
Sweet pepper	200	300	4
Cucumber	100	400	5
Eggplant	100	350	4

that the size of the leaves and, possibly, hair on their surface have a role in determining air movement around them, thus the boundary layer resistance. As explained in Box 6.1, the boundary layer resistance of a greenhouse crop can be taken as constant. Typical values calculated for some greenhouse crops are given in Table 6.2.

Weighing 'trays' to measure transpiration of greenhouse crops (Box 6.5) are increasingly available as optional 'sensors' with climate control systems. They provide useful additional insights that are much valued by growers, although the information is not (yet) used on-line by the controller.

Box 6.5. How can we measure crop transpiration?

Usually transpiration is determined by a lysimeter: that is enclosing the root zone, so that the water balance (irrigation and drain) can be determined. At equilibrium, irrigation water can either percolate or be transpired. As the amount of water in the root zone changes during a day, such a water balance lysimeter gives reasonable estimates of transpiration on period longer that one day.

Weighing a whole crop (including root medium) and measuring percolation is needed to determine transpiration on short intervals, as irrigation will also be weighed.

Of course water is also stored into the biomass (growth) and may be stored/depleted in/from the root zone. So, a system as the one shown here, that weighs independently the root medium and the crop, gives much more information (e.g. De Koning and Tsafaras, 2017).

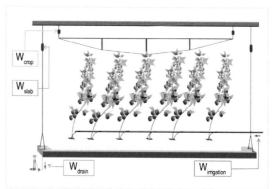

Drawing: Ad de Koning, Ridder Hortimax

6.4 Humidity in greenhouses

The presence of a transpiring crop has a huge impact on greenhouse climate: transpiration converts (sun) energy into latent heat, and thus mitigates the warming up of the greenhouse. On the other hand, as far as there is energy for the process, the production of vapour determines humidity in the greenhouse which, in turn, determines transpiration, which makes management of humidity quite tricky.

The humidity (amount of vapour in air) in greenhouses has to be managed mainly to prevent diseases, that grab their chance at high humidity. As could be inferred from Figure 5.6 the vapour that transpires from a greenhouse crop is not yet 'far away', when it has crossed the boundary layer of the leaves. Indeed, before it is really away, it has either to go through the vents, or condense somewhere. Management is easier when one understands the processes that affect greenhouse humidity, and how.

Let's start with writing the vapour balance of the greenhouse air at equilibrium:

$$E = V + C \qquad\qquad \text{g m}^{-2}\text{ s}^{-1} \quad (6.13)$$

where E, V and C are the vapour flux density of, respectively crop transpiration, ventilation and condensation. Taking into account Equation 5.5, we can write:

$$V = g_V (\chi_a - \chi_{out}) \qquad\qquad \text{g m}^{-2}\text{ s}^{-1} \quad (6.14)$$

Condensation will take place only if the saturated vapour concentration at the condensation surface (usually the cover) is lower than the vapour concentration of the greenhouse air, therefore:

$$C = \max \{0, g_C (\chi_a - \chi^*_{T_{cover}})\} \qquad\qquad \text{g m}^{-2}\text{ s}^{-1} \quad (6.15)$$

The conductance for vapour condensation g_C can be determined on the basis of the theory of heat and mass transfer. Papadakis *et al.* (1992) showed that, for relatively small slopes, the theory of transfer to and from horizontal planes could be applied to greenhouse covers without much loss of accuracy. Therefore Stanghellini and De Jong (1995) proposed:

$$g_C \cong \max \left\{ 0, \frac{A_{cover}}{A_{soil}} 1.64 \times 10^{-3} (T_a^v - T_{cover}^v)^{1/3} \right\} \qquad\qquad \text{m s}^{-1} \quad (6.16)$$

Where the superscript v indicates the virtual temperature, in order to account for the effect of vapour concentration gradients on the density (and thus buoyancy) of air, see, for instance, Monteith and Unsworth (2014: p15). Using instead actual (dry bulb) temperature would result in a slight underestimate of the condensation flow.

It is handy to re-write the transpiration of a crop, Equation 6.9, in the form of a transfer equation, as Equation 6.14 and 6.15. Indeed, by defining:

$$\chi_{crop} \equiv \chi_a^* + \varepsilon \, \frac{r_b}{2LAI} \frac{R_n}{L} \qquad\qquad \text{g m}^{-3} \quad (6.17)$$

and:

$$g_E \equiv \frac{2LAI}{(1+\varepsilon)\, r_b + r_s} \qquad\qquad \text{m s}^{-1} \quad (6.18)$$

And substituting into Equation 6.9, it gets the desired form:

$$E = g_E \, (\chi_{crop} - \chi_a) \qquad\qquad \text{g m}^{-2}\,\text{s}^{-1} \quad (6.19)$$

Finally, substituting Equation 6.14, 6.15 and 6.19 into the mass balance, Equation 6.13 and solving for equilibrium humidity of air inside the greenhouse, we get:

$$\chi_a = \frac{g_E \, \chi_{crop} + g_V \, \chi_{out} + g_C \, \chi_{T_{cover}}^*}{g_E + g_V + g_C} \qquad\qquad \text{g m}^{-3} \quad (6.20)$$

In the case that there is no condensation ($\chi_a < \chi_{cover}^*$) the terms containing g_C disappear from Equation 6.20.

In fact, usually is the cover the coldest element in the greenhouse, and if g_C were very high, the dew point of the greenhouse air would be the cover temperature, as all vapour in excess would condense there. In reality the dew point (the humidity of the air) will be higher than cover temperature as far as the transfer coefficient for condensation is small.

What Equation 6.20 shows is that the equilibrium humidity in the greenhouse is high in the measure that: there is crop transpiration, the humidity outside is high and the temperature of the cover is high. The relative importance of these factors depends on the transfer coefficients: obviously, if there is no ventilation, the humidity outside does not play a role, nor does the temperature of the cover have an effect when condensation is prevented, for instance by an impermeable screen. Indeed, in a highly insulated greenhouse, for instance with double cover, condensation does not contribute much to the removal of vapour, as the inside cover surface usually is hardly colder than the air. So, all the rest being the same, a given humidity of the air is maintained by ventilating away the vapour that would otherwise condense on the cover. That is the experience of many growers, who found out that the energy saving after placing a double cover was much less than they would expect from the decreased U-value of the cover. Indeed, the potential energy saving that can be obtained by increasing insulation of the cover of heated greenhouses is fully attained only when dehumidification methods other than ventilation are applied (Campen, 2009), possibly with regain of sensible heat (Kempkes *et al.*, 2017b).

6.4.1 Condensation on the cover

A cold cover warrants a dry air. However, there are two issues: a cold cover cools the greenhouse air and the condensed water will drip down. Cooling of the greenhouse air is usually undesired: it costs energy in heated greenhouses and it cools unheated ones. The one exception may be tropical environments (or temperate environments in summertime), where radiative cooling may contribute to limit average temperature.

Condensed water forms droplets that merge and grow while sliding down, until they fall. The dripping down is always undesired, since it creates humid spots were pathogens will be happy to settle. The conditions for a drop to detach depend on the slope of the surface and its properties. Usually droplets will not detach from glass with a slope higher than about 25° so condensed water can usually be recollected in small profiles placed under the gutters in Venlo glasshouses. On the other hand, one of the angles in the cover of a round tunnel must be the detachment angle, so that dripping will happen, and always around the same (wet) spot. Besides gothic arches, well-tensioned flat covers with various inclinations have been investigated for low-tech greenhouses (e.g. Montero *et al.*, 2017). Inclination, however, should be higher than is usual in low-slope flat covers, such as the parral greenhouse typical of Almeria or the Canarian greenhouse of Morocco. Often a transparent film is applied, to collect condensed water (and to increase thermal insulation) in the cold months (see Figure 6.2).

Figure 6.2. Transparent film placed below the greenhouse cover in the cold months, both to collect dripping condensed water and to increase insulation. Left, in a parral greenhouse with cucumber, in Almeria and right, a wooden greenhouse with tomato near Valparaiso (Chile). (photos: Cecilia Stanghellini)

Polyethylene for plastic covers is often marketed with 'anti-condensation' treatment. Of course, condensation will happen anytime the cover is below the dew point of the air, and there is no treatment that can prevent this. What such treatments do is to ensure that condensed water forms a film on the surface, instead of droplets, so a more adequate name is 'film-forming treatment'. Besides limiting dripping, film forming also helps in preventing light loss caused by droplets (Stanghellini *et al.*, 2012a). Unfortunately most treatments available today seldom last more than one or two seasons.

6.4.2 Temperature of the crop and dew point

Although there are more (see Section 7.4), the main purpose of humidity control is to prevent fungal diseases that are known to thrive in warm, wet places. These are often the 'wounds' of a plant (such as where leaves and side shoots are removed), the places where water condensed on the cover drops down and the places where condensation takes place. Hereafter we analyse the process of condensation, which factors affect it, and what can be managed to decrease the chance of condensation happening on elements of the crop.

Condensation happens on any surface that is below the dew point of the air. Although the 'big leaf' representation that we have adopted cannot, by definition, tell us anything about temperature distribution in a crop, it is worthwhile analysing the processes that determine the temperature of the 'big leaf', thus the chance that it may be cooler than the dew point of the air. Of course the crop energy balance Equation 6.1, the heat and vapour transfer of the crop Equation 6.6 and 6.7 can be also solved, together with the hypothesis of saturation Equation 6.4, for the crop temperature:

$$T_{crop} - T_a = \frac{(r_b + r_s)\dfrac{R_n}{2LAI} - L(\chi_a^* - \chi_a)}{\rho c_p \left(1 + \varepsilon + \dfrac{r_s}{r_b}\right)} \qquad °C \quad (6.21)$$

All symbols have been defined after Equation 6.9. What we see here, is that there are two conflicting processes determining the temperature of the crop. A positive net radiation would make it warmer than the air, whereas evaporation (driven by the humidity deficit of the air) would make it cooler. Drying the air lowers the dew point, but also increases the deficit (thus crop transpiration), thus it also cools somehow the crop, which may end up only slightly farther from the new dew point. This process is sometimes called 'the hydraulic feed-back'. That is, in a semi-closed environment a change in humidity affects transpiration, which in turn affects the humidity.

Indeed, it may be that the practice of Dutch growers (now infamous for its cost in energy) of maintaining a night-time 'minimum pipe temperature', whatever the conditions, attained its purpose mainly thanks to the (radiative) energy released by the pipe heating system, rather than by the fact that heating coupled to a low night-time temperature set-point was bound to result in ventilation.

In a unheated (and unscreened) greenhouse at night, the crop will be inevitably colder than the air. High humidity means a small dew point depression (the difference between air temperature and dew point), then there is a good chance that at least some part of the crop will be below the dew point (Box 6.6). An aluminum screen or a double cover limits thermal radiation losses (the crop-facing layer of a double cover is relatively warm), so that, all the rest being the

---

Box 6.6. Is ventilation always the best humidity control?

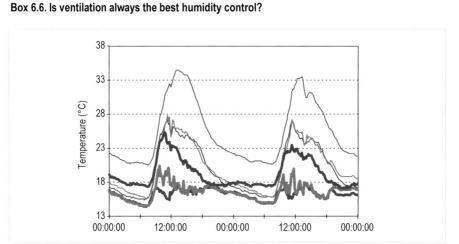

The figure shows the temperature measured, for a two-day period, in two identical un-heated multi-tunnels in Sicily (Southern Italy), one (green) was ventilated as the grower meant it to be, the other (purple) was not, due to a failure of the vents' motors. Blue represent conditions outside. Thin lines represent air temperature and thick lines dew point.

The well-ventilated house could not have been dryer: its humidity (dew point) was nearly always the same as outside. However as most of the time the difference between air temperature and dew point was small, it is unlikely that at night all crops elements managed to be colder than the air (which they must have been) and yet warmer than dew point. That is, high chance of condensation on at least some parts of the crop.

The dew point depression in the unventilated compartment was much larger, and the chance of condensation much smaller. As there was heating nowhere, the non-ventilated compartment at night was warmer than the other one by keeping inside the heat released from the soil, that had been warmed by the (admittedly high) daytime temperature, combined to a relatively high radiation load of the soil, only partly covered by a young crop.

Nevertheless, although this particular example may be rather extreme, in relatively warm and humid conditions it may be better to limit ventilation to allow the temperature to raise, rather than keep ventilating, cooling unnecessarily the greenhouse.

---

same, the crop temperature is raised, and a smaller dew point depression (higher humidity) might be allowed. High insulation also warrants an increased uniformity of temperature (air and crop), which allows for smaller 'safety margins' on humidity (Stanghellini and Kempkes, 2004). Increased homogeneity of temperature (both crop and air) can [additionally] be gained also through air circulation (Box 6.7). The latter, less intuitively, also raises crop temperature, whenever it is colder than the air.

Equation 6.21 gives the temperature of the crop at equilibrium, that is: under steady conditions and after the time required for the temperature of the crop to adapt after a variation in conditions. Leaves have little mass and are efficient heat exchangers. Stanghellini (1987, p 82) deduced

---

Box 6.7. Air circulation.

(photo: Cecilia Stanghellini)

Equation 6.20 gives a 'mean' temperature of a crop: in a tall crop there will be vertical gradients (the net radiation available will vary with depth in the crop) and there may be horizontal variations as well, particularly in poorly insulated and poorly air-tight greenhouses. So, there may be parts of the crop colder than dew point, even if the mean crop temperature is not. In such cases fans that circulate the air inside the greenhouse, without air exchange with outside, may be helpful. Obviously, air circulation reduces horizontal and vertical gradients, so the chance that some part of the crop be colder than dew point. In addition, by reducing the boundary layer resistance, an increased air movement lowers the temperature difference between the crop and the air, thus it warms up the crop, whenever it is colder than the air. More intuitively, it cools the crop whenever it is warmer than the air (as we humans nearly always are).

---

from her experimental data that the time constant (the time it takes to reach a new equilibrium after a change) of leaves of a greenhouse tomato crop is about one min (daytime) up to two (night-time), which is short enough to be neglected. However, stems and fruits have more mass and take much longer to adapt. This is an issue, particularly in the morning: as soon as sun rises, transpiration increases, pushing vapour into the greenhouse air, whose dew point, in turn, goes up. Montero *et al.* (2017, Figure 3) have shown that the dew point [almost] inevitably overtakes the temperature of the tomato fruits, that are slowly warming up. Condensation on the fruits is the consequence. In greenhouses with climate control this is prevented by raising the heating set-point well before sunrise. In the absence of a heating system, ventilation may be the only choice, at least if outside air is drier than inside.

Whenever there is a net radiation gain (such as sunshine or artificial light), there is little chance that the crop may be colder than dew point, so there is little use for humidity control aimed at preventing fungal diseases. One may want to control humidity for other reasons, of course, such as maintaining a minimum transpiration rate to prevent nutrient deficiencies, particularly Ca, see Figure 13.1. Stanghellini and Kempkes (2008) developed an indicator of 'crop well-being', that could be applied for the dynamic selection of humidity set-point in a greenhouse, accounting for net radiation.

6.5 Concluding remarks

We have seen that very basic physical principles (energy balance, vapour and heat transfer) make it possible to write a general equation identifying the factors driving crop transpiration. In particular, we have seen which crop-specific factors are relevant, such as the fraction of available radiation that is absorbed and the behaviour of the stomata in relation to light, and which ones are less significant is a greenhouse environment, such as the boundary layer resistance.

Thereafter, the balance of vapour in a greenhouse has been used to identify the factors that determine 'humidity' inside, which gives insight into ways to steer humidity. Finally we have discussed humidity management in relation to its purpose and prevailing conditions.

Chapter 7
Crop response to environmental factors

Photos: Frank Kempkes, Greenhouse Horticulture, Wageningen University & Research and Cecilia Stanghellini (right, bottom)

Environmental factors have a significant impact on plant growth and development. The principles of plant and crop growth analysis, processes such as light interception, photosynthesis and respiration, as well as biomass partitioning were discussed in Chapter 2. The influence of light, temperature, CO_2, humidity, drought and salinity on crop growth, development and yield is the subject of this chapter. A grower can influence these factors and can thus influence yield, quality of the product and quality of the crop. For additional reading, the book *Plant Physiology in Greenhouses* (Heuvelink and Kierkels, 2015) provides an easy-to-read overview of external influences on the plant.

7.1 Light

Photosynthetic active radiation (PAR; see also Chapter 3), which is often simply called 'light', provides the energy for photosynthesis. There is almost never too much PAR in a greenhouse. A rule of thumb often applied states that 1% less light means 1% less yield, hence yield is assumed to be proportional to the amount of light (see also Eq. 2.23). This is not always entirely correct, but for fruit-producing vegetables it is fairly accurate. For ornamentals such as chrysanthemum, poinsettia or ficus, the yield reduction per 1% less light is around 0.6% (Marcelis *et al.*, 2006). Potted plants are often grown under low light conditions to avoid leaf damage (Box 7.1). The quantitative response of crop yield to light is important, because, for example, measures taken to save energy (e.g. installing screens, coating on the glass, double glass) often reduce the light level in the greenhouse. Therefore, what is saved by cutting the costs of energy could be nullified or even reversed by the reduction in yield. The impact of energy-saving measures on net return depends on many factors, such as the reduction in energy use, the price of energy and cost of investment, but also light reduction and its consequences for (financial) yield per m^2.

In addition to the effects on growth and production, light also influences product quality (Figure 7.1). At a higher light level, *Kalanchoe blossfeldiana* produces more biomass, more inflorescences and more flowers per inflorescence (Carvalho *et al.*, 2006). The height of the plants, however, is not affected – this aspect is more influenced by temperature. The development rate is also influenced: the flowers open earlier at higher light levels. This effect of light is more of an indirect effect: higher light levels mean more radiation on the crop, which

Box 7.1. Potted plants love light too.

Many potted plants, like *Dracaena* and *Calathea*, are grown under low light conditions to avoid leaf damage.

Most potted plants originate from the understory of tropical rainforests and as such they are adapted to low light conditions. Growers of those plants are slightly afraid of light, and daily light integrals as low as three to five mol PAR m^{-2} d^{-1} are typically applied. Growers use screens for many hours per day and apply sun-blocking coatings on the greenhouse (such as whitewash). If sun-intolerant plants are exposed to too much radiation, photo oxidation can occur, resulting in yellow leaves, tip burn or even necrosis of parts of the leaves. This is, of course, unacceptable for ornamental plants. It is, however, clear that this heavy shading carries a production penalty, since potential crop growth is directly related to the amount of light that can be captured and efficiently used.

It has been shown that several shade-tolerant potted plant species can be grown faster when more light is allowed. Additionally, a significant reduction in energy use for heating can be achieved if more natural irradiance is allowed to enter the greenhouse. On the other hand, the use of more light requires higher levels of relative humidity to avoid light damage (Kromdijk *et al.*, 2012), and high humidity keeps the stomata from closing. Allowing more light in can also be achieved by diffusing the light (Li *et al.*, 2015) or blocking near infrared (NIR) radiation (Box 3.5). Under these circumstances, growers can allow more light, resulting in a faster growing plant which they can sell earlier.

(photo: Frank Kempkes, Greenhouse Horticulture, Wageningen University & Research)

often results in a higher organ temperature and temperature is the main factor determining the development rate. In tomato plants, this is exclusively the case (De Koning, 1994), while in other crops (e.g. cucumber; Schapendonk *et al.*, 1984) other factors, such as light, also have an influence (even when the indirect effects of light on organ temperature are excluded). Light sum determines the production time of young plants (Box 7.2) which suggests that plant weight rather than developmental stage alone determines delivery date.

Figure 7.1. Effect of photosynthetic photon flux density (μmol m^{-2} s^{-1}) on growth of *Kalanchoe blossfeldiana* 'Anatole', grown in a climate room at 21 °C (Carvalho *et al.*, 2006). Effect of temperature is shown in Figure 7.8.

Box 7.2. Light sum determines the production time of young plants.

The table shows the sowing and delivery dates of young tomato plants from a nursery in a Mediterranean climate. The plant cycle duration varies between 39 and 58 days. To determine whether the longer cycle in winter is explained by lower light in winter, we will determine the total light integral for all batches.

The average outside global radiation (MJ m^{-2} d^{-1}) per month in November until March is as follows: 9.3, 8.1, 8.5, 10.6 and 14.6, respectively. Greenhouse transmissivity is 70%, fraction PAR in global radiation equals 47% and 1 MJ PAR = 4.6 mol. Therefore, the light integral for the first cycle (17 days in November, 31 in December and 8 in January) is:

$$(17 \times 9.3 + 31 \times 8.1 + 8 \times 8.5) \times 0.47 \times 0.7 \times 4.6 = 724 \text{ mol m}^{-2}$$

The table shows that light sum is rather constant for all cycles (on average 759 mol m^{-2}, with standard deviation 56 mol m^{-2}). Note that in the middle of March, whitewash was applied to the greenhouse cover, resulting in a transmissivity of only 40%. This is not taken into account in our calculations, therefore the light sum of batches 14 to 16 is increasingly overestimated in the above table.

Cycle	Sowing	Delivery	Length of cycle (days)	PAR sum (mol m^{-2})
1	13 Nov	8 Jan	56	724
2	20 Nov	17 Jan	58	741
3	27 Nov	24 Jan	58	732
4	4 Dec	28 Jan	55	692
5	11 Dec	4 Feb	55	709
6	18 Dec	10 Feb	54	719
7	25 Dec	16 Feb	53	730
8	2 Jan	24 Feb	53	759
9	9 Jan	28 Feb	50	734
10	16 Jan	4 Mar	47	732
11	23 Jan	9 Mar	45	753
12	30 Jan	14 Mar	43	773
13	6 Feb	20 Mar	42	797
14	13 Feb	25 Mar	40	794
15	20 Feb	31 Mar	39	814
16	27 Feb	7 Apr	39	934

Note: If only developmental stage (number of leaves) determines delivery date a constant temperature sum would be expected rather than a constant light sum (see Chapter 2).

7.1.1 Supplementary light

Growers in regions with dark winters increasingly use supplementary light (Heuvelink *et al.*, 2006), which is needed to avoid dramatic drops in winter production. In these regions, no winter production is possible for crops such as tomato and sweet pepper without supplementary light. By using supplementary light, growers are able to raise both crop production and product quality, in addition to having better control over yield and quality. Made possible by supplementary light, producing year-round meets market demands, improves the efficiency of investments and gives a more regular labour demand. In the Netherlands, supplementary light is mainly used in the production of ornamentals and tomato. An example of calculating the contribution to supplementary light to the total light integral in a greenhouse is given in Box 7.3. In many situations (crops or greenhouse locations), the use of supplementary light is not economically feasible. The high investment and operational costs are not compensated for by a high-enough increase in yield (in financial terms, also influenced by better quality). The technical aspects of supplementary light are outlined in Chapter 10.

High pressure sodium lamps (HPS) are most commonly used for supplementary light, however there is increasing interest in light emitting diodes (LEDs) (Figure 7.2). The advantages of LED lighting are listed in Box 10.1. Since growers are used to HPS lamps, they sometimes feel that the transition to LEDs is not advantageous because the head of the plant requires additional heating to maintain an acceptable development rate. On the other hand, the 'coldness' of LED light has the advantage that the addition of light is separated from the addition of heat, which makes it possible to bring the lamps between crops for interlighting. The application of LEDs in horticulture has been reviewed by Mitchell *et al.* (2015) and Bantis *et al.* (2018). The main

---

Box 7.3. Contribution of supplementary light to total light integral.

Compared to the intensity of solar radiation, supplementary light often represents only a small contribution for most of the year. On an average day in mid-February in the Netherlands, the global radiation outside is 4 MJ m^{-2} d^{-1}. Half of this is PAR and 70% (greenhouse transmissivity) reaches the crop. In terms of sunlight, 1 J PAR equals 4.6 µmol (Langhans and Tibbitts, 1997). This means the crop gets 1.4 MJ m^{-2} d^{-1} = 1.4×4.6 = 6.4 mol m^{-2} d^{-1} of natural light (PAR). In a tomato crop, 200 µmol m^{-2} s^{-1} is a reasonable intensity for supplementary lighting (some growers use lux, i.e. more or less 15,000 lux. Note: lux is the wrong measure for light on plants, as it is based on the spectral sensitivity of the human eye, not on the number of photons; see Box 3.2). If supplementary light is supplied for 16 hours per day, this supplementary light provides 200×16×3,600/10^6 = 11.5 mol m^{-2} d^{-1}. Therefore, the additional light is roughly twice as much as the natural light in this period. In May, however, when natural light intensity and day length is much higher, the supplementary light only contributes slightly.

Dueck *et al.* (2012) reported that during a tomato experiment with 170 µmol m^{-2} s^{-1} supplementary light in the Netherlands, cumulative incident light (mol m^{-2}) from the lighting system (HPS and/or LED; details in the section on crop interlighting) was more or less equal (87%) to that of the sun during the lighting season (October 15th – May 20th). The maximum day length was 18 h and in March and April the use of the assimilation lighting depended on solar radiation.

Figure 7.2. Overhead lighting with high pressure sodium (HPS) lamps (left) and light emitting diodes (LED, right). (photos: HPS Cecilia Stanghellini, LED Tom Dueck, Greenhouse Horticulture, Wageningen University & Research)

limitation of using LED lighting in greenhouses is the high investment costs (Persoon and Hogewoning, 2014; Nelson and Bugbee, 2014).

Many experiments have been conducted to determine the effect of supplementary light on a wide range of crops. In an experiment with sweet peppers, raising the light level from 125 to 188 µmol m^{-2} s^{-1} resulted in a higher yield, but the fruit weight stayed constant. Therefore, the additional light resulted in a better fruit set and thus a higher yield. Supplementary light shortens the production time of young plants (Box 7.4).

Light control

An important question remains: when should the lights be on? Several aspects must be taken into account: is the extra revenue higher than the additional running costs (see Chapter 12)? Is it physiologically possible to maintain a longer day? Several plants like chrysanthemum and poinsettia are short-day plants, they need long nights to flower, hence, there are limited hours in which supplementary light should be provided. Besides the flowering response, there can be another reason why the lights should not be on too long each day. For example, a tomato plant needs at least six hours of darkness, as it develops a detrimental leaf injury when grown under continuous light. Additionally, eggplant shows chlorosis under continuous light (Murage and Masuda, 1997). Recently, a gene was suggested that would make it possible to supply light for 24 h (Velez-Ramirez *et al.*, 2014). Introgressing this tolerance into modern tomato hybrid lines resulted in an up-to 20% yield increase under continuous light, showing that limitations on crop productivity, caused by the adaptation of plants to the terrestrial 24 h day/night cycle, can be overcome. In roses it has been shown that continuous light during cultivation is detrimental for vase life (Mortensen *et al.*, 2007).

The effect of supplementary lighting on plant growth depends on the balance between assimilate production in source leaves and the overall capacity of the plants to use assimilates (sink demand). Sink organs are, for example, fruits, flowers, young leaves and roots. If the source is larger than the sink, it does not make sense to turn on the light; there is already too

Box 7.4. Supplementary light shortens the production time of young plants.

In Box 7.2 we saw that the light sum remains rather constant over cultivation cycles (different sowing dates) of young tomato plants. On average, a light sum of 759 mol m^{-2} was needed between sowing and delivery date. It can now be calculated, for each cycle, how much shorter the necessary cultivation time would be if supplementary light were applied. Let us assume we apply 150 µmol m^{-2} s^{-1} over 16 h every day; this means 150 × 16 × 3,600 / 1,000,000 = 8.64 mol m^{-2} each day.

For the first cycle, 17 days in November result in a light sum of 17 × 14.1 + 17 × 8.64 = 387 mol m^{-2} and in December one day provides a light sum of 12.3 + 8.64 = 20.94 mol m^{-2}. To obtain 759 mol m^{-2}, we therefore need an additional (759 – 387) / 20.94 = 18 days in December. Total cultivation time with supplementary light is 17+18 = 35 days, compared to the previous 17 + 31 + 8 = 56 days without supplementary light, meaning a reduction of 21 days, or 21/56 × 100% = 38%. The table below shows that for sowing dates in January or February, the result of supplementary lighting is a shortening in days (12-16 days). This represents a 29-36% reduction in time between sowing and delivery date.

Cycle	Sowing	Delivery	Length of cycle without light (days)	n days 2nd month with light	Total mol m^{-2} per cycle with light	Days gained	Reduction in time (%)
1	13 Nov	8 Jan	56	18	764	21	38
2	20 Nov	17 Jan	58	25	751	23	40
3	27 Nov	24 Jan	58	33	759	22	38
4	4 Dec	28 Jan	55	9	759	19	35
5	11 Dec	4 Feb	55	16	763	19	35
6	18 Dec	10 Feb	54	23	767	18	33
7	25 Dec	16 Feb	53	29	749	18	34
8	2 Jan	24 Feb	53	5	747	19	36
9	9 Jan	28 Feb	50	12	770	16	32
10	16 Jan	4 Mar	47	18	768	14	30
11	23 Jan	9 Mar	45	24	766	13	29
12	30 Jan	14 Mar	43	30	764	12	28
13	6 Feb	20 Mar	42	7	760	13	31
14	13 Feb	25 Mar	40	13	771	12	30
15	20 Feb	31 Mar	39	18	751	13	33
16	27 Feb	7 Apr	39	24	763	14	36

much assimilate production to meet the demand. In this situation, supplementary light would only cost money and could even damage the crop.

Li *et al.* (2015) calculated the pattern of source/sink ratio over time during a greenhouse experiment for three tomato cultivars: a cherry type (small fruit), a normal tomato and a cultivar with large fruit (Figure 7.3). Initially, for all three cultivars there was a period in which the source/sink ratio was above one, which means assimilate production was higher than

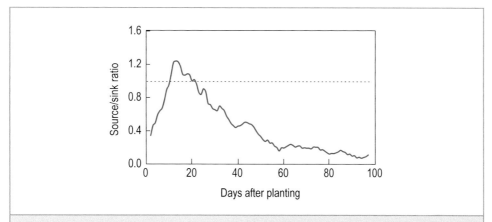

Figure 7.3. Calculated source/sink ratio over time for tomato cv. Komeett (a mid-sized tomato variety). The solid line represents the moving average over five days. The dashed horizontal line represents a source/sink ratio of one. The fruit set started between 20-30 days after planting. The data is taken from Li *et al.* (2015) and based on cultivation in the Netherlands (Wageningen), planted on 16 August.

the demand for assimilates. This occurs because in the young crop there are only a few sink organs: roots, young leaves and perhaps one truss with small fruit, whereas the crop already intercepts a lot of the light (a relatively high leaf area). In this period, additional supplementary light is not useful. For most of the growing season, however, the source/sink ratio is (much) lower than one, therefore supplementary light will stimulate growth and yield. Whether it is profitable depends on the cost of lighting and the financial return of the extra production (see Section 12.4.1).

In the Netherlands, the difference between the received natural light in the middle of summer and winter is a factor of ten: the light intensity in mid-summer is five times higher and the day length is twice as long as in mid-winter. If a grower does not adapt to this, and we assume a constant light use efficiency (LUE), plant growth would also be reduced by a factor of ten. Note that this factor will, in reality, be somewhat lower, as LUE is expected to be higher under low light in winter than under high light intensity in summer (Figure 2.13). *Chrysanthemum* growers mitigate the effect of seasonal light sum differences to some extent by adapting the planting density (Lee *et al.*, 2002). In winter, without supplementary light, a typical planting density would be 40 plants per m^2, whereas in summer this would increase to at least 66 plants per m^2. Tomato growers do not have this option as the same plants are grown for 11 months however these growers retain additional stems to increase stem density in summer (see Chapter 2; Figure 2.8).

Crop interlighting

LEDs provide so-called 'cold' light: these lamps produce relatively little radiative heat. This makes it possible to place the lamps in-between the crops, so-called interlighting (Figure 7.4). When grown under overhead lighting, leaves lower in the canopy have a low photosynthetic

Figure 7.4. Light emitting diodes (LED) interlighting in a cucumber crop (photo Cecilia Stanghellini).

capacity because they acclimate to low light conditions (Trouwborst *et al.*, 2011). However, with interlighting there is less acclimation and the lower leaves stay more productive.

In countries with very low natural light levels in winter, such as Norway, Finland or Iceland, interlighting has been shown to be more profitable than just overhead lighting. A combination of overhead lighting and interlighting (e.g. 75:25, or even 50:50) can raise yield by 10% compared to only overhead lighting at an equal total light intensity (Hovi-Pekkanen and Tahvonen, 2008). The total amount of light (as well as the running costs) stays the same in both situations.

In order to take advantage of both HPS and LED lighting, hybrid systems using both type of lights have shown great potential for greenhouse crops. For example, an HPS and LED hybrid system resulted in a 20% higher yield for two tomato cultivars compared with top HPS lighting only (Moerkens *et al.*, 2016). However, the tomato production under HPS alone was highest when four lighting treatments (photosynthetic photon flux density (PPFD) 170 µmol m^{-2} s^{-1}) were compared: (1) overhead lighting with HPS; (2) overhead lighting with LED; (3) overhead hybrid lighting with HPS and LED lighting (50:50) above the crop; and (4) HPS above the crop in combination with LED between the canopy (interlighting) with a proportion of 50:50 (Dueck *et al.*, 2012).

To optimise the intra-canopy lighting system, the positioning of LED lamps should also be considered. By using a 3D light model (ray tracing) in combination with a 3D model of a tomato crop to simulate the three-dimensional plant structure and light distribution for an interlighting system, De Visser *et al.*, (2014) showed that LEDs shining slightly upward (20°) increase light absorption and LUE relative to horizontal beaming LEDs, which could prevent loss of light to the ground.

Light spectrum

Besides light intensity and day length (the period during which the plant receives natural or artificial light), the light colour (spectrum) is also important. With LEDs, the grower can determine the exact colour of the additional light. HPS lamps, on the other hand, have a high component of yellow-red light (Figure 10.4). There is minor variation in the photosynthetic effectiveness of photons of different wavelengths, though red light has the highest quantum efficiency (Figure 7.5). Note that Figure 7.5A shows the so-called 'McCree curve' (see also Box 3.2). This curve should not be confused with the absorption spectrum of chlorophyll in a solvent, which has strong peaks in blue and red light. Per μmol of absorbed light, green leaves are more efficient than red leaves, which is in accordance with the higher content of anthocyanins in red leaves. Anthocyanins function as a light filter, reducing the quantum flow rate absorbed by chlorophyll (Paradiso *et al.*, 2011). Green light is better utilised at canopy level relative to leaf level (Figure 7.5B), since at leaf level green light shows lower absorbance, while at crop level this reflected light is (partly) absorbed by other leaves in the canopy. Hence, findings at leaf level should not automatically be extrapolated to crop level without considering interactions within the crop.

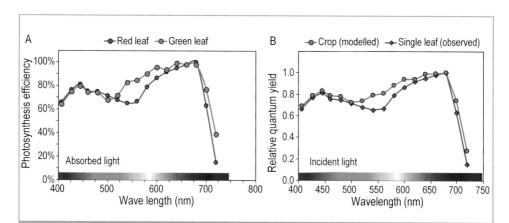

Figure 7.5. (A) Spectral quantum yield (μmol CO_2 per μmol absorbed photons) of photosynthesis in green and reddish leaves of rose cv. 'Akito', and (B) relative spectral quantum efficiency simulated at crop level and observed at leaf level (see A) for green leaves. Values presented are relative to the maximum value at 680 nm for each type of leaf. Redrawn from Paradiso *et al.* (2011).

It is tempting to use Figure 7.5A to estimate the differences in efficiency of different light spectra (e.g. comparing red/blue LED with HPS), however, care is needed when drawing conclusions from instantaneous effects. Leaves acclimate to the growth light spectrum (Hogewoning *et al.*, 2012) and hence short-term (instantaneous) spectral effects on leaf photosynthesis (like in Figure 7.5) may differ from the long-term response. Hogewoning *et al.* (2012) show quantitatively that leaves acclimate their photosystem composition to their growth light spectrum and how this changes the quantum yield for CO_2. Furthermore, the light spectrum influences many aspects of crop physiology and morphology under prolonged illumination (Hogewoning *et al.*, 2007).

Light colour effects on the efficiency of photosynthesis are limited (Figure 7.5), therefore its main effects on plant growth are through morphological changes. Ouzounis *et al.* (2015) reviewed the impact of light spectra on plant physiology and on secondary metabolism in relation to greenhouse production. These authors indicated the possibility of a targeted use of LEDs to shape plants morphologically, increase the quantity of protective metabolites to enhance food quality and taste, and potentially trigger defence mechanisms of plants. Often, the ratio between light colours is important e.g. a low red (R; 640-680 nm)/far-red (FR; 710-750 nm) ratio results in elongated plants, one aspect of the so-called shade-avoidance syndrome (SAS; Pierik and De Wit, 2014). Plants exhibit increased shoot elongation rates as well as several other responses, including hyponasty (i.e. more upward-oriented leaves) and early flowering, to avoid shading by neighbours. The major light signal triggering these shade avoidance responses is the ratio of R to FR light, which is lowered by selective absorption of red light by chlorophyll (Pierik *et al.*, 2004). It has also been shown that at the end-of-day; the last 20-30 min of the light period) the R:FR ratio is decisive for its effect on elongation (Lund *et al.*, 2007).

A higher fraction of blue light results in more compact plants (Hernández and Kubota, 2016). However, 100% blue light results in very elongated plants (Hernández and Kubota, 2016), maybe because it results in a phytochrome equilibrium comparable to that obtained with a low R/FR ratio (shade avoidance syndrome, see above). The blue part of the light spectrum has been associated with leaf characteristics which also develop under high irradiances. The increase in photosynthetic capacity with blue light percentage (0-50%) is associated with an increase in leaf mass per unit leaf area, nitrogen (N) and chlorophyll content per area and stomatal conductance (Hogewoning *et al.*, 2010b).

The role of green light (Figure 7.6) in plant physiology and morphology has long been ignored. The mechanisms of the action of green light on plant growth and development, as well as the nature of a green light photoreceptor, have rarely been studied (Golovatskaya and Karnachuk, 2015). There is a growing pile of evidence that green wavelengths enhance stem elongation, CO_2 assimilation and biomass production and yield. These results suggest that a certain amount of green light may be needed in LED light recipes.

Figure 7.6. Young tomato plants grown under red and green light emitting diodes (LEDs). (photo: Xue Zhang, Wageningen University & Research)

The importance of light spectrum for plant morphology is also illustrated by Hogewoning *et al.* (2010a; Figure 7.7). These authors grew cucumber plants at the same light intensity but under different spectra. The total dry weight of plants grown under artificial solar light (AS; sulphur plasma lamp with the addition of quartz-halogen lamps) was 2.3 and 1.6 times greater than that of plants grown under fluorescent tubes and HPS, respectively. The height of the AS plants was 4-5 times greater, while plants under HPS or fluorescent tubes remained very short. This could be interesting for potted plants, which are appreciated for their compact spherical outline. This striking difference appeared to be related to a more efficient light interception by the AS plants, characterised by longer petioles, a greater leaf unfolding rate and a lower investment in leaf mass relative to leaf area. Photosynthesis per leaf area was not greater for the AS plants. The effects of light spectrum can be very large in young plants, whereas in an adult crop they are more moderate but still present. For plant growth, the spectrum of solar light is superior to that of artificial light.

A positive effect of FR radiation on tomato yield has recently been shown. The addition of FR increases total biomass production due to the larger leaf area and longer internodes

High pressure sodium   Fluorescent tubes   Artificial solar

Figure 7.7. Cucumber plants grown at 100 μmol m⁻² s⁻¹ under a high-pressure sodium lamp (left), fluorescent tubes (middle), and an artificial solar spectrum (right) 13 days after planting the seedlings (bar = 10 cm). The lower three images are of plants different from those on the upper image. These three images are not scaled; the leaf colour appears unnatural due to the growth light environment (Hogewoning *et al.*, 2010a).

(better light distribution in the crop). Furthermore, FR addition improves the partitioning towards the fruits (Kim *et al.*, 2019). The favourable effect of FR radiation on leaf expansion and subsequent dry mass accumulation has also been reported for lettuce (Li and Kubota, 2009) and several ornamental plants (Park and Runkle, 2017). Park and Runkle (2017) also observed that inclusion of FR during seedling growth promoted subsequent flowering in one long-day species (snapdragon), whereas in other species flowering was not affected by FR. The acceleration of flowering by low R:FR is one characteristic of the shade-avoidance responses (Pierik and De Wit, 2014).

Adventitious rooting is a critical process in the vegetative propagation of ornamental plants. Christiaens *et al.* (2015) studied spectral light quality to improve the adventitious rooting of cuttings in three ornamental species: *Chrysanthemum × morifolium*, *Lavandula angustifolia* and *Rhododendron simsii* hybrids (azalea). Different combinations of R and blue (B) LEDs were tested: R:B 100:0, 90:10, 80:20, 50:50, 10:90 and 0:100 at a light intensity of 60 μmol m⁻² s⁻¹ for chrysanthemum and lavender and 30 μmol m⁻² s⁻¹ for azalea. No natural light was supplied. For all three species, rooting was highly efficient under 100% red light. For chrysanthemum, rooting was most inhibited under 10:90 R:B, while for azalea the inhibition of rooting was highest under 50:50 R:B.

7.1.2 Diffuse light

On a cloudy day it looks as if the light comes from all angles; we call this diffuse light. On a cloudless day there is more direct light, producing shadows, but still a considerable part is diffuse (Figure 3.2). In southern countries, the ratio between direct and diffuse light is higher than in countries such as the Netherlands, Canada or Sweden. In the 1960s, researchers already knew that diffusing the light increases production, but only recently has it been possible to diffuse the light more without giving much in to total light intensity. Nowadays, both diffusing glass and coatings scatter the light but barely affect the intensity, because an anti-reflection coating is simultaneously applied (Box 3.4).

Diffuse light is more equally distributed in the crop, both horizontally and vertically. The light penetrates deeper into the canopy, meaning the extinction coefficient is lower. Diffusing greenhouse covers result in no sun flecks or shaded spots. This has a significant impact: the leaves at the top of the canopy absorb less light, however these leaves are usually close to light saturation, and the leaves in the middle of the canopy can photosynthesise more. Overall, it results in greater production of assimilates, resulting in a yield increase of 5-10% (e.g. Dueck *et al.*, 2012; García Victoria *et al.*, 2012). The higher production of tomato plants under diffusing glass has been shown to result from four factors: a better horizontal light distribution, a better vertical light distribution, a higher leaf photosynthesis capacity (of lower leaves) and a slightly higher leaf area index (Li *et al.*, 2014). All these studies originated in The Netherlands, where the reference is clear glass. Please be aware that nearly all plastic covers used elsewhere already are somewhat diffusive (page 62), so results cannot simply be extrapolated to other conditions.

As mentioned previously, heavy shading is commonly applied during the production of potted plants in order to avoid damage caused by high light intensities. The daily light integral (DLI) is usually limited to 3-5 mol m^{-2} photosynthetically active radiation (PAR). However, shading carries a production penalty, as light is the driving force for photosynthesis. In greenhouse experiments with two Anthurium cultivars and two Bromeliads, Van Noort *et al.* (2013) diffused the light by using a diffusing screen or diffusing glass. During the summer, the growth of all crops increased substantially by allowing more (diffuse) light, and leaf damage never occurred. The anthurium plants were marketable within 16 weeks, whereas this normally takes 22 weeks; these plants were also 25% larger. Applying light integration and decreasing the set-point for heating, saved 25% energy (Van Noort *et al.*, 2013). Li *et al.* (2015) concluded from the same experiment with two *Anthurium andreanum* cultivars that less shading under diffuse light not only stimulated plant growth but also improved plant ornamental quality (i.e. compactness). A higher DLI (10 instead of 7.5 mol m^{-2}) under diffusing glass led to more flowers, leaves and stems; more compact plants were obtained without light damage in leaves or flowers in both cultivars.

7.2 Temperature

Temperature is the main influencing factor on development. The relationship between temperature and development is an optimum response (see Chapter 2). Where the optimum lies depends on plant species and specific (enzymatic) processes. At sub-optimal temperatures, there is a range of temperatures where development rates increase more or less linearly with temperature (Figure 2.2). In this linear part, a heat sum (degree days) can be used to predict development. In young plants, temperature can have a significant effect on the build-up of biomass. For example, in an experiment where young tomato plants were grown at a constant temperature of 18 or 24 °C (Heuvelink, 1989), the fresh weight at 24 °C was almost four times higher, just like the leaf area. This effect is remarkable, because leaf photosynthesis in a broad temperature range is barely affected by temperature, at least at the PPFD (125 µmol m^{-2} s^{-1}) of this experiment (e.g. Figure 7.13). At a higher temperature the plants developed more leaves, which were thinner, resulting in a much larger leaf area, and hence light interception and plant growth rate were greater. Hence, the faster build-up of biomass at a higher temperature is mainly due to morphological effects.

Temperature can also influence quality, e.g. in potted plants. *Kalanchoe blossfeldiana* plants were grown under 200 µmol m^{-2} s^{-1} light and four different temperature levels: 18, 21, 23 and 26 °C (Carvalho *et al.*, 2006). At higher temperatures, crop cultivation time was reduced, which is positive, but a higher temperature also led to fewer flowering side shoots, fewer flowers per inflorescence and much taller plants (Figure 7.8). All of which are negative effects: a compact, abundantly flowering plant fetches a higher price. Therefore, producing a nicely shaped plant requires some time.

Figure 7.8. Effect of temperature on the growth of *Kalanchoe blossfeldiana* 'Anatole' (Carvalho *et al.*, 2006). Effect of light is shown in Figure 7.1.

Temperatures that are too high or too low can lead to stress. In tomato plants, the fruit set is the most vulnerable process. In a day/night regime of 32/26 °C (moderately high temperature stress) the number of fruits per plant drops drastically compared to a regime of 28/22 °C (Sato *et al.*, 2002). The reason for this is poor pollination, which also results in a poor fruit set. Both the pollen quality of the stamen and pollen release from the stamen are negatively affected by elevated temperatures. Remarkably, total plant growth stayed the same in both regimes, and photosynthesis was not influenced.

On the other hand, too-low temperatures can also have negative effects. We see that in sweet peppers: the fruits are too small and are misshapen (Tiwari *et al.*, 2007). Such fruits are often seedless, once again because of poor pollen quality. Too-low temperatures also inhibit the transport of sugars from the leaves to the sink organs. Since photosynthesis continues at roughly the same rate (hardly temperature-dependent in a wide range of temperatures) the sugars need to be stored as starch in the leaves.

Low root temperatures result in slower growth and smaller tomato fruits (Cooper, 1973). The critical root temperature for tomato growth is about 15 °C (Martin and Wilcox, 1963). The impact of shoot temperatures on plant development and growth is much larger than that of root temperatures. Cucumber plants, grafted on *Cucurbita ficifolia*, were less sensitive to low-root temperatures than cucumber plants with their own roots (Ahn *et al.*, 1999). In roses, the root temperature is important for the formation of bottom breaks (new shoots). In an experiment with different root temperatures and the same air temperature, the number of bottom breaks increased at higher root temperatures (Figure 7.9; Dieleman *et al.*, 1998).

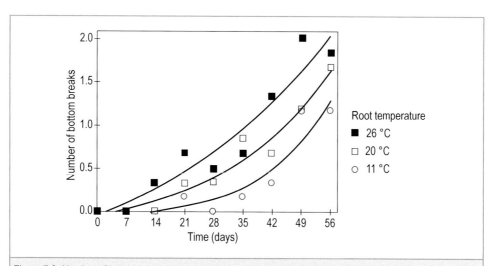

Figure 7.9. Number of bottom breaks per rose plant as affected by root temperature of 11, 20 and 26 °C. Each point represents the average value of six plants (Dieleman *et al.*, 1998).

Vertical temperature gradients are typical for closed and semi-closed greenhouses (see also Chapter 9). Qian (2017) investigated the possible differences in crop growth and yield, comparing cooled and dehumidified air blown into a greenhouse from below or above for tomato crops grown year-round in semi-closed greenhouses. Cooling below the canopy induced vertical temperature gradients. The temperature at the top of the canopy was over 5 °C higher than at the bottom, whenever solar radiation exceeded 700 W $^{-2}$. Despite the occurrence of vertical temperature gradients, plant growth and fruit yields were mostly unaffected. Leaf and truss initiation rates did not differ between treatments, since air temperatures at the top of the canopy were comparable. The only observed response of plants to the vertical temperature gradient was the reduced rate of fruit development in the lower part of the canopy. This resulted in a longer time between anthesis and fruit harvest in the treatment with a vertical temperature gradient, and a slight increase in the average fruit weight in summer. However, total fruit production over the whole season was not affected (Qian, 2017).

7.2.1 DIF and DROP

DIF

In addition to the average temperature, the temperature regime (or temperature pattern) is also important. It makes a difference whether plants are kept constantly at, for example, 21 °C, in a day/night regime of 25/17 °C (day length of 12 hours; average 21 °C) or even in the reverse regime 17/25 °C. The difference between day and night temperatures is called 'DIF' (short for DIFFERENCE), and the reverse regime is called a 'negative DIF'. Growers, especially of potted plants, use negative DIF to reduce elongation and obtain more compact plants.

Many plants, including tomato plants, are significantly affected by negative DIF (Figure 7.10). A negative DIF of 10 °C results in a more compact plant, which also exhibits reduced growth and shorter side shoots. The leaf angle is more upright: this has a big impact on light interception. It is no surprise that the biomass of the plant on the right (Figure 7.10) is much higher, as at a negative DIF of 10 °C the tomato plant also produces thicker leaves. Heuvelink (1989) conducted a plant growth analysis (see Chapter 2) to determine why plant growth rate is much lower in an inversed temperature regime. Young tomato plants were grown at different day and night temperature combinations with an average of 21 °C. Day (T_d) and night (T_n) temperature varied between 16 and 26 °C, while the day length was 12 h. Light intensity was about 125 µmol m^{-2} s^{-1} PAR. An inversed temperature regime (T_d lower than T_n) reduced plant growth, which was caused by a reduction in leaf area ratio (LAR). This decrease in LAR in an inversed temperature regime was caused mainly by a decrease in specific leaf area (SLA), while net assimilation rate (NAR) was not influenced by the temperature regime. For young, widely spaced plants, a lower SLA (thicker leaves) results in less light interception and thus in growth reduction. The upright leaf angle enhances this effect.

It is now possible to explain the different reactions to temperature regimes at the same temperature integral between young plants and closed canopies. In a closed canopy (a producing crop), differences in leaf area index, brought about by differences in leaf thickness,

Figure 7.10. Tomato plants grown for 40 days at 12-h day length with a day temperature of 16 °C and night temperature of 26 °C (left) or day temperature of 20 °C and night temperature of 22 °C (right). Despite the same average temperature of 21 °C, large differences are visible between the two plants (Heuvelink, 1989).

have very little influence on light interception because most light is already intercepted anyway. This explains why a producing crop, in contrast to young plants, shows a limited reaction to temperature regime at the same temperature integral.

Carvalho *et al.* (2002) showed for *Chrysanthemum* that DIF can predict the final internode length only within a temperature range where the effects of day temperature and night temperature are equal in magnitude and opposite in sign (18-24 °C).

DROP

Not only the day/night temperature regime is important, but also the regime during the day. In particular, a drop in temperature during a brief period (up to a few hours; called DROP) can reduce plant elongation. The point during the day at which this DROP occurs plays a significant role. In the first hours of the light period, elongation growth is most sensitive to temperature: a low temperature during those hours reduces internode length. In the last hours of the light period, DROP has no effect on elongation (Cockshul *et al.*, 1998). Potted plant growers use this phenomenon to keep their plants compact (Carvalho *et al.*, 2008). In a greenhouse, DROP is easier to create than a negative DIF (in late spring and summer), as day temperature only needs to be lower than night temperature during the first hours of the light period, and not throughout the whole light period. DROP in the early-morning hours often has a similar effect to negative DIF.

7.2.2 Temperature integration

It has been demonstrated above that temperature regime, at the same average temperature, can be important, especially for young plants and fast-growing potted and bedding plants, but in general full-grown crops react more to the long-term average temperature than to the precise regime (De Koning, 1990). This provides the opportunity to work with temperature integration: in the longer term (e.g. 24 h, several days or one week) the average temperature must meet the setpoints. Besides the integration interval, the amplitude between lowest and highest temperature is limited. Most crops will allow for the compensation of a 24 h temperature of 19 °C with 21 °C the next day to obtain 20 °C on average. However, 24 h of 5 °C cannot be compensated by 35 °C the next day. Temperature integration is only possible within the temperature range where the rate of the relevant process (e.g. photosynthesis or flowering) depends linearly on temperature. Only then one constant temperature sum is needed (see also Section 2.1). For example, if flower development shows an optimum at 20 °C and is substantially slower at 16 and 24 °C, we cannot compensate 16 °C with 24 °C later. Temperature integration has been investigated and successfully applied in many crops, e.g. cucumber (Hao *et al.*, 2015), potted plants (Rijsdijk and Vogelezang, 2000), sweet pepper (Bakker and Van Uffelen, 1988), roses (Buwalda *et al.*, 1999), tomatoes (De Koning, 1990) and minor greenhouse crops such as freesias, radishes, lettuce and endives (as mentioned in Grashoff *et al.*, 2004). Greenhouse temperatures can thus be regulated with a view to saving energy (Box 12.3). When the sun heats up the greenhouse, the energy is free and the temperature can be allowed to rise slightly above daytime setpoint. If the grower must use the heating the next day (e.g. a very windy or cloudy day), a lower temperature can be maintained such that the long-term average temperature is realised. This integration is also possible within 24 hours (compensating a higher day temperature with a lower night temperature) or over a week (Elings *et al.*, 2006).

If the price of energy is variable, a grower can choose to use it, especially at times where it is cheap, and apply less heating when energy is expensive, as long as the average long-term temperature setpoint is realised. Modern climate computers allow for this temperature integration, enabling the grower to provide a long-term average temperature setpoint and an accepted bandwidth for the temperature variation. A combination of screens, ventilation and heating leads to the desired average temperature at the lowest possible energy costs. This may result in energy savings of 3% at a bandwidth of 2 °C to 13% at a bandwidth of 10 °C (Buwalda *et al.*, 1999).

7.3 Carbon dioxide (CO$_2$)

Raising the CO$_2$ concentration in the greenhouse air stimulates crop growth considerably, as without a CO$_2$ supply the greenhouse air CO$_2$ concentration often limits crop growth and yield. Gross photosynthesis shows a saturating response to radiation, and at all radiation levels a higher CO$_2$ concentration improves photosynthesis (Figure 7.11).

In lower CO$_2$ ranges the effect is large, but an increase from 600 to 700 µmol mol^{-1} does not contribute a lot. Above 1000 µmol mol^{-1}, there is no further improvement. Therefore, the relationship between CO$_2$ concentration and canopy gross photosynthesis shows a saturating response. Raising the CO$_2$ level significantly, e.g. to 1,200 or 1,500 µmol mol^{-1}, can be risky, especially when CO$_2$ is supplied via flue gases from a heater or co-generator (see Section 11.3 and Table 11.2). Additionally, CO$_2$ itself can cause undesired effects at elevated levels: the stomata may close and crop temperature can become too high.

CO$_2$ enrichment is very common in modern greenhouses. It increases growth rate and yield by 15-30% and also leads to better plant quality. A compilation of experiments with a variety of vegetable crops conducted in unheated greenhouses in Southern Spain showed yield increases between 12 and 22% (Table 7.1). Elevated CO$_2$ causes higher production of assimilates, resulting in more branches, thicker leaves, less blind shoots (e.g. in roses) and a better fruit set. Poorter and Pérez-Soba (2001) conducted a meta-analysis of literature data and found

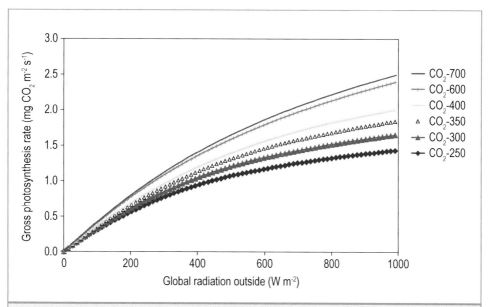

Figure 7.11. Simulated effects of global radiation outside and greenhouse air CO$_2$ concentration on canopy gross photosynthesis rate.

Table 7.1. Increase in yield resulting from CO_2 supply for several vegetable crops grown in unheated greenhouses in Spain (Sanchez-Guerrero *et al.*, 2010).

Crop	Greenhouse type	CO_2 max (vpm)	CO_2 consumption (kg m^{-2})	ΔYield (%)
Cucumber – Autumn	multi-tunnel	700	2.2	19
Bean – Spring	parral	600		12
Bean – Autumn	parral	600		17
Pepper – Spring	multi-tunnel	750		22
Pepper – Autumn	multi-tunnel	750	4.7	19
Tomato – Autumn	multi-tunnel	800	3.3	19

that low soil nutrient supply or sub-optimal temperatures reduced the proportional growth stimulation of elevated CO_2.

When leaves are exposed to elevated CO_2 there is an immediate increase in photosynthesis due to decreased photorespiration, which is negatively affected by an increase in atmospheric CO_2/O_2 ratios. A stimulation of net photosynthesis can be observed even when plants are grown for extended periods at high CO_2 levels, however, in that situation photosynthetic capacity can decrease, a process known as 'acclimation' (Foyer *et al.*, 2012). The degree to which 'acclimation' occurs depends on the source/sink balance which is, for example, influenced by environmental conditions (see Chapter 2). An elevated CO_2 concentration in a semi-closed greenhouse did not cause feedback inhibition in a high-producing tomato crop, because the plants had sufficient sink organs (fruits) to utilise the extra assimilates (Qian *et al.*, 2012). In (semi)closed greenhouses, production is higher than in open greenhouses (Table 7.2). This production increase is due to the higher assimilation (dry matter production), i.e. higher total

Table 7.2. Cumulative tomato fruit production until week 29 after planting and total amount of supplied CO_2 in greenhouses with 700 W m^{-2}, 350 W m^{-2} and 150 W m^{-2} cooling capacities, respectively, and in the conventional greenhouse. Values in parentheses indicate the relative yield increase compared to a conventional greenhouse (Qian *et al.*, 2011).

Cooling capacity (W m^{-2})	Yield (kg m^{-2})	Supplied CO_2 (kg m^{-2})
700	28 (14%)	14
350	27 (10%)	30
150	26 (6%)	46
0	24	55

crop photosynthesis. Dry matter partitioning to the fruits did not differ between treatments. An analysis of climate and plant growth data with a crop growth model shows that the differences in dry matter production could be explained by the realised higher CO_2 concentration in (semi) closed greenhouses (Figure 7.12; Qian, 2017).

In this chapter, crop responses to light, temperature and CO_2 have been presented more or less separately, however, these responses depend on the level of the other factors. For example, the optimum temperature for leaf photosynthesis is higher at an elevated CO_2 level than an ambient level (Taiz and Zeiger, 2010). The interaction among light, CO_2 and temperature effects on tomato leaf photosynthesis is shown in Figure 7.13, whereas an example of the same interacting effects on tomato crop photosynthesis is shown in Figure 7.14. With low light levels, photosynthesis decreases at temperatures above 25 °C, whereas with high light levels, photosynthesis increases with temperatures above 25 °C. Also note the different responses for leaves (Figure 7.13) and crops (Figure 7.14), mentioned earlier in Chapter 2 (Figure 2.10).

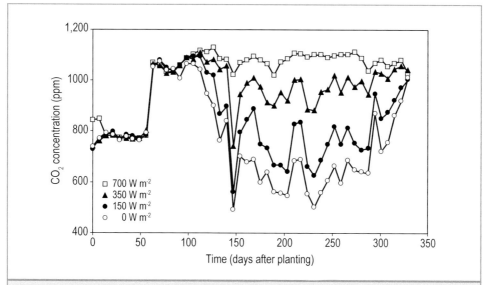

Figure 7.12. Weekly average day-time CO_2 concentrations in greenhouses with 700, 350, and 150 W m^{-2} cooling capacities, and in a conventional greenhouse (Qian *et al.*, 2011).

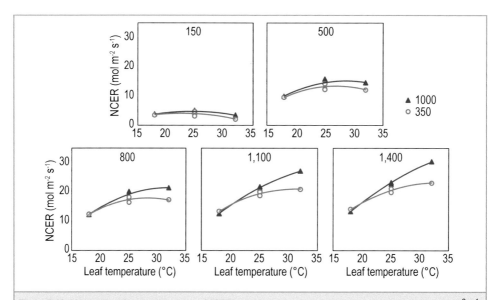

Figure 7.13. Mean net carbon exchange rate (NCER) as a function of irradiance (top of each panel; μmol m^{-2} s^{-1}) and prevailing CO_2 concentration (legend; μmol mol^{-1}) and a 1.1 (T=18 and 25 °C) or 2.2 (T=25 and 30 °C) kPa vapor pressure difference. Mean values of plants grown at 350 and 700 μmol CO_2 mol^{-1}, leaf temperature 18, 25 or 32 °C obtained by adapting air temperature (re-drawn from the data set described by Stanghellini and Bunce, 1993).

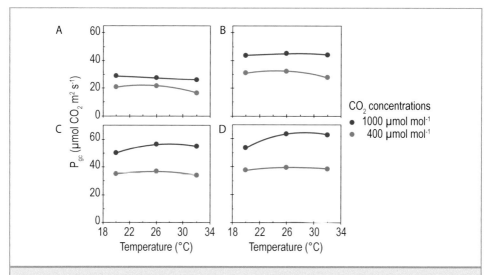

Figure 7.14. Measured gross crop photosynthesis (P$_{gc}$) values for tomato as a function of temperature at two CO_2 concentrations (400 or 1000 μmol mol^{-1}) at four photosynthetic photon flux densities (PPFD) (A: 300, B: 600, C: 900 or D: 1,200 μmol m^{-2} s^{-1}). Data were fitted using quadratic functions. Standard errors were smaller than the symbols used. (Körner *et al.*, 2009).

7.4 Humidity

Air humidity can be expressed as partial vapour pressure, vapour concentration or relative humidity (RH; see Table 4.1). Air humidity at saturation increases significantly with air temperature. Humidity is an important physical aspect of the greenhouse air: the crop puts water vapour into the greenhouse air via transpiration, which plays a significant role in the energy balance of the greenhouse (see Chapter 6). Humidity represents latent heat: preventing too much humidity, for example through natural ventilation, costs energy, because humid warm air is removed from the greenhouse (see Chapter 5). In a wide range, humidity is not very important for the physiological plant processes such as photosynthesis or biomass partitioning, except in extreme situations (a very high or very low humidity).

7.4.1 High humidity

Extremely high humidity (a vapour pressure deficit below 0.2 kPa [<1.5 g m^{-3}], which means a RH above 94% at 25 °C) can cause problems. The plant cannot transpire enough and therefore the water flow in the plant is almost absent. This means there is almost no calcium (Ca) transport (Figure 13.1), leading to a Ca shortage in the leaves, resulting in smaller leaves, therefore less light interception and less crop photosynthesis. Furthermore, at a high RH the pollen will not release from the anthers and fertilisation is inhibited. High humidity is often prevented in greenhouses, as it increases the risk of diseases (e.g. botrytis), especially in combination with a heterogeneous temperature distribution such that water vapour condenses on the coldest parts. While transpiration stagnates, water uptake will to some extent continue. This combination leads to problems such as cracks in tomato fruit (the fruit 'explodes' because of too much water) and glassiness in lettuce (the leaves contain water in the intercellular spaces).

High humidity may also reduce yield; for example, 18-21% reduction in tomato yield at a vapour pressure deficit (VPD) of 0.1 kPa (high humidity) compared with a VPD of 0.5 kPa, which may be related to a diminution of the leaf area due to a calcium deficiency in the foliage (Holder and Cockshull, 1988). Cultivation at high humidity may negatively influence the vase life of cut flowers, e.g. roses. The severity of this effect depends on the cultivar (Mortensen and Gislerød, 1999). The main reason for a shorter vase life for roses cultivated at high humidity is a compromised stomatal closing ability, resulting in desiccation (Fanourakis *et al.*, 2016).

7.4.2 Low humidity

The other extreme is very low humidity: a deficit more than 1 kPa (>7.5 g m^{-3}) which means a RH below 70% at 25 °C. This dry air creates a high evaporative demand and leads to stress. The plant partly closes the stomata to reduce transpiration, but this also means higher stomatal resistance for CO_2 influx. Hence, leaf photosynthesis is reduced. In addition, cell elongation is reduced, leading to smaller leaves that cannot intercept as much light. Both effects of low humidity reduce crop photosynthesis and consequently growth and yield.

In tomato plants, prolonged low humidity results in a lower fruit fresh weight. Lu *et al.* (2015) studied the effects of VPD controlled by a fogging system on greenhouse environment and tomato plant growth. The VPD was effectively reduced by the fogging system from 1.4 to 0.8 kPa on average at midday during the entire winter season. Maintaining a lower VPD at midday increased the tomato leaf stomatal index and stomatal conductance during the majority of the day, which led to an increase in the net photosynthetic rate. Furthermore, maintaining a lower VPD increased the mean tomato biomass and yield by 17.3 and 12.3%, respectively. Under Mediterranean conditions, a VPD higher than 2 kPa during the hottest hours of the day reduced fruit growth and fruit fresh weight, probably due to a reduction in the fruit-stem water potential gradient. Under those conditions, total yield is also reduced by a decrease in fruit size (higher dry mass content; Leonardi *et al.*, 2000). Ehret and Ho (1986) found comparable results when studying the effects of salinity on dry matter partitioning in tomato fruits. Even when fruit dry weight is not influenced, fruit fresh weight is reduced. While this can result in a better taste, it certainly reduces yield.

There is yet another effect of humidity. The big leaves continue to transpire and attract all the xylem water flow in the plant. Consequently, there is no water left for the non-transpiring parts, such as the fruits, leading to local calcium deficiency because calcium is exclusively transported passively with the water flow. Even if there is enough Ca in the plant as a whole, local deficiencies can occur. Calcium is not redistributed in the plant (it stays where it arrives with the xylem flow), whereas most other ions can be redistributed. In tomatoes and sweet peppers, a shortage of Ca in the tip of the fruit causes blossom end rot (Figure 7.15; Ho and White, 2005). When plants are grown in saline conditions, the same occurs due to the difficulties of absorbing enough water. Furthermore, at saline conditions, the concentration of Ca relative to the other cations decreases.

Figure 7.15. Blossom end rot (local Ca deficiency) in tomatoes (left) and peppers (right). (photos: Wim Voogt, Greenhouse Horticulture, Wageningen University & Research)

7.5 Drought and salinity

7.5.1 The Maas-Hoffman model

Early responses to water and salt stress are essentially identical (Munns, 2002): the result of both is that an uptake of water by the plant is limited. For a general description of yield response to drought or salinity, the Maas-Hoffman model (Maas and Hoffman, 1977 cited in Grieve *et al.*, 2012) is very useful. Yield is at its maximum (plateau) for a range of drought or salinity levels (the optimum range), however it drops when levels are above a certain threshold. This drop is linearly related to the factor on the x-axis and is characterised by its slope: yield gradually falls to zero (Figure 7.16).

The shape of the curve for drought and salinity (expressed by the electrical conductivity (EC; dS m^{-1}; Box 13.1) of the solution) is similar. Differences are seen in the factor on the horizontal axis, the position of the threshold level and the slope of the linear decrease. The horizontal axis shows $1 - \theta$ (θ being the volumetric water content of the root zone), or the EC of the solution in the root environment. If $1 - \theta$ approaches zero, the situation is called water logging. The soil or substrate is saturated with water, but the plant cannot use it. Due to an oxygen shortage, the roots cannot function well. This explains the low yield in the first section of the graph (Figure 7.16): excess water can lead to decreased yield. This is followed by an ideal section for $1 - \theta$. After the threshold harvest decreases due to increasing drought, the form of the curve remains the same, the levels and thresholds are crop specific. At very low EC levels, there are simply not enough nutrient ions available for good growth; yield is

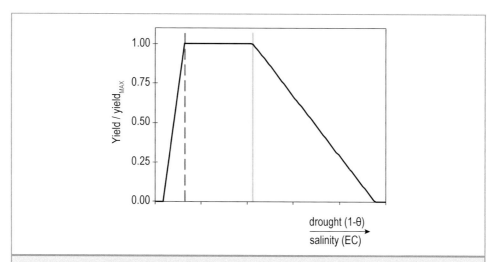

Figure 7.16. Crop response to drought and salinity. To the left of the "optimal plateau" is either water logging or nutrient deficiency; to the right there is yield decrease due to drought or salinity. The two parameters of the Maas-Hoffman model (see Figure 7.17) are the threshold (green line here) and the slope of the decrease thereafter.

suboptimal, then follows the optimal EC range. After the threshold, increasing salinity affects production negatively (Figure 7.16). Crops differ in sensitivity to salinity (Grieve *et al.*, 2012): both the threshold and the declining slope are crop specific (Figure 7.17). Asparagus is very salt-resistant, radishes, carrots and lettuce are very sensitive and tomatoes are moderately sensitive. Besides plant species, the genotype within a species (cultivar) also determines the sensitivity to salinity.

7.5.2 Spatial and temporal variation in electrical conductivity

If the EC in the root environment shows a spatial variation, the plant response corresponds to the lowest EC level, not the average level (Sonneveld, 2000). To combine the advantage of a higher EC (better taste) with the advantage of a low EC (higher yield), a split root system could be a solution. Half of the roots receive a high dose of minerals, while the other half a very low dose (Box 7.5). Mulholland *et al.* (2002) observed that for a split-root, high EC (2.8/8.0 dS m^{-1}) treatment enhanced the tomato fruit quality by increasing the concentrations of sugar, acid and total soluble solids, and reducing the incidence of visual defects such as uneven ripening and gold-spot.

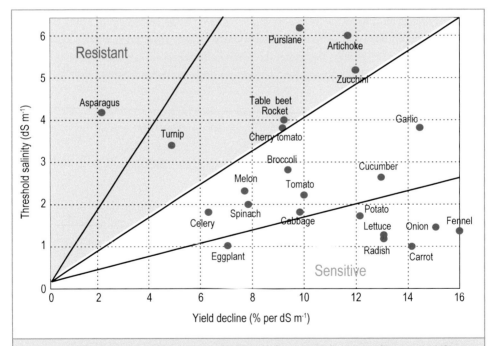

Figure 7.17. Crop-specific threshold and slope in the Maas-Hoffman model for salinity effects on yield (Shannon and Grieve, 1998). Note that cherry tomato is more resistant to salinity than tomato. Within species cultivars can differ in sensitivity. Furthermore, radish is indicated as sensitive, whereas Marcelis and Van Hooijdonk (1999) report radish as resistant to salinity, which may be explained by differences in other environmental conditions.

Box 7.5. Poor nutrient distribution in root substrate is not always a disaster.

Whichever substrate you use, there will always be small areas with more or less minerals. The question is, is this bad for plant growth? Cees Sonneveld, former head of plant nutrition and substrates at the research station in Naaldwijk, the Netherlands, carried out research during the 1990s with interesting results (Sonneveld, 2000).

Sonneveld split the root system of a tomato plant. One half was grown in one stone wool block, the other half in another block. In this way he could supply both blocks with a nutrient solution of different electrical conductivity (EC) (see illustration).

Remarkable results were obtained. Water was preferably taken up by the plant from the block with the lowest EC, so from where it was easiest to withdraw water. Nutrient uptake preferably was from the block with the highest EC. Yield was not reduced if the EC increased, even to 10 dS m^{-1}, as long as the other block remained low in EC. So even with a big difference in EC the yield stayed the same as at a normal EC. Local salt accumulation in the slab therefore does not seem to cause a problem. The researcher also noticed that plants respond mainly to the EC in the irrigation water and less to the average EC in the slab. The explanation: irrigation water is applied where the most active roots have accumulated. Cucumber exhibited the same behaviour, but is more sensitive to areas with a high EC. These trials also clearly show that it is possible to separate the uptake of water and nutrients.

In several trials, half of the roots a plant were grown in a stone wool block with a low EC, the other half were grown in a block with a high EC. The plant drew water from the block with a low EC and the nutrients from the block with a higher EC. (illustration: Wilma Slegers)

A more practical method for varying the EC is to utilise a different EC during day and night (temporal pattern). When the sun is shining and transpiration is high, a low EC makes it easy for the plant to absorb a lot of water. However, a constantly low EC leads to poor fruit taste and nutrient deficiencies. Therefore, at night time when transpiration is low, the lower daytime EC should be compensated for with a high night EC. For example, in tomato plants, day time EC = 1 and night time EC = 9 dS m^{-1}. This method leads to a low stress level and easy water and nutrient uptake, and, in tomatoes and sweet peppers, it drastically reduces blossom end rot (Van Ieperen, 1996). A quick change in EC is only possible in a cultivation system without a substrate (e.g. nutrient film technique; NFT , Box 13.6), or a system with little water buffer in the root zone (Box 13.3). However, when the cultivation is in a substrate, e.g. stone wool or perlite, it helps to irrigate during the day with a low EC and make this EC dependent on the radiation (more radiation, lower EC of nutrient solution). Most growers are already applying such a strategy. When growing in soil you do not have such possibilities, because you cannot change the EC quickly enough (large buffer) (see Section 13.1.2).

7.5.3 Electrical conductivity and fruit quality

Tomato shelf life and sugar content improve when the EC in the root environment is higher (Table 7.3). A higher EC can be obtained by adding NaCl to the nutrient solution or by increasing the concentration of all macro nutrients. When adding NaCl, yield decreases by 10%, whereas increasing EC by raising the concentration of macro nutrients hardly influences yield. However, in the latter case, positive effects on fruit quality are also much smaller. The negative effects of NaCl on tomato yield depend on the greenhouse climate and the cultivar. For example, Li *et al.* (2001) compared yield of tomato at 9 dS m^{-1} obtained by either adding NaCl to the standard nutrient solution or supplying a concentrated solution. The small (negative) effect of NaCl could be fully explained by the slight effect on osmotic pressure in the root zone. A higher EC stimulates the content of healthy substances in tomato fruit, such as vitamin C and lycopene (Fanasca *et al.*, 2007). Petersen *et al.* (1998) observed in tomato plants that increased salinity in the root zone increased the concentrations of dry matter, sugars, titratable acid, vitamin C and total carotene in the fruit, and in all cases but one, these effects were independent of salinity source.

7.5.4 Salinity effects on yield – underlying processes

Salinity creates difficulties in water absorption. There are three possible processes via which salinity leads to a lower yield (Figure 7.18). First, SLA is reduced because the leaves cannot expand enough due to the difficulty in water uptake, therefore, the leaves get thicker. Second, the plant will try to reduce transpiration and in doing so closes the stomata, which reduces the influx of CO_2 and thus photosynthesis decreases. Third, the fruits cannot be diluted with enough water, making their dry matter content (DMC) relatively high, resulting in a lower fresh yield.

Table 7.3. Effect of electrical conductivity (increase of either macro-nutrients or by adding sodium chloride) on tomato fruit quality (Verkerke *et al.*, 1993).

Electrical conductivity (dS m^{-1})	Shelf life (days)	Yield	Sensory panel (scale 0-100)	Sugar content (% FW)
3	24	100%	53	2.9
6	28	98%	54	3.1
6 NaCl	31	90%	60	3.5

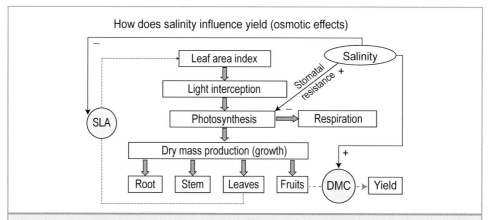

Figure 7.18. Schematic presentation of how the osmotic effects of salinity influence crop yield. Salinity can increase stomatal resistance, reduce leaf area extension (lower SLA) and/or increase fruit dry matter content, all resulting in a lower fruit fresh yield.

In a simulation study (Heuvelink *et al.*, 2003), the relative contribution of these three processes have been weighed up. DMC is 5% at EC = 2 dS m^{-1} and increases linearly by 0.2% per dS m^{-1}. SLA decreases 8% per dS m^{-1} from a threshold of 3 dS m^{-1}. Stomatal resistance is estimated at 50 s m^{-1} if the stomata are wide open and increases over the range of 1 to 10 dS m^{-1} by a factor of 2 or 4.

Increased fruit DMC or stomatal resistance, as a result of salinity, only have moderate effects on tomato fruit yield, however the effect via SLA is dramatic (Figure 7.19). At high salinity, the leaves are very small and thick, which lowers yield considerably. What can a grower do? Keep as many leaves on the plants as possible by delaying leaf picking and increasing plant density.

7.5.5 Salinity stress mitigated at high air humidity

Three cultural techniques that have proven useful in tomato plants to overcome, in part, the negative effects of salinity are: (1) treatment of seedlings with drought or NaCl, which ameliorates the adaptation of adult plants to salinity; (2) mist (increasing humidity) applied to tomato plants, improving vegetative growth and yield in saline conditions; and (3) grafting tomato cultivars onto appropriate rootstocks (Cuartero *et al.*, 2006). An elevated CO_2 concentration may also alleviate the negative effects of salinity on photosynthesis and plant growth, probably owing to the improvement in oxidative stress, as well as the water status through stomatal closure at a high CO_2 concentration (Takagi *et al.*, 2009).

It may be expected that at high greenhouse air humidity, salinity is less of a problem since the plant water demand is lower (reduced transpiration). This hypothesis was investigated by Li *et al.* (2001, 2004) in a tomato crop. These authors combined contrasting root-zone salinity treatments (EC 2, 6.5, 8 and 9.5 dS m^{-1}) with two climate treatments – a reference

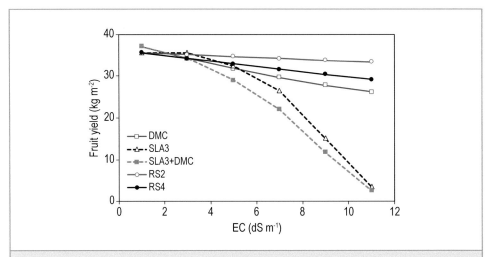

Figure 7.19. Total fruit fresh yield as affected by salinity with 5 different underlying assumptions: dry matter content (DMC), fruit dry matter content increases by 0.2% per dS m^{-1}; SLA3, specific leaf area decreases with 8% per dS m^{-1} starting from a threshold of 3 dS m^{-1}; DMC+SLA3, combination of both former effects; RS2 and RS4, stomatal resistance increases with a factor of 2 or 4 over the salinity range 1 to 10 dS m^{-1} (Heuvelink *et al.*, 2003).

(high transpiration, HET0) and a 'depressed' transpiration (low transpiration, LET0), obtained through misting. Marketable fresh-yield production efficiency decreased by 5.1% for each dS m^{-1} above 2 dS m^{-1} (Figure 7.20). The number of harvested fruits was not affected; yield loss resulted from reduced fruit weight (3.8% per dS m^{-1}) and an increased fraction of unmarketable harvest. At the LET0 treatments, yield loss was only 3.4% per dS m^{-1} in accordance with the reduction in fruit weight. Li *et al.* (2004) concluded that control of the shoot environment in a greenhouse to manipulate the fresh weight of the product may mitigate the effects of poor quality irrigation water without affecting product quality.

7.6 Concluding remarks

Light intensity and light integrals primarily influence photosynthesis and growth, whereas light spectrum (light quality) mainly influences plant morphology. The average temperature primarily influences development; diel temperature pattern influences elongation and therefore growth of young plants, while fully-grown crops allow for 'temperature integration'. CO_2 enrichment increases photosynthesis. Plant shape is influenced by light, temperature and CO_2 via the source/sink ratio, which is the ratio between assimilate production and assimilate demand. The most dramatic effects of salinity on yield result from reduced leaf expansion. This chapter provides many examples of the influences of environmental factors on production, quality of the product and quality of the crop. In the following chapters we will examine which actuators are used for climate control and how crop knowledge is used in control strategies.

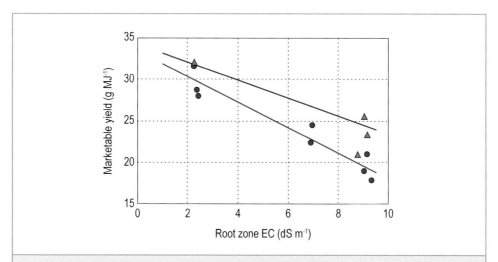

Figure 7.20. Efficiency of production of marketable yield (tomato) vs mean electrical conductivity (EC) in the root zone. Each symbol is the mean of a full-cycle experiment, red with 'standard' climate management, blue for experiments with humidity raised through misting, but same temperature (re-drawn from Li *et al.*, 2001).

Photo: Bert van 't Ooster

In a well-equipped greenhouse the climate is often controlled by an interacting set of measures, taking into account the vapour, energy and CO_2 balances. The goal is to adjust the climate with a minimum of resources; however, if climate factors for crop growth are not well tuned, the use of these resources may be suboptimal. In Section 5.2 (minimum) ventilation is described. The present chapter focusses on heating requirement and actuators for heating. Both the theory and incorporation in energy balances are discussed to evaluate their role from the perspective of climate control and resource use efficiency.

In temperate climate regions a greenhouse is heated to control the interior air temperature or better to prevent cold stress of the crop. Since the crop is the actual target, the focus of temperature control should be on crop temperature. In steady-state conditions the absolute temperature difference between leaves and air is normally less than 2 °C, but a sudden rise in solar radiation can cause differences up to 5-6 °C; this, however, will only last until the stomata have responded to the sudden rise in energy input. In practice most growers only measure air temperature, but for adequate control more sensors are relevant (Box 8.1). It is tempting to think (Jongschaap *et al.*, 2009) that instead of heating the air it would be better to heat the crop directly by means of radiative heating. However, as leaves are very good heat exchangers, they immediately release the energy to the air, so that in the end there is hardly any difference in the energy used to obtain a desired leaf temperature through radiative heating or through hot water pipes heating the air and the leaves.

8.1 Heat demand of a greenhouse

We use the energy balance of the greenhouse to determine the heating requirement. It is necessary to distinguish between the required combined capacity of energy sources and the actual heat requirement at any given moment in time, both in ($W\ m^{-2}$). Using this information it is possible to select the most feasible actuator.

Box 8.1. Basics of climate control.

The figure below shows the basic setup of a climate control system with some relevant sensors to measure temperature, humidity, radiation, wind speed and wind direction. An inner feedback loop supports automatic control by the process computer and the outer feedback loop is based on observations by the grower who makes modifications in the inner control loop (that is: selects the set-points for the 'target' variables, Box 12.1). The climate control system needs temperature sensors inside and preferably outside the greenhouse as well. Temperature sensors alone give a poor picture of the climatic situation and the necessary measures to reach a desired climate inside the greenhouse. Other useful sensors are humidity sensors (Box 4.1), photosynthetic active radiation (PAR) or quantum sensors (Box 3.3), radiation sensors (pyranometer for outside global radiation I_s (Box 3.3) and a pyrgeometer for net longwave radiation $R_{n,TIR}$ (Box 3.7)), anemometer for wind speed and a wind direction sensor.

Sensor data are logged to deliver feedback signals to the controller. Wind speed and direction affect leakage ventilation and heat transfer and therefore overall energy loss. Wind speed is mainly used for safety; wind direction to choose leeward and windward side (Box 12.1). A controller can use these signals to respond to changes in heat loss. All sensors (except radiation sensors) should be within a ventilated white box containing a thermometer, a humidity sensor and optionally a CO_2 sensor. The white box protects the sensors against direct solar radiation and longwave radiation effects , and the ventilation ensures air properties representative of the wider environment. There is one disadvantage to clustered sensors: temperature and humidity are measured at only one spot, while there can be big horizontal and vertical gradients in the greenhouse. To solve this problem, networks of wireless sensors have been successfully applied to map heterogeneity of temperature and humidity within greenhouses (e.g. Balendonck *et al.*, 2014). Wired and wireless sensors together provide a good picture of the variation within the greenhouse climate. Wireless temperature and humidity sensors are increasingly popular in climate control but have less reliability, need more maintenance and are in general not ventilated what can have significant effect on the accuracy of the measurement during daytime. Use of multiple sensors in climate control routines is limited by the number of actuators (for instance, one heating valve per compartment).

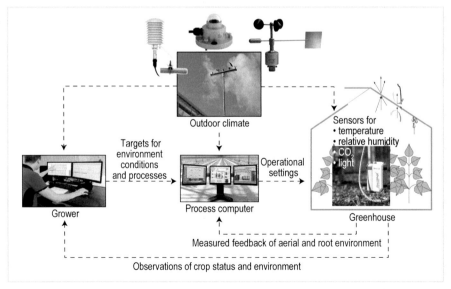

Basic setup of a climate control system: Boxes with amongst others shaded and ventilated temperature and humidity sensors inside and outside, light, wind speed and direction sensors outside. An inner control loop for automatic control by the process computer and an outer control loop based on observations by the grower.

8.1.1 Required heating capacity

A quick approach to finding the required total heat capacity is to balance the supplementary heat flux $H_{heating}$ with the net heat loss, i.e. the heat loss of the greenhouse $H_{in,out}$ (see also Equation 3.7) minus the heat gain resulting from solar radiation H_{sun}:

$$H_{heating} = max\left(H_{in,out} - H_{sun}, 0\right) \hspace{3cm} \text{W m}^{-2} \quad (8.1)$$

$$H_{heating} = max\left(U \cdot \frac{A_c}{A_s} \cdot \left(T_{iH} - T_{out,D}(t_{th})\right) - (1-f) \cdot \tau \cdot (1-\rho) \cdot I_s, 0\right)$$

Where the *U*-value in building physics is the unit thermal conduction through construction elements in W m^{-2} K^{-1}, also called the effective heat transfer coefficient. It combines the effects of convection, radiation and conduction (Chapter 3). Details of conversion are given in Box 8.2 and Box 8.3. A_c / A_s is the specific greenhouse cladding area (-), A_c is the area of the cover and the sidewalls (m^2), A_s the soil or floor area (m^2), T_{iH} is the target greenhouse air temperature

---

Box 8.2. Calculation of the U-value by summing thermal resistances.

Method 1: The total thermal resistance of a construction element is $R_o = R_i + R_c + R_e$ where R_c is the thermal resistance of the construction (m^2 K W^{-1}), R_i and R_e are the thermal resistance of the inner and outer boundary layer of the construction (m^2 K W^{-1}). Index i = inside, c = construction, e = outside (external). The U-value is the reciprocal value of R_o: $U_c = 1 / R_o$.

The thermal resistance components are the construction R_c and the boundary layers inside R_i and outside R_e. The resistance of the construction R_c is the sum of thermal resistance of its layers:

$$R_c = \sum R_{c,nh} + \sum_{l=1}^{m} \frac{d_l}{\lambda_l} + R_{cav}$$

which may consist of non-homogeneous layers $\sum R_{c,nh}$ like the duct layer(s) in twin wall panels, homogeneous layers $\sum d_l / \lambda_l$ like plastic or glass sheets with λ_l the thermal conductivity of the material in layer l (W m^{-1} K^{-1}), and cavities R_{cav} like in double glazing (m^2 K W^{-1}). The thermal resistance of the boundary layers may be defined as:

$$R_i = \frac{1}{\alpha_{ci} + \alpha_{Ri}^*} \hspace{2cm} R_e = \frac{1}{\alpha_{ce} + \alpha_{Re}^*}$$

Where α_{ci} and α_{ce} are the convective heat transfer coefficients of the inner and outer boundary layer. These are functions of dimensions, temperature difference and air velocity; α_{Ri}^* and α_{Re}^* are the apparent linearised heat transfer coefficients for long wave radiation of the inner and outer boundary layer (apparent because radiant energy is exchanged with environmental 'objects' and not with air). Therefore α_{Ri} and α_{Re} are temperature corrected:

$$\alpha_{Ri}^* = \alpha_{Ri} \frac{T_{Rin} - T_{iS}}{T_{air} - T_{iS}} \text{ and } \alpha_{Re}^* = \alpha_{Re} \frac{T_{eS} - T_{Rout}}{T_{eS} - T_{out}}$$

Where α_{Ri} and α_{Re} are the heat transfer coefficients inside and outside resulting from linearisation of the Stefan-Boltzmann's law (Box 3.6); T_{iS}, T_{eS}, T_{Rin}, T_{Rout}, T_{air}, and T_{out} are (best estimates of) inner surface temperature of the cover, outer surface temperature, radiation reference temperature inside and outside, and air temperature inside and outside (°C). This approach is closest to the definition of a U-value used in building physics (heat loss equals U times area times temperature difference between indoor and outdoor air space), but requires surface temperatures of the cover.

Box 8.3. Alternative for calculating the heat loss of dry and thin construction layers.

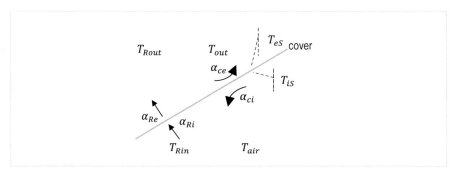

Overview of symbols and position.

Method 2: Three heat flows, through (1) the inner boundary layer, (2) the construction and (3) the outer boundary layer, are all equal at equilibium, and defined as:

$$H_{in,out} = \alpha_{Ri} \cdot (T_{Rin} - T_{iS}) + \alpha_{ci} \cdot (T_{air} - T_{iS}) \qquad (1)$$

$$H_{in,out} = \lambda / d \, (T_{iS} - T_{eS}) \qquad (2)$$

$$H_{in,out} = \alpha_{Re} \cdot (T_{eS} - T_{Rout}) + \alpha_{ce} \cdot (T_{es} - T_{out}) \qquad (3)$$

Substitution and separation of variables delivers:

$$H_{in,out} = U_c^* \cdot (T_{air}^* - T_{out}^*)$$

Where $U_c^* = \left(\dfrac{1}{\alpha_{ci}+\alpha_{Ri}} + \dfrac{d_c}{\lambda_c} + \dfrac{1}{\alpha_{ce}+\alpha_{Re}} \right)^{-1}$; $T_{air}^* = \dfrac{\alpha_{Ri} \cdot T_{Rin} + \alpha_{ci} \cdot T_{air}}{\alpha_{Ri} + \alpha_{ci}}$; $T_{out}^* = \dfrac{\alpha_{Re} \cdot T_{Rout} + \alpha_{ce} \cdot T_{out}}{\alpha_{Re} + \alpha_{ce}}$.

U_c^* is defined with use of the real heat transfer coefficients, and inside and outside air temperature are the weighted average of the air temperature and the radiation reference temperature. The inner and outer surface temperature of the cover are eliminated.

Example:
Method 1 and Method 2 compared for a single glass cover and 4 m s^{-1} wind speed.

Data: $\alpha_{Ri} = 6$; $\alpha_{Re} = 6$; $\alpha_{ci} = 3.7$; $\alpha_{ce} = 20$; $T_{air} = 20\ °C$; $T_{Rin} = 22\ °C$; $T_{Rout} = 0\ °C$; $T_{out} = 10\ °C$.
$\dfrac{A_c}{A_S} = 1.2$; $d_{cov} = 4\ mm$; $\lambda_{glass} = 0.84\ W\ m^{-1}K^{-1}$.

Answer: *Method 1:* $H_{in,out} = \dfrac{A_c}{A_S} \cdot U_c \cdot (T_{air} - T_{out}) = 1.2 \times 9.26 \times (20 - 10) = 111\ W\ m^{-2}$

Method 2: $H_{in,out} = \dfrac{A_c}{A_S} \cdot U_c^* \cdot (T_{air}^* - T_{out}^*) = 1.2 \times 6.83 \times (21.2 - 7.7) = 111\ W\ m^{-2}$

Exercise: Find U_c, U_c^*, T_{air}^*, T_{out}^* yourself.

for heating (°C)[1], $T_{out,D}$ (t_{th}) (°C) is the design outdoor air temperature. This is the threshold where in the long term for t_{th} % of the time outdoor air temperature T_{out} is less than or equal to $T_{out,D}$. The parameter t_{th} is either set in standards or chosen by the designer of the system. f is the estimated fraction of solar radiation in the greenhouse that is converted to latent heat under normal operation of the system, τ is transmissivity of the greenhouse and ρ light reflection from within the greenhouse. In Equation 8.1 solar irradiation I_s (W m^{-2}) is present because the night is not necessarily the time with the highest heat demand. The standard U-value is defined as the energy flow through the building envelope, when temperature difference between inside and outside is 1 °C. The *U*-value in Equation 8.1 is defined more generically as it includes the effects of screening and leakage ventilation:

$$U' = U_{tr} \cdot \frac{A_c}{A_s} + U_{vleak} \qquad\qquad \text{W m}^{-2}\text{ K}^{-1} \quad (8.2a)$$

$$U_{tr} = U_c \cdot (s_{screen} \cdot (1 - p_{screen}) + (1 - s_{screen})) \qquad\qquad (8.2b)$$

Where U_{vleak} is the heat loss contribution from leakage ventilation. Leakage ventilation contributes an additional loss of 0.4 to 1 W m^{-2} K^{-1}. Specific leakage ventilation $g_{v,leak}$ (m^3 m^{-2} s^{-1}) depends on the greenhouse structure and wind. U_c is the U-value of the greenhouse cover construction, s_{screen} is a binary switch function (0 when open; 1 when closed), $p_{screen} \in [0 ; 1]$ is the energy-saving fraction by the screen. In fact, Equation 8.1 is not very accurate because both f in Equation 8.1 and U_c in Equation 8.2b are assumed constants, which is not quite accurate. Nevertheless, it gives a good and quick indication of the magnitude of the required heating capacity. In fact U_c depends on several factors (Figure 8.1): long wave radiation from the inside to the walls and cover, convection in the boundary layer of the cover/wall inside, (condensation to the cover), thermal conduction through the construction, convection in the boundary layer of the cover/wall outside, and long wave radiation from the cover/wall to the outside environment.

In a well-insulated building the resistance of the construction is much larger than the resistance of the boundary layers. The effect of R_i and R_e is then small and the use of standardised values of 0.13 and 0.04 m^2 K W^{-1} for R_i and R_e respectively is allowed. Greenhouses, however, are usually poorly insulated and in that case $R_c \ll R_i$ and $R_c \ll R_e$. As a result the effects of drivers of convection and radiation are prominent. A greenhouse construction is normally a single layer of glass or plastic. In that case the U-value is U_c (Box 8.2) or the alternative U_c^* (Box 8.3):

$$U_c = \left(\frac{1}{\alpha_{ci} + \alpha_{Ri}^*} + \frac{d_c}{\lambda_c} + \frac{1}{\alpha_{ce} + \alpha_{Re}^*} \right)^{-1} \qquad\qquad \text{W m}^{-2}\text{ K}^{-1} \quad (8.3)$$

$$U_c^* = (R_i + R_c + R_e)^{-1} = \left(\frac{1}{\alpha_{ci} + \alpha_{Ri}} + \frac{d_c}{\lambda_c} + \frac{1}{\alpha_{ce} + \alpha_{Re}} \right)^{-1}$$

---

[1] Temperature should be expressed in Kelvin but in case of temperature difference you may use degrees Celsius instead.

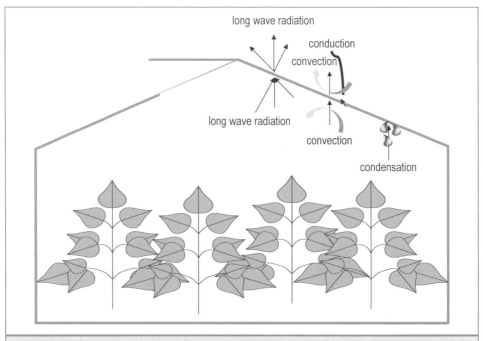

Figure 8.1. Thermal processes at the greenhouse cover: long wave or thermal radiation, convection, conduction and condensation (possibly supplemented with water cooling on the outside, see Chapter 9, Section 1.3).

Where U_c is closest to the formal definition of a U-value. The heat loss must be calculated from:

$$H_{in,out} = A_c / A_s \cdot U_c \cdot (T_{air} - T_{out}) \text{ or} \qquad \text{W m}^{-2} \quad (8.4)$$

$$H_{in,out} = A_c / A_s \cdot U_c^* \cdot (T_{air}^* - T_{out}^*) \text{ (alternative)}$$

Despite all the above objections, U-values for heat loss calculation and dimensioning of heating equipment are generally assumed to be constant or proportional to wind speed for a given construction and are presented in tables. Table 8.1 is an example indicating upper range values depending mainly on greenhouse type, cover material and heating system (Von Zabeltitz, 1986).

The α's in Equation 8.3 are influenced by air movement inside the greenhouse, wind speed and sky temperature. That's why the values in Table 8.1 are worst-case values, valid at a wind speed of 4 m s^{-1}, low sky temperature and sub optimal air movement inside the greenhouse. So, U' values are subject to constraints. They can vary roughly 1 point around the mean under different conditions.

Table 8.1. Indication of maximum U_c values (see Equation 8.2a) for different greenhouse cover materials at wind speed 4 m s^{-1} (Von Zabeltitz, 1986).

Cladding material	U_c (W m^{-2} K^{-1})
Single glass	8.8
All double glass	5.2
Double acrylic	5.0
Twin wall polycarbonate	4.8
Single polyethylene film	8.0
Double polyethylene film	6.0

Yet another loss factor is neglected, namely condensation. It delivers heat to the cover. The cover warms up and this increases the sensible energy loss through convection and long wave radiation. Stanghellini *et al.* (2012a) measured an increase of total energy loss from 12 to 16% for a wet compared to a dry cover.

Box 8.4 shows a simple U-value based calculation of the required heating capacity.

Box 8.4. Example of calculating heat capacity.

A wide-span (12 m span width) greenhouse with gross dimensions 96 m × 105 m (10,000 m^2) has an eaves height of 3.5 m and roof angle 25°. What heating capacity is needed at a design temperature of -10 °C for a single glass house and a twin wall polycarbonate house without an energy saving screen at night when the desired internal temperature is 18 °C?

The actual temperature is required to be higher than the design temperature for 99.5% of the time $t_{th} = 0.5\%$.

$$H_{heating} = U_c \cdot A_c / A_s \cdot (T_{air,t} - T_{out,D}(t_{th})) - (1 - f) \cdot \tau \cdot (1 - r_e) \cdot I_s$$

$H_{heating} = 8.8 \times 1.28 \times (18 -- 10) - 0 = 315$ W m^{-2} \qquad single glass (worst case)

$H_{heating} = 4.8 \times 1.28 \times (18 -- 10) - 0 = 172$ W m^{-2} \qquad polycarbonate twin wall panel (worst case)

$$\frac{A_c}{A_s} = \frac{2 \cdot (L + W) \cdot H_{eaves} + \frac{A_s}{\cos \alpha} + \frac{1}{2} \cdot n \cdot W_s^2 \cdot \tan \alpha}{A_f}$$

$$= \frac{2 \times (105 + 96) \times 3.5 + \frac{96 \times 105}{\cos 25} + \frac{1}{2} \times 8 \times 12^2 \times \tan 25}{96 \times 105} = 1.28 \ (-)$$

L is length of the greenhouse (m), W width of the greenhouse (m), H_{eaves} is eaves height (m), W_s is span width (m) and n is number of spans.

For single glass the capacity needed is 315 W m^{-2}, for twin wall polycarbonate 172 W m^{-2}. Usually the grower will use an energy screen at night, in which case the heating capacity is less, depending on the properties of the screen p_{screen}. Also, heat buffers help reduce the required heating capacity, as shown in Section 8.8.

8.1.2 Actual heating demand

If the actual heating demand is required rather than the capacity of the equipment, a more detailed equation is needed to account for minimum ventilation $g_{v,min}$ and heat conversion to latent heat H_{cnv}. The actual heating demand can be calculated using the following steady state balance:

$$H_h = \left(U_{tr} \cdot \frac{A_c}{A_s} + g_{v,min} \cdot \rho c_p\right) \cdot (T_{iH} - T_{out}) - (1-f) \cdot (1-\rho) \cdot \tau \cdot I_s$$
$$+ H_{cnv} + H_{AiSo} \qquad \text{W m}^{-2} \quad (8.5)$$
$$H_h = max(H_h, 0)$$

Or, in the event that air conditioning is applied at the inlet of a forced ventilation system:

$$H_h = \left(U_{tr} \cdot \frac{A_c}{A_s} + g_{v,leak} \cdot \rho c_p\right) \cdot (T_{iH} - T_{ext1})$$
$$+ max\big((g_{v,min} - g_{v,leak}),0\big) \cdot \rho c_p \cdot (T_{iH} - T_{ext2})$$
$$- (1-f) \cdot (1-\rho) \cdot \tau \cdot I_s + H_{cnv} + H_{AiSo} \qquad \text{W m}^{-2} \quad (8.6)$$
$$H_h = max(H_h, 0)$$

Where H_h is the actual heat demand at a given time, H_{cnv} is the heat flux (W m^{-2}) resulting from sensible heat conversion to latent heat. Both f and H_{cnv} fulfil a similar function. When $f \geq 1$, then H_{cnv} should be selected to account for the latent heat flux. Alternatively use $f = 0$ and let H_{cnv} account for the full latent heat flux. In Equations 8.5 and 8.6 ρ in ρc_p is air density and ρ in the radiative heat flux is light reflection from within the greenhouse. Air conditioning means that incoming air is pre-conditioned by a heat recovery system or a cooling pad which results in incoming air temperature T_{ext2}, °C. T_{ext1} is normally T_{out}, °C. In manual calculation, the interaction with the soil (H_{AiSo}) is often given a pre-set value or neglected, in a model evaluation H_{AiSo} is computed. U_{tr} can have a day and night value when the energy screen is open or closed according to Equation 8.2b. When the heating demand is calculated for every hour of the year using a climate dataset and Equation 8.5 with wind speed and sky temperature dependent U_c-value, a fluctuating graph as shown in Figure 8.2A results.

The cumulative energy use for a Dutch greenhouse is given in Figure 8.2B. In winter the line of cumulative energy use rises steeply. In summer it is almost flat. Curve a is the cumulative curve of the graph in Figure 8.2A. In curve b the thermal screen is removed, in c the leakage ventilation is 0.5 h^{-1}, and in d a 75% temperature efficient heat recovery system is applied to ventilation. The total energy use E_h was respectively 1.10, 1.63, 1.16, and 1.05 GJ m^{-2}. For conversion to fuel use please refer to Table 8.2.

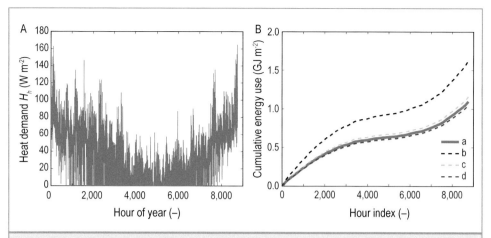

Figure 8.2. (A) Example of simulated hourly heat demand of a Dutch greenhouse with A_c / A_s = 1.12, U_c = mean 5.6, with standard deviation SD 0.78 W m^{-2} K^{-1} during the day and mean 3.1, SD 0.32 W m^{-2} K^{-1} during the night. T_{iH}=20 °C day and night. Average height was 6 m and leakage ventilation rate while heating was 0.25 h^{-1}. Greenhouse transmissivity 0.7, (1-f)=0.45, thermal screen with p_{screen}=0.4. Initial time t_0 is midnight January 1st. (B) Typical cumulative energy use of a Dutch greenhouse during the year (a), the same greenhouse without a thermal screen (b), with double leakage ventilation (c), with a 75% temperature efficient heat recovery system (d).

8.1.3 Critical outdoor temperature

The time variate critical outdoor temperature T_{coh} points out at what outside temperature (°C) the heating system should be activated. The heat balance may be used to decide when heating is needed. For a greenhouse ventilated with outside air (Section 5.3) this follows from:

$$T_{coh} = T_{iH} - \frac{\tau \cdot (1 - \rho) \cdot I_s \pm H_{AiSo} - g_{v,min} \cdot \Delta \chi \cdot L}{\frac{A_c}{A_s} \cdot U_{tr} + g_{v,min}\rho c_p} \qquad °C \quad (8.7)$$

Where T_{iH} is the target temperature (°C) for heating and T_{coh} the critical outdoor temperature for heating. If $T_{out} < T_{coh}$ then heat demand is positive. For heating the controlled minimum ventilation, either natural or forced, is the starting point. When heating is necessary, ventilation must be as low as possible. As Equation 8.7 shows, T_{coh} depends on greenhouse design, ventilation, evapotranspiration, the soil flux and radiation.

Table 8.2. Gross and net calorific values also called higher and lower heating value of some common fuels and hydrocarbons (engineering toolbox, several sources).

Fuel (phase)	Formula	Molar mass (g mol⁻¹)	Mol% methane (%)	Unit	Density (kg l⁻¹)	Gross calorific value (MJ unit⁻¹)	Net calorific value (MJ unit⁻¹)
Methane (g)	CH_4	16.043	100	kg		55.53	50.05
Methanol (l)	CH_4O	32.042	-	kg	0.792	22.66	19.92
Ethane (g)	C_2H_6	30.070	-	kg		51.90	47.52
Ethanol (l)	C_2H_6O	46.069	-	kg	0.789	29.67	26.81
Propane (l)	C_3H_8	44.097	-	kg	0.493	50.33	46.34
Butane (l)	C_4H_{10}	58.123	-	kg	0.579	49.15	45.37
Natural gas (g)[1]			83	m³		35.17	31.65
Natural gas (g)[2]			97	m³		40.30	34.00
LNG (Nigeria) (g)			91	m³		44.00	
Gasoline (l)	$C_nH_{1.87n}$	100-110	-	kg	0.72-0.78	47.30	44.00
Light diesel (l)	$C_nH_{1.8n}$	170	-	kg	0.78-0.84	46.10	43.20
Heavy diesel (l)	$C_nH_{1.7n}$	200	-	kg	0.82-0.88	45.50	42.80
Bio-oil[3]			-	kg		30.1-39.7[5]	
Biomethane			-	m³		38.30	
Woody biomass[4]			-	kg			18.0-21.0

[1] Low calorific Dutch natural gas.
[2] High calorific Russian natural gas.
[3] Bio-oil derived from hydrothermal treatment of microalgae.
[4] Based on 0% water content, palm oil residues (Paul *et al.*, 2015).
[5] Values by Yang *et al.* (2011) and Demirbas (2009).

8.1.4 Dimensioning the heating system

So far we have seen that the heat demand per hour can be estimated by solving the steady state sensible heat balance. The steady state energy balances may be solved for day and night time, including the management effects of screening, light dependent set points for heating and ventilation. Then the hourly data of the calculations are sorted descending to result in what is called in practice a cumulative frequency curve of the heating demand (Figure 8.3). This is however not a frequency distribution curve in the true sense of the word. At the same time the top t_{th}% values are removed to prevent peak loads with very short duration.

The blue curve in Figure 8.3 indicates that in a year there are 5,000 hours with a heat demand of at least 30 W m⁻²; 6,397 h have a heat demand, so there are 8,760-6,397 = 2,363 h with no heat demand. The graph in Figure 8.3 is a tool for making the right decisions about dimensioning

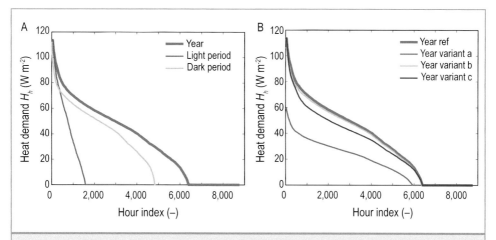

Figure 8.3. (A) Cumulative frequency curve of the heating demand of a greenhouse for a one-year simulation with climate reference data (Figure 8.2A), t_{th}=0.5%. The curves indicated with 'Dark Period' and 'Light period' separate heat demand at night and during daytime. The area under the graphs represents total heat demand (Wh m^{-2}) overall (Year), at night (Dark Period), daytime (Light period). (B) Effects in the year curve (ref) of (a) a double cover (better insulation, R_c=0.18 m^2 K W^{-1}); (b) application of a 75% temperature efficient heat recovery system; and (c) improved thermal screen with p_{screen}=0.5 instead of 0.4.

heating systems. Is a boiler of a certain capacity sufficient or is it smart to choose a set of different energy sources? Is a 60 W m^{-2} heating circuit in the greenhouse sufficient and what percentage of overall heat demand will it cover when used as the primary heating source?

Also, the effect of several management measures that affect the course of the lines can be evaluated, including:
> improved greenhouse insulation;
> application of heat exchanger units;
> improved screening;
> alternative climate control strategies.

The outcomes of these evaluations support decisions about energy demand and the layout of the energy supply system. The peak value (at the y-axis) indicates the capacity of the heating system. The area under the curve is the total energy needed yearly. If the capacity of an energy source is less than the maximum heat demand, Figure 8.3 can be used to find the number of hours this source can be used effectively and how much of the total yearly energy demand can be delivered by that source.

8.2 Heating systems of greenhouses

There are many systems for transporting heat to (and distributing heat inside) the greenhouse. This section deals with relevant actuators inside the greenhouse. Sections 8.3 to 8.8 deal with the energy sources. Several systems are used inside the crop space: convective heaters that work mainly through forced or free convection. Forced convection units are fan driven, often with fabric tubes or perforated plastic film tubes to distribute warm air. When poorly designed these units cause temperature inequality. Free convectors use thermal buoyancy to transport air along the heater. Section 8.2.1 deals with direct and indirect convective heating systems. In section 8.2.2 a second group, heating systems using water filled pipes, is discussed. Several pipe designs exist that contain more or less water. A smaller diameter results in a quicker reaction to changes but also to a higher flow resistance and less heat capacity.

8.2.1 Direct and indirect heating systems

A direct heating system burns a fuel inside the greenhouse and heats the greenhouse air directly; it may also discharge the flue gases for carbon dioxide enrichment. The capacity, the fuel and the fan(s) that distribute the air, determine the uniformity of heat distribution in the greenhouse and, possibly also flue gases (CO_2, H_2O). The coefficient of heat transfer increases with the velocity of the air passing through the system, but in most cases fuel supply and air supply are attuned. Such systems are most commonly controlled by a thermostat (on/off), with poor control of greenhouse temperature.

Indirect heating systems have a remote flame chamber with a water-based heat exchanger or a flame chamber with an air-to-air heat exchanger on location, but flue gases are diverted into a separate flue gas transport unit. Flue gases are normally dispersed in the outdoor air. In some cases flue gases can be diverted to the greenhouse for CO_2 enrichment (Chapter 11). Figure 8.4 gives examples of a direct and an indirect air heating system.

Convective heating systems normally use a water-air heat exchanger or a finned coil. Air is forced along a heating element and discharged freely into the greenhouse, but it is preferable to distribute it evenly by means of ducts to prevent undesirable temperature gradients. Another option with air heating systems is a central air conditioning unit branched onto a network of ducts located under benches, under hanging substrate gutters or between crop rows (Kempkes *et al.*, 2017b). Convective water-air heating systems respond fast to control actions and can handle a large range of water temperatures.

Figure 8.4. Direct heating systems where a fuel is burnt in a flame chamber and flue gases are added directly to greenhouse air (A) and an indirect heater (B) where fuel is burnt inside but flue gases are dispersed in outside air. (https://www.winterwarm.nl/producten/agri/dxadxb_heater.aspx; www.gvzglasshouses.co.uk). Convective heaters, Fiwihex heat exchangers (https://www.fiwihex.nl) (C), and perforated tubes with water filled heating tubes or pipes (D).

8.2.2 Pipe heating systems

Pipe heating systems are commonly used in greenhouses at latitudes above 45°; in mild winter climates with low heat demand growers tend to avoid pipe heating because of the cost (Baille, 2001). There are several systems, including high pipes, pipe rail system, bench heating and special systems for warming the root environment or intercrop solutions. A pipe heating system is a stretched heat exchanger with a predesigned layout (Box 8.5) to reduce spatial temperature gradients in the greenhouse.

Water-filled heating pipes release heat to the greenhouse by means of (free) convection and thermal radiation (TIR, Stefan-Boltzmann, see Section 3.2). The higher the temperature the larger the TIR-fraction. At lower temperature ranges T_{pipe} <60 °C, the overall heat transfer (convection and radiation) α_{pipe} in W per square metre pipe per K temperature difference between pipe and air for a 51 mm round pipe can be described by (De Zwart, 1996):

$$\alpha_{pipe} = a \cdot (T_{pipe} - T_{ambient})^{b}; \qquad a = 1.99, b = 0.32 \qquad\qquad \text{W m}^{-2}\text{ K}^{-1} \quad (8.8)$$

Where T_{pipe} is pipe temperature K or °C, $T_{ambient}$ is air temperature around the pipe in K or °C.

Box 8.5. Heating circuits within the greenhouse.

The water inside the heating pipes releases energy and cools down as it travels through the pipes. It is important how the pipes are arranged. In a parallel system the water travels a shorter distance through the first pipe than through the last pipe, resulting in different cooling curves and therefore in horizontal temperature gradients. Albert Tichelmann (1861-1926) solved this problem by designing circuits that ensure a uniform heat distribution. This is achieved by a so-called ring transfer in which everything is congruent. The sum of the lengths of the supply line must the same as the length of the return line for each element, whereby all pipes must correspond to the same pipe diameter. Thus, all elements are subject to the same pressure losses, so they all have the same volume flows. The pressure loss over the heating circuit is controlled by valves.

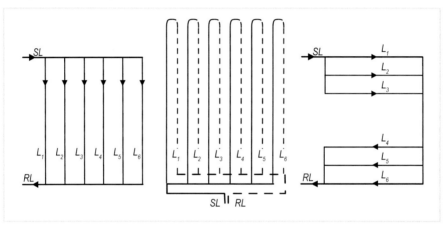

Here above, three possible designs of a heating circuit are shown: Parallel (left), Tichelmann floor (mid) and Tichelmann side walls (right). The heating pipes L_1 to L_6 are normally folded in a so-called hairpin shape, thus bringing the supply *SL* and return line *RL* to one side.

The effective heat release to the greenhouse per square metre soil area can be calculated with:

$$H_{pipe} = \alpha_{pipe} \cdot A_{s,pipe} \cdot (T_{pipe} - T_{ambient}) \qquad\qquad \text{W m}^{-2} \quad (8.9)$$

Where $A_{s,pipe}$ is the specific pipe area per m^2 floor area. Specific pipe area may be calculated as:

$$A_{s,pipe} = \pi D_{pipe} \cdot S \cdot L_{s,pipe} \,; \qquad\qquad L_{s,pipe} = \frac{L_{pipe}}{A_s}$$

Where D_{pipe} is pipe diameter, S is a shape factor if the pipe surface is increased by welded-on fins, $L_{s,pipe}$ is the pipe length per m^2 floor area, L_{pipe} is total pipe length in the greenhouse, A_s is soil or floor area in m^2.

8.3 Transport units servicing compartments, different greenhouses

The heat has to be transported from the source to the greenhouse and distributed equally. For direct and indirect heaters the fuel has to be transported safely to the units. For convective heaters and pipe heating systems warm water is transported. A transport system exists between the energy centre and greenhouse heating circuits in one or more greenhouse compartments. It has to deliver heat fast and on demand with minimal loss. A transport system consists of different subsystems in which water flows if necessary are mixed. The greenhouse itself is heated with pipe-fed water-air convectors, heat exchangers or heating pipes like the pipe rail system. Figure 8.5 shows the basic principle of heat distribution to greenhouse heating circuits.

The heat H_h in (W) delivered by the transport group is calculated from:

$$H_h = \phi_w \, (\rho c_p)_w \, (T_s - T_r) \qquad\qquad \text{W} \quad (8.10)$$

Where ϕ_w is the water flow in the transport group l s^{-1}, $(\rho c_p)_w$ is the specific heat of water J l^{-1} K^{-1}, T_s is supply temperature and T_r is return temperature.

Often circuits are supplied with heat by means of a four-way valve (Figure 8.5). The benefit of a four-way valve, compared to a two-way valve is control over a large temperature range, a constant flow rate in the pipes, and quick response of the heat source when there is (no) heat demand. Figure 8.6 shows an example of the temperatures in both the transport group and the heating circuit in the greenhouse.

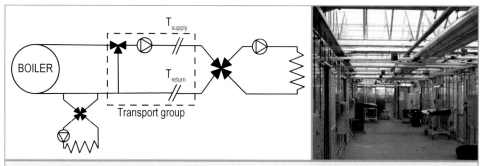

Figure 8.5. Source heat (boiler) is transported directly to a heating circuit nearby or via a transport group to circuits further away from the boiler (left). A transport group in the main aisle of a greenhouse (right). The big pipes transport the water to and from the greenhouse heating circuits.

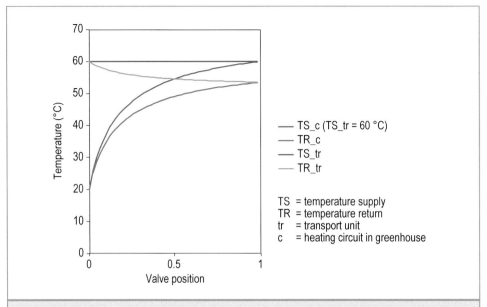

Figure 8.6. Temperatures of the branches of the four-way valve, supply and return of the transport unit, and supply and return of the heating circuit in the greenhouse. Supply temperature in the transport unit is 60 °C, that of the heat circuit ranges from 20 to 60 °C depending on the valve position. The heat delivered is not linear with valve position.

8.4 The energy centre of the greenhouse

Figure 8.7 gives an impression of an energy centre for a greenhouse. The distributor supplies transport groups or heating circuits with heat. The collector takes back the return water to the heating centre. In this example the heating centre consists of boiler 1 and boiler 2, a CHP and a heat buffer tank. Only boiler 1 is used to supply the greenhouse with carbon dioxide.

Growers often combine co-generators with two boilers to ensure backup is available at times of maintenance of a heat source without losing the ability to heat the greenhouse and to allow 'peak-shaving'. Boilers have a much higher thermal efficiency than co-generators; therefore they produce more heat with less fuel. This allows the grower to limit connectivity costs which depend on the peak demand for fuel delivered from a grid or fuel storage.

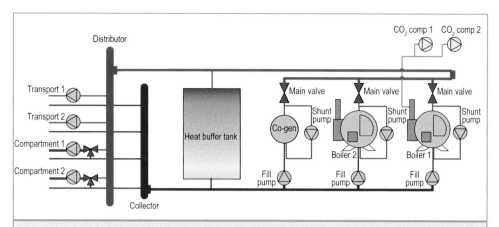

Figure 8.7. An example of combined heat sources. The distributor brings hot water to the transport lines and the circuits of greenhouse compartments. The collector accepts the return water. Between the main return and the main supply a heat buffer tank, a co-generator and two boilers are connected to a Tichelmann circuit. One or more units supply the heat demand. Boiler 1 is also used for CO_2 enrichment (see also Chapter 11, Section 3).

8.5 Energy sources

Several energy sources can be used to heat a greenhouse using equipment such as a central hot water boiler with flue gas condenser, a co-generator (CHP), a geothermal heat well, solar collectors, wind power, residual energy from industrial plants, and heat pumps. Solar energy, wind energy and geo-thermal heat are considered sustainable energy sources. A drawback of sustainable energy is that CO_2 for enrichment (Section 11.3) has to be bought elsewhere. Boilers and co-generators run on fossil fuels, biomass or biomass-derived oil.

A suitable heat source for a greenhouse has to meet several requirements. The temperature level has to fit the design of the heating grid and vice versa. The temperature level has to be sufficiently high at the right time, i.e. the heat must be available when needed. The return temperature has to be relatively low so that a maximum amount of energy can be delivered to the greenhouse. On the other hand the return temperature is a typical design issue. Normally a ΔT of 10 to 15 degrees is used. In case you increase this ΔT too much in one heating system, for instance a pipe-rail, the heat release of one meter of combined forward and backward pipe is not the same over the full length causing horizontal temperature differences.

If a source produces more than one resource simultaneously (heat, electricity, CO_2), then all resources must be used efficiently and not go to waste. Choosing the right device requires knowledge of the system and answers to questions like: Does it produce a single resource or multiple resources simultaneously? How can simultaneously produced resources be processed? What device or combination of devices is most feasible or most resource-use efficient? How controllable is the device? What are the environmental effects?

8.5.1 Central hot water boiler and flue gas condenser

The most common device is the central hot water boiler. An example of a boiler is shown in Figure 8.8. It produces heat and CO_2 at the same time. A fuel, for instance natural gas, is burned within the boiler. The large flame heats up water inside heat exchanging tubes. The heated water is transported to the greenhouse. Hot water can reach 105 °C and return water to the boiler should not be cooler than 65-70 °C to prevent corrosion in the flame tubes as a result of condensation and damage by thermal tension in materials. The flue gas temperature after the combustion chamber is 400-500 °C and is cooled inside the boiler by means of a flue gas cooler to about 120 °C. This flue gas contains sensible energy as well as latent energy kept in water vapour. This vapour is mainly produced by fuel combustion, so the energy originates from the fuel energy content. A flue gas condenser, a heat recovery device (Box 8.6) 'harvests' this energy. Depending on its design this device cools the gas down to 40 °C, which is below the dew point temperature of the flue gas (50 to 60 °C); the harvested heat is transferred to water in a heat exchanger to feed a secondary heating system.

The hot water available for greenhouse heating is maintained at 70-90 °C by constantly mixing extra hot water from the boiler or the condenser if necessary. The most efficient use of the condenser heat is in a secondary, low temperature heating network (30-40 °C), namely inter-crop heating, floor heating or root zone heating (Figure 8.9).

The flue gas that has been through the condenser can be used for CO_2 injection (Section 11.5) if it is clean and cool enough.

Figure 8.8. A twin central hot water boiler with (internal) flue gas condenser (left). Insulated distributor and collector pipes and flue gas unit (right). (photos: Bert van 't Ooster)

Box 8.6. The heat exchanger.

A central element in many energy processes in the greenhouse is the heat exchanger. It consists of two separated circuits that exchange energy through a large and highly conductive contact area. The water in both circuits is never mixed. One side is heated up, while the other is cooled down.

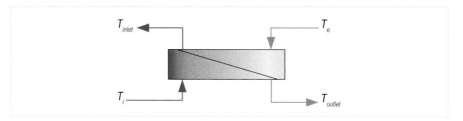

T_e = Temperature of the cold input flow (°C)

T_i = Temperature of the warm input flow (°C)

T_{inlet} = Output temperature of the heated medium (°C), normally inlet to the system of interest

T_{outlet} = Output temperature of the cooled medium (°C), normally output of the system of interest

The exchange efficiency (η_T) is design dependent and can be calculated through:

$$\eta_T = \frac{T_{inlet} - T_e}{T_i - T_e}$$

The efficiency of a typical heat exchanger is about 70-80%.

The heat flux delivered by a source is defined as:

$$H_{hd} = \frac{\phi_{fuel} \cdot E_{fuel} \cdot \eta}{A_s} \cdot f_{cnv} \qquad\qquad \text{MW ha}^{-1} \text{ or W m}^{-2} \quad (8.11)$$

With H_{hd} the heat flow delivered by the source, in this case the boiler, in (MW ha^{-1}) or in (W m^{-2}), ϕ_{fuel} the fuel flow into the boiler in units (m^3, l, kg) per hour, E_{fuel} is the combustion heat of the fuel either for the gross or the net calorific value[2] in MJ per unit, η is 'water sided' efficiency of the boiler relative to the gross or net calorific value of the fuel, A_s is the soil area (m^2), f_{cnv} is $10^4 \times 3,600^{-1}$ for output H_{hd} in (MW ha-1) and $10^6 \times 3,600^{-1}$ for output H_{hd} in (W m^{-2}).

For instance, when the boiler uses 100 m^3 h^{-1} (at 100% efficiency relative to gross calorific value), how much heat is available for a one hectare greenhouse? The gross calorific value of Dutch natural gas is 35.17 MJ m^{-3} (Table 8.2):

$$H_{hd} = \frac{100 \ (\text{m}^3 \ \text{h}^{-1}) \times 35.17 (\text{MJ m}^{-3}) \times 1}{10,000 \ (\text{m}^2)} \cdot 10^6 \times 3600^{-1} = 97.7 \qquad\qquad \text{W m}^{-2}$$

---

[2] For a definition of the gross and net calorific value of a fuel see Wikipedia or Hydrogen Centre.

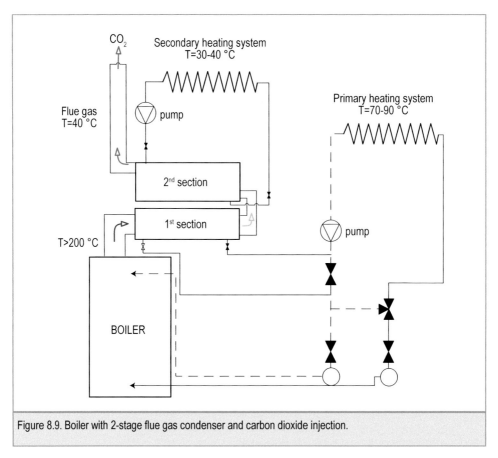

Figure 8.9. Boiler with 2-stage flue gas condenser and carbon dioxide injection.

This is very close to a 1:1 ratio. A practical rule of thumb says that $1 \text{ m}^3 \text{ h}^{-1} \text{ ha}^{-1}$ natural gas use by a boiler contributes 1 W m^{-2}. This rule is often used in practical calculations when 100% accuracy is not required. Equation 8.11 is more accurate and generic for any heat source.

The difference between the gross and net calorific value, also called the higher and lower heating value in Table 8.2 is the latent heat of the vapour in the flue gas.

8.5.2 Combined heat and power

Combined heat and power (CHP) mainly found its way to greenhouse horticulture in northwest Europe (the Netherlands and Belgium) for good reasons. CHP units consist of an engine that drives a generator and it is therefore also called a co-generator. The CHP process produces heat, electricity and CO_2 simultaneously. The electricity that is produced can be put to internal use, e.g. supplementary lighting, or sold to the grid.

Decentralised production of electricity in co-generators connected to greenhouses is more efficient (Box 8.7) than a central power plant, because the heat and flue gas can be used locally. The central power plant also produces a lot of heat and flue gas but this usually goes to waste because heat is difficult to transport, although there are some exceptions. Figure 8.10 shows a co-generator in context with heat, power and CO_2 delivery to greenhouse or grid. The cooling water of the motor and of the flue gas is stored in the heat buffer for heating the greenhouse. Motors with internal combustion produce NO (nitrogen oxide) and NO_2 (nitrogen dioxide), CO and un-burnt hydrocarbons. Catalysts scrub the exhaust gases of these pollutants with an efficiency of up to 95% (Clarke Energy). The exhaust gas, treated with metered urea reactant solution, passes through the fine cell honeycomb-patterned converters, reducing the oxides of nitrogen to water and nitrogen, in a process known as selective catalytic reduction. CO and

---

Box 8.7. Co-generator more efficient than central power plant.

When a co-generator uses 100 units of fuel to produce 40 units of electricity and 50 units of heat, its overall energy efficiency is 90%. When the electricity and heat are produced separately the grower has a boiler and buys the electricity from the grid. With an assumed thermal efficiency η_{th} of 95%, the boiler uses 52.5 units of fuel to produce the units of heat. The central power plant has an electrical efficiency η_e of roughly 40%: it uses 100 units of fuel to produce 40 units of electricity, and the heat gets lost to the environment. So separate production costs 100 + 52.5 = 152.5 units of fuel. At society level co-generation therefore saves (152.5-100)/152.5 = 34% energy. Large co-generators have a higher electrical efficiency than smaller ones.

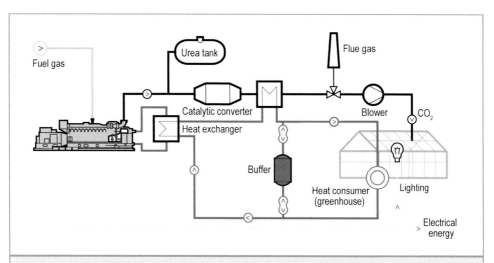

Figure 8.10. Co-generator in context: a single stage cooling system with two heat exchangers and a buffer to store heat; a catalytic converter purifies the flue gas. Electricity is used in the greenhouse or delivered to the grid (www.gepower.com).

un-burnt hydrocarbons are removed by catalytic oxidation. In this process, the pollutant gases diffuse through to the surface of a ceramic honeycomb, coated with noble metals, causing a reaction to form hydrogen and carbon dioxide. If flue gas scrubbing catalysts are in place (such as in Figure 11.1), the exhaust gases can be used for carbon dioxide injection.

The co-generator usually runs only at 100% capacity. The graph in Figure 8.11 is a sketch of a cumulative frequency curve for heating with an imaginary relative heat demand of the greenhouse. The figure shows how many running hours are required from the co-generator at full load and how many at partial load. Partial load is not common due to efficiency issues. It is avoided as much as possible by running it at 100% with a part of the heat stored in the buffer tank. When CO_2 enrichment originating from the co-generator is included, the system can be active for more hours a year. CHP systems require several thousand operating hours to be economically feasible.

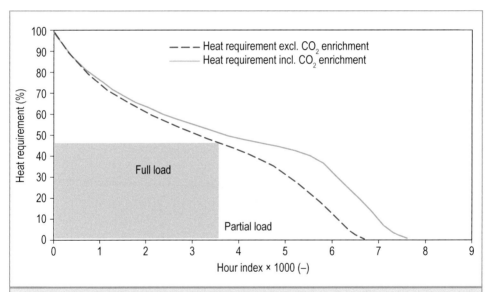

Figure 8.11. Fictive indication of the running hours and energy delivery of a co-generator unit by means of a cumulative frequency curve for heating. If the co-generator produces 48% of the maximum heat demand it can run until the heat requirement curve and the full load line intersect (3,500 hours in this example). Then it has to stop or run at a partial load for 1,400 hours up to hour 4,900. The lower line is without CO_2 enrichment, the higher line with enrichment. In the space between the two lines the engine runs to produce CO_2 while there is no heat demand. Surplus heat can be stored in a buffer tank for later use. Heat indicated by the space between the dashed and solid line can thus be migrated in time to 'fill' the unused space under the dashed line.

8.5.3 Geothermal heat

At varying depth (from soil surface to some 5.5 km) the earth's rigid mantle (thickness 5-50 km) holds geothermal heat which can be mined when this heat is present in water conducting porous layers. Limberger *et al.* (2018) presented maps indicating the geothermal potential of various world regions. A geothermal source will not deliver hot water forever, it will eventually cool down. Nevertheless, technical lifetime estimates range from 30 to 100 years (Sullivan *et al.*, 2010).

The hot water (60-100 °C) is pumped up and led through a heat exchanger. It transfers its heat partially and the cooled water (20-50 °C) is pumped back into the soil to the same depth but in another place. The hot and cold pit, also referred to as a doublet, have to be at a certain distance from each other, otherwise the cold pit may cool down the hot pit. The water circuits are closed: there is no direct contact between the soil water and the water used in the greenhouse. As in Equation 8.10, the heat retrieved from a deep geothermal pit is defined by:

$$H_{gt} = \phi_w \, (\rho c_p)_w \, (T_{gt} - T_r) \hspace{4cm} \text{MW} \quad (8.12)$$

Where H_{gt} is the capacity in MW, ϕ_w is the pit yield in $m^3 \, s^{-1}$, $(\rho c_p)_w$ is the specific heat of water which is around 4.186 MJ m^{-3} K^{-1} but depends on salinity of the pit water, T_{gt} is the water temperature delivered by the geothermal source K or °C, T_r is the return temperature of the water to the geothermal source K or °C.

Except in 'blessed' regions (such as Southern Hungary, central Turkey, central Italy or Iceland, to name a few) geothermal heat is to be found at depths often in excess of 1 km. Drilling is therefore the major investment cost, and there is also the drawback of uncertainty about actual thermal yield. In the Netherlands, for instance, depth exceeds 2 km and the initial investment ranges from 4 to 10 million euros for an estimated hot water yield 150 m^3 h^{-1}. If a grower (or a consortium of growers) can stomach the investment (with the help of subsidies) the operational costs are low compared to the variable costs of boilers and co-generators, as will be shown in Section 9. Cooperation is often necessary (and stimulated) to match the extraction and reinjection area of one well with a large area of greenhouses. The heat production of a pit is flat, while greenhouse heat demand fluctuates. This is balanced by both short- (daily) and long-term (seasonal) buffers.

In the system a lot of heat exchangers are included to transfer the energy efficiently. Heat pumps (Figure 8.12) may also be useful because they cool down the return water to roughly 20 °C, so more geothermal energy can be harvested. Electrical energy is still needed to drive the water pumps and the heat pump. The ratio between generated heat and electrical energy input, the apparent coefficient of performance (COP; Section 9.2.3), is estimated at 8 (-) for a system with a heat pump and 12 (-) without a heat pump.

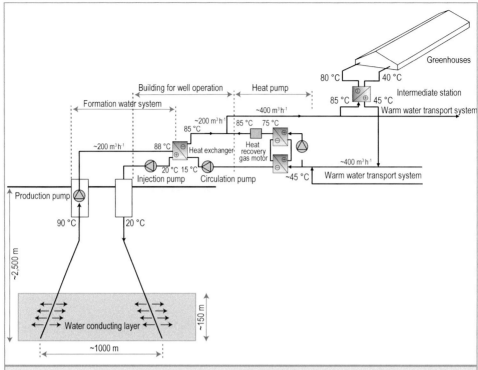

Figure 8.12. Geothermal installation with heat buffers and one or more greenhouses connected. A heat exchanger separates the water circuits. A heat pump may be added to minimise the return temperature.

8.6 Industrial residual heat

Industrial plants, especially power plants, and energy intensive industry such as fertilizer factories, often waste energy by sending heat into ambient air by means of cooling towers or into water. In some cases this waste energy can be made available for use in greenhouses. This mainly depends on network infrastructure, distance, available capacity and temperature constraints. In many cases the supply water is around 80 °C and the return temperature should not be higher than a ceiling value of 25 or 40 °C. In dimensioning a residual heat supply system, the source, the transport system, and the heat demand side must be matched. Growers still need a boiler for backup and need to find different solutions for CO_2 enrichment. The maximum investment costs for an industrial residual heat supply system are estimated based on the fixed and variable costs of the heating system for the greenhouses serviced in € GJ^{-1}.

8.7 Heat pump

Heat pumps are increasingly used by growers in combination with long-term low temperature heat storage. Heat pumps are mostly used in combination with cooling, hence their working principle and performance is explained in Chapter 9. Heating by heat pumps is also possible, however, other techniques are applied when heating alone is necessary. It is not recommended to satisfy the full heat requirement of a greenhouse by means of heat pumps (Ruijter, 2002). Therefore, the heat pump capacity is designed to cover about 70-80% of total heat demand. The rest of the required heat demand is then supplied by the (gas) fired boiler to prevent the COP of the heat pump from being at times too low.

8.8 The role of buffer tanks

Buffers are necessary to store surplus heat, which can be used later in the day (short-term heat buffer) or later in the year (long-term heat buffer) depending on the size and use of the buffer. Horizontal and vertical tank buffers are mainly used for a short or midterm storage period. Compared to horizontal buffers, vertical buffers have a much larger volume, belong to large companies but must be made on site, have a good vertical temperature gradient and have equal volume sections. The horizontal buffer is limited in capacity because of transport limitations. Aquifers (see also Section 9.4.1) and basement buffers below the greenhouse are used for long-term storage (Figure 8.13).

A buffer is intended to minimise heat waste and to avoid peak loads in fuel consumption. Therefore the buffer capacity has to be tuned to sources and sinks to keep it manageable at all times. When heat (and electricity) and CO_2 are produced at the same time the buffer allows for decoupling of their use.

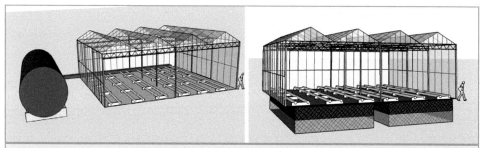

Figure 8.13. Tank buffer (left) is used as a short-term buffer that is filled and emptied daily or in a cycle of several days. These buffers allow matching of supply to demand at a high use efficiency of both heat and carbon dioxide. The basement buffer (right) is a long-term buffer to carry solar heat from one season to another. (Pictures: Janssen *et al.*, 2006)

Another important role of the buffer is to lower the required capacity of boiler or CHP. If you can store 25 W m^{-2} and the capacity of the boiler is 50 W m^{-2}, you can supply the greenhouse with 75 W m^{-2}. A lower capacity of the boiler is relevant because the costs of the connection to the gas grid are normally based on the peak capacity on a monthly and annual basis.

In applying geothermal heat, buffers enable a higher degree of utilisation of the source, result in better coverage of greenhouse heat demand, increase maximum heating capacity of the source and help manage utilisation of the heating power.

The maximum total energy that can be stored in a tank buffer is:

$$E_{max} = V_b \cdot (\rho c_p)_w \, (T_h - T_b) \qquad\qquad \text{MJ} \quad (8.13)$$

Where V_b is the volume of the buffer in m^3, $(\rho c_p)_w$ is the specific heat of water 4.186 MJ m^{-3} K^{-1}, T_h is the fill temperature of the water delivered to the buffer in K or °C, T_b is the base temperature of a cooled buffer, normally return temperature of cooled water to the buffer in K or °C.

Besides filling, a buffer must also be emptied in terms of energy. In other words, there must be enough hours with heat demand. If we consider that there is no radiation at night, and we neglect both crop transpiration and energy exchange with the soil, the average heat demand of the greenhouse at night equals:

$$H_h = U' \cdot A_s \cdot \Delta T \qquad\qquad (W)$$

Box 8.8 gives a numerical exercise on buffer management.

8.9 Economic feasibility

To decide which method of heating is the most economical, both the capital costs and operational costs play a role. The capital costs can be estimated as

$$C = \frac{a \cdot I}{H_s} \qquad\qquad \text{€ GJ}^{-1} \quad (8.14)$$

Where a is the annuity, from the chosen interest rate and the investment lifetime, I the total investment cost (€), H_s the heat annually produced by the system (GJ yr^{-1}). Capital costs result in a cost parameter € MWh^{-1} or € GJ^{-1}. There may be a limit to I because of the financial climate (how large a loan one wants/can get). Variable costs include the fuel and purchase of resources if not generated by the company itself.

Box 8.8 Buffer capacity and energy storage.

A buffer tank with volume V_b equals 100 m^3 ha^{-1} has a supply temperature of 95 °C and a return temperature of 45 °C. What maximum amount of thermal energy E_{max} can be stored and how much fuel can be burnt in a boiler with η_{th}=0.85 and a co-generator with η_{th}=0.53?

Answer: E_{max}=20,930 MJ ha^{-1}, this is 595 gas equivalents at (35.2 MJ m^{-3}). Gas boiler: 700 m^3 ha^{-1}, cogenerator 1,123 m^3 ha^{-1}.

The greenhouse has a U-value, U' equal to 8 W m^{-2} K^{-1}, and the duration of the night is 14 hours. Is it then possible to empty the buffer during the night?

Answer: The energy requirement per 1 K temperature difference is 80 kW K^{-1} per ha. The buffer, as we saw above, contains 20,930 MJ ha^{-1}. The average heating power available is:

$$\frac{E_{max}}{t} = 20,930 \text{ MJ}/(14 \text{ h} \times 3,600 \text{ s h}^{-1}) = 0.4153 \text{ MW} = 415.3 \text{ kW},$$

$$\Delta T = \frac{H_h}{U' \cdot A_s} = 415.3/80 = 5.2 \text{ °C}.$$

So without thermal losses a full buffer will maintain a temperature difference with outside of 5.2 °C for 14 hours.

If that is more than needed (or the night is shorter), the buffer may not be emptied fully.

A central boiler has relatively low fixed costs because of a long lifetime and high energy efficiency. Variable costs depend on the fuel price. A co-generator requires a higher investment but has shorter payback period if spark spread is good. The spark spread is the difference between the price received for electricity produced by a generator and the cost of the fuel needed to produce that electricity. It is typically calculated using daily spot prices for fuel and power at various regional trading points. The spark spread is expressed in € MWh^{-1}. In general, total costs during the life span are lower for CHP than a boiler, but that largely depends on the spark spread.

An example of the type of calculation supporting the choice of heating system is given in Figure 8.14.

Despite the very high investments, costs per MWh are lowest for geothermal heat. The more it can be used, the lower the costs per MWh. Short- and long-term heat buffering support maximised use of a geothermal source. The variable cost of geothermal heat ranges from 5-14 € MWh^{-1} in the Netherlands (Vermeulen, 2013). This mainly depends on the pump installed and the energy yield of the geothermal doublet. Energy yield depends in turn on water yield and the difference between production and re-injection temperature of the doublet water. Fixed costs depend on the investment, the energy yield of the geothermal doublet, the life-time of the source, the number of operational hours per year and maintenance requirement. Roughly, these costs range from 20-40 € MWh^{-1} (Vermeulen and Van Wijmeren, 2013).

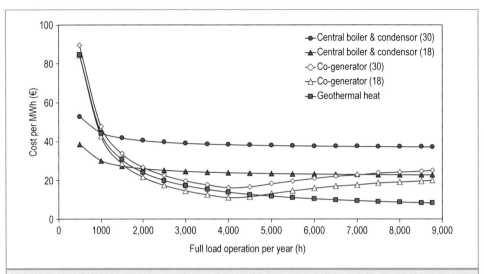

Figure 8.14. Economic comparison of three heat sources: boiler, CHP and geothermal, for a 10 ha greenhouse with a crop cycle starting early December and ending the third week of November. Target temperature day/night is 20.5 and 17-19 °C respectively. The heating system has a total capacity of 13.6 MW: 2 central boilers of 5.8 MW_{th} each, 2 co-generators of 4.4 MW_{th} and a short-term heat buffer of 2,640 m^3. The buffer allows separation in time of use of heat, CO_2 and electricity. The greenhouse is equipped with a 40% energy saving screen. The legend shows between brackets two fictive average price levels of natural gas, 18 and 30 in €cents m^{-3}, and for the co-generator two price levels for power delivered to the grid 60 € GWh^{-1} between 18 pm and 7 am and 100 € GWh^{-1} between 7 am and 18 pm. The cost per MWh includes the capital cost and maintenance of the systems, $\alpha \cdot I$ were estimated at 9,750, 141,000 and 405,000, for the central boiler, the co-generator and the geothermal heat source respectively.

8.10 Concluding remarks

For a good understanding of greenhouse heating one needs to distinguish between heat demand and meeting this demand using energy sources and heat distribution systems. Working principles of energy sources for heating greenhouses differ with respect to combined production of the resources (heat, electricity and CO_2), to temperature levels and constraints. The demand for resources and the type of resources needed determine what is the best solution. This is region dependent. Therefore, instead of presenting a best solution, a method was presented to design and evaluate heating systems for any relevant geographical region. As a result of the ongoing transition to sustainable energy, growers can differentiate more in the selection and management of (multiple) energy sources and heat distribution systems in (and between) greenhouses. New solutions, such as residual heat, geothermal heat and solar heat, are struggling in terms of viability. To harvest solar heat and use this heat at times of heat demand requires heating and cooling functions in the greenhouse (Kempkes *et al.*, 2017b) as well as buffering of heat. The integration of heating, cooling and dehumidification that will be needed is the subject of Chapter 9.

Chapter 9
Cooling and dehumidification

Photo: Barend Gehner

Natural ventilation is a cost effective way to cool and dehumidify a greenhouse (Chapter 5). However, as the air leaves the greenhouse, solar heat and water vapour and in some cases also carbon dioxide, are lost to the environment. Active cooling is needed if ventilation capacity is not sufficient to maintain the required temperature range or if the otherwise lost resources are to be put to viable use. Dehumidification is required to prevent high humidity, which is a cause of crop diseases and yield reduction (Section 7.4). Humidification is relevant in the case of ventilation with warm and dry air. In high-tech greenhouses and specifically in (semi-)closed greenhouses the goal is optimal control of the climate combined with smart use of resources. In this chapter active cooling and (de-)humidification are discussed from the perspective of resource use.

The greenhouse air temperature can be lowered by ventilation or active cooling. Ventilation, at best, can bring the greenhouse temperature down to the outside temperature, ignoring the cooling effect of crop transpiration (see Section 6.4.2 and Section 4.2.3). External shading screens to reduce incoming solar radiation are also unable to reduce greenhouse air temperature to levels below the outside temperature (Box 5.3). Furthermore, screening reduces the light absorbed by the crop and therefore potential yield (Chapters 2 and 7).

If the target for greenhouse air temperature is lower than the outside temperature, active cooling is required. In a similar way to heating (Equation 8.7), we can define the critical outdoor temperature for active cooling:

$$T_{coc} = T_{iC} - \frac{\tau \cdot (1 - \rho) \cdot I_s \pm H_{AiSo} - g_{v,100\%} \cdot \Delta\chi \cdot L}{\frac{A_c}{A_s} \cdot U_{tr} + g_{v,100\%}\rho c_p} \qquad °C \quad (9.1)$$

Where T_{iC} is the target temperature (°C) for cooling and T_{coc} the critical outdoor temperature for cooling. When 100% ventilation is reached and it turns out to be insufficient to obtain the indoor target temperature for cooling T_{iC}, then active cooling (and no ventilation) is preferred. If $T_{out} > T_{coc}$ then a demand for active cooling exists since full ventilation capacity $g_{v,100\%}$ is insufficient. For symbols not explained see Equation 8.7. T_{coc} depends on greenhouse design, ventilation, evapotranspiration, the soil flux and radiation (Equation 9.1). T_{coc} indicates for a given target temperature and maximum ventilation at what outdoor temperature active cooling is needed. The options for active cooling are:

> ➤ evaporative cooling by evaporation of water; this is sometimes called nature's way of cooling because crop transpiration is a natural form of this type of cooling;
> ➤ roof cooling: a layer of water runs down the roof and takes up heat and partly evaporates;
> ➤ heat pumps;
> ➤ stored cold water from an aquifer;
> ➤ other cold sources and desiccant systems.

9.1 Evaporative cooling

The cheapest way of active cooling is adiabatic (no external input of energy) cooling which is based on evaporation of water and therefore also called evaporative cooling. The physics of the process are described in Section 4.3.2. It is obvious that evaporative cooling is possible only with a large wet bulb depression, i.e. in hot and dry climate regions. A drawback is that evaporative cooling uses a lot of water. In a review paper, Misra and Ghosh (2018) describe systems of three main kinds: direct evaporative cooling, indirect or roof evaporative cooling and two-stage or mixed mode cooling systems.

9.1.1 Misting and fogging systems

Mist and fogging systems are two examples of direct evaporative cooling. These systems spray water directly into the greenhouse air. The amount of water sprayed is a trade-off between the desired cooling capacity and the target range of humidity in the greenhouse. For direct evaporative cooling, the only external energy consumption is electricity to drive the pumps. This is a relatively small amount and these systems therefore have low running costs. The effectiveness of this system depends on the ventilation capacity which depend on the wind forces and less on temperature forces (Section 5.2). As the greenhouse is being cooled the air temperature inside the greenhouse may be lower than outside. The buoyancy effect which normally contributes to the ventilation therefore works negatively on the ventilation. So wind is the driving force for the ventilation. If there is no wind this type of cooling will not work as desired.

Misting systems deliver a fine spray of water to greenhouse air to provide relief in very hot and/or dry conditions. Mist systems operate at a pressure of 200-500 kPa (Bottcher *et al.*, 1991; Li and Willits, 2008). Droplets volume median diameter (VMD) ranges from 100 to 200 µm; fine droplets result in better cooling than coarse droplets. Depending on pressure and droplet size, spray rate, ventilation rate and vapour pressure deficit (VPD) 10-60% of sprayed water evaporates (evaporation efficiency) and converts sensible heat into latent heat. This results in an increase in relative humidity (RH) as a result of cooling and humidification as shown in Figure 4.12. Crop transpiration tends to fall because of lower VPD. In misting systems, a fraction of the droplets is too big to evaporate before they reach the crop or the greenhouse soil. This may lead to reduced product quality and an increased incidence of diseases as a result of a wet canopy.

Fogging systems operate at a high pressure of 2-7 MPa (Li and Willits, 2008; Lu *et al.*, 2015), with small diameter steel lines. Fogging line pressure and fogging nozzles need to be well maintained for proper functioning. Droplet size (VMD) is typically within a range of 2-60 μm (Li and Willits, 2008). When the very fine 'atomised' droplets are suspended into the greenhouse air, the water particles evaporate within seconds. However, fogging systems also show overall evaporation efficiency of 60-100% depending on spray rate, ventilation rate and VPD (Li and Willits, 2008). A well-operated system has to ensure that no water falls on the crop, and that the air is sufficiently dry to absorb the moisture (Hanan, 1998). For this reason fogging systems normally use fogging cycles with on/off time cycles (a few seconds on, in cycles of a few minutes). Increasing ventilation enhances the evaporation of fog droplets because air is kept drier (higher VPD); the trade-off is that more warm air passes through the greenhouse and more fog droplets will drift out of the greenhouse without contributing to cooling. If humidity is low and extreme heat loads to indoor climate are avoided, then the indoor temperature can fall 3 °C below outdoor temperature in Mediterranean areas (Katsoulas *et al.*, 2006) and 10 °C in hot dry regions (Davies, 2005) (see Figure 4.12). The drier the air, the bigger the cooling effect. Abdel-Ghany and Kozai (2006) defined the mean value of the cooling efficiency η_{mC} as:

$$\eta_{mC} = \frac{1}{t_{tot}} \int_0^{t_{tot}} \eta_{iC}(t) \cdot dt \quad \text{with} \quad \eta_{iC}(t) = \frac{T_{db,u} - T_{db,i}}{T_{db,u} - T_{wb,i}} \tag{9.2}$$

Evaporation of fog cools the air inside the greenhouse from the uncooled conditions ($T_{db,u}$, x_u) to the cooled conditions ($T_{db,i}$, x_i). $T_{wb,i}$ is the wet bulb temperature of the air inside. The instantaneous cooling efficiency η_{iC} is integrated over an operational time period t_{tot} to obtain mean cooling efficiency. Using Equation 9.2 they found η_{mC} in the range 0.2-0.3. Öztürk (2003) used a more practical but less accurate efficiency parameter for the fog system η_{fs} (%) based on outside dry bulb temperature $T_{db,out}$:

$$\eta_{fs} = \frac{T_{db,out} - T_{db,i}}{T_{db,out} - T_{wb,out}} \cdot 100 \tag{9.3}$$

With $T_{db,i}$ the dry bulb temperature inside, $T_{wb,out}$ the web bulb temperature of the air outside or better of the ventilation air entering the greenhouse. Equation 9.3 is less accurate than Equation 9.2 because it focus is the cooling of the inlet air and not that of the greenhouse air.

9.1.2 Pad and fan cooling

Pad and fan is the most popular direct cooling system in hot, dry climate regions. This system cools and humidifies inlet air (Figure 9.1). The air is cooled at the location of the wet pad. Fans pull the cooled air through the greenhouse (negative pressure, the most common system by far) or push it into the greenhouse (positive pressure). The system relies on well-maintained fans and a well permeable pad not blocked by salt. Horizontal temperature differences inevitably occur with this system, causing unequal growth and development of the plants. The coolest area of the greenhouse is adjacent to the pads and temperature increases towards the fans as a result of solar heat load. This temperature gradient can be reduced by applying misting or

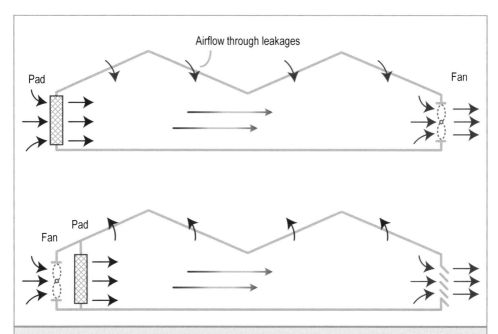

Figure 9.1. The pad and fan system (redrawn from Von Zabeltitz (1986)). The pad is an open cross-flow heat exchanger meant to evaporate water. The fan generates the required pressure for air flow through the pad. In the negative system (top) the fan generates negative pressure in the greenhouse with a flow of warm air through leakages. In the positive system (bottom) the fan generates positive pressure in the greenhouse to keep warm air out, but there is loss of cool air through leakages.

fogging at one or some points in the pathway of the ventilation air. The water in the pad will assume the wet bulb temperature of the air (to be found in the psychrometric chart), unless it is actively cooled. In this case, the air temperature could cool down to below wet bulb temperature. A water temperature below the external dew point would to a limited extent dry external air passing through the pad, which may seem to be a way to operate pad and fan cooling in humid regions. In that case the pad system operates as a sensible cooling system and some water may be gained. However to work with cooled water in the pad is (in most cases) not economical. A downside of the pad and fan system is that in most cases you are also cooling the surroundings. For example, if the outside temperature is for example 45 °C, RH 10%, air after pad wall is 25 °C and the air leaving the greenhouse is around 30 °C. So already in the normal operation a lot of cooling power is wasted. If one is to cool the incoming air even further this would increase the losses. In that case it is advised to cool the inside of the greenhouse without ventilation with outside air.

The saturation efficiency (Equation 9.3) of a pad and fan system depends on the thickness of the pad and the air velocity through the pad (Figure 9.2). The pressure drop in the pad also depends on thickness of the pad and velocity of the air. In order to keep temperature gradients

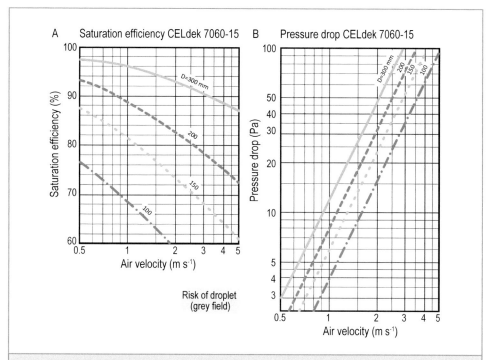

Figure 9.2. Performance specifications of the CELdek 7060-15 pad systems (Munters). The saturation efficiency (A) and pressure drop (B) for four pads of different pad depth D as a function of air velocity in the pad. How humid and cool the air is when it leaves the pad depends on two main factors: pad depth/thickness and air velocity. A thin pad is less effective than a thick one. Only at 100% efficiency can the wet bulb temperature be reached, but normally a higher temperature than wet bulb temperature results.

in the greenhouse low, a pad and fan system normally has a high ventilation rate, with hourly exchange rates of up to 20 h^{-1} (Misra and Ghosh, 2018). However, a low ventilation rate would limit water use and attain higher saturation efficiency. Therefore, the ventilation rate is a trade-off between water use efficiency and acceptable horizontal temperature gradients. Greenhouses are advised not to exceed 50 meters in the direction of the air flow.

In principle evaporated water could be regained in a cooling element at the outlet of the greenhouse. This has been proposed as a mitigating measure in water-scarce regions. For instance, Paton and Davies (1996) and Bailey and Raoueche (1998) presented a 'seawater greenhouse' system that could be applied in coastal and arid regions to produce a suitable microclimate for plant production and also provide irrigation water. The system uses two pads: the inlet is wet by seawater (in a large stream, to wash out the salts left behind by evaporation) and vapour is recovered at the outlet heat exchanger, which is cooled by deep seawater. Indeed, simulations, for instance Sablani *et al.* (2003), have often shown that such a system, possibly coupled to cover materials with low near infrared (NIR) transmission, could allow crops to be grown and fresh water to be generated in hot and arid regions. Unfortunately, prototypes have been marred by higher than expected costs and technical 'hitches'. One of these drawbacks is the difficulty of gaining environmentally friendly and cost-effective access to deep seawater cold enough to be used to recover vapour. The most recent commercial version of a 'seawater greenhouse'-style complex is Sundrop Farms. Here they use a solar power plant to produce electricity to power the greenhouse and to desalinate seawater to produce fresh water for irrigation and optionally also for cooling in a pad and fan system.

Indirect evaporative cooling as shown in Figure 9.3 (right) is an alternative use of evaporative cooling if an increase in air humidity is undesirable. In this system two air streams are separated by a heat recovery system. On the dry side hot primary air is cooled against a cold surface and on the wet side secondary air is cooled by evaporative cooling. The moist secondary air is not used in the greenhouse.

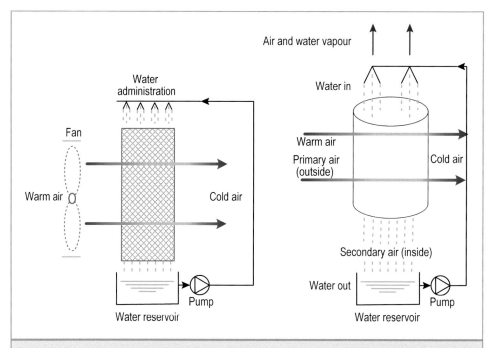

Figure 9.3. Pad and fan is an example of direct evaporative cooling (left). With indirect evaporative cooling a heat exchanger is added to transform to a sensible cooling system with no effect on humidity (After Misra and Ghosh, 2018).

9.1.3 Roof cooling

Roof cooling is a simple method of evaporative cooling that is actually a combination of convective cooling and evaporative cooling. In this system, cooling is achieved by spraying or trickling water on the roof of the greenhouse or on external shade screens. Because the greenhouse cover is relatively large, it may well be used for effective cooling at more than 4 °C under outside air temperature (Ghosal *et al.*, 2003; Sutar and Tiwari, 1995). When cold water flows over the deck it will both cool the deck and cool the external boundary layer partly by convection and partly by evaporative cooling. The cooler air in the boundary layer above the cover will add to the cooling effect of ventilation, whenever ventilation is used. When roof cooling is applied, outside air temperature and the partial evaporation of the water film must be taken into account (Campen, 2009). This cooling system is effective when it is used in single layer glass or plastic film greenhouses (Campen, 2009). With double glazing the effect is much smaller because of the increased thermal resistance between the hot greenhouse air and the cool outer surface.

9.2 Air conditioning through heat pumps

Growers increasingly use a heat pump in combination with long-term low temperature heat storage. A heat pump can best be compared to a reversed refrigerator. In the case of a refrigerator, the heat pump working principle is used to cool, but it can also be used to dehumidify or heat.

9.2.1 Heat pump types

There are two types of heat pumps: a compression heat pump (Box 9.1) and an absorption heat pump (Box 9.2). A compression heat pump is either driven by an electromotor or a gas-fuelled internal combustion engine (De Zwart, 2004). Both processes (compression and absorption) involve condensation and evaporation of the refrigerant within the system. However, an absorption heat pump uses a thermo-chemical process whereas a compression heat pump relies on mechanical energy. The absorption heat pump is less energy efficient, but a useful characteristic is that it uses heat to generate cold and one option is to use solar heat. Absorption heat pumps have proven to be ideal in places where power is unreliable, unavailable or costly and where high temperature waste heat or solar heat is available and otherwise not (fully) used.

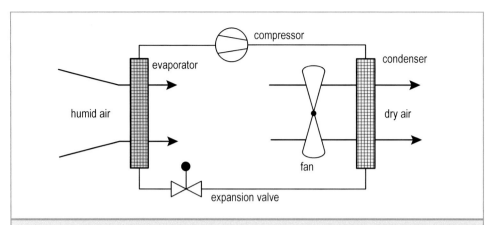

Figure 9.4. The operation scheme of a compression heat pump used as a dehumidifier and sensible heater. Effect on the external air that passes through: the evaporator cools and dehumidifies the air and latent plus sensible heat cooled out is returned to the passing air as sensible heat at the condenser. An application of this is presented in Kempkes *et al.* (2017b). Internal process: the refrigerant is vaporised in the evaporator. The heat required for this process is withdrawn from the passing external air which thus drops in temperature. The refrigerant gas is subsequently compressed resulting in much increased pressure and temperature. Due to this high pressure it is possible to liquefy the vapour at a higher temperature in the condenser. The condenser will release the heat absorbed in the evaporator together with the work delivered by the compressor.

9.2.2 Heat pump functions

Although a heat pump can be used for heating, cooling and dehumidification, in greenhouse horticulture a heat pump is mostly used in combination with cooling, as other techniques are applied when heating alone is necessary. When an air-based heat pump like the one shown in Figure 9.4 is used for dehumidification and heating, inlet air enters the evaporator where it cools down and loses vapour by condensation. This heat is taken up by the evaporator to evaporate the cooling medium. The sensible and latent heat gained plus the work delivered by the compressor is released at the condenser side where heat is released in the condensation process of the cooling medium. This sensible heat is (partly) added to the greenhouse air. Thus warm and dehumidified air is produced. The airflows at the evaporator and the condenser are separated if only dehumidification or only heating is required. The airflow along the evaporator and the condenser is often replaced by water as transport medium to allow spatial separation with heat exchangers at the heat source and sink locations combined with large central heat pump(s).

Box 9.1. Compression heat pump.

In a compression heat pump, a liquid called refrigerant is vaporised in the evaporator. The heat required for this process is withdrawn from the medium to be cooled, either water or air. This medium thus drops in temperature. The refrigerant gas is then subsequently compressed by a compressor so that the pressure and temperature is much higher. Due to this high pressure it is possible to liquefy the vapour at a higher temperature in the condenser. The condenser will release the heat absorbed in the evaporator together with the energy supplied by the compressor. Then, when the liquid refrigerant leaves the condenser it passes through an expansion valve where the pressure drops. Then the refrigerant flows back to the evaporator and the process can start over again.

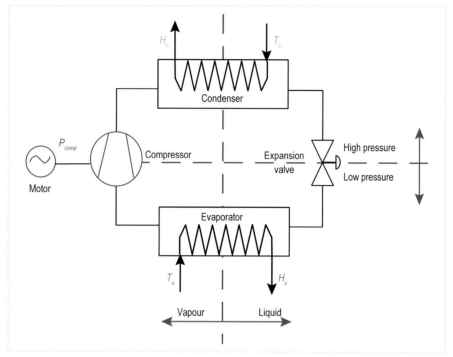

A schematic representation of a compression heat pump with P_{comp} the work added to the system by the compressor, H_c the heat flow delivered at the condenser, T_c temperature of the condenser, H_e the heat flow (sensible and latent heat) taken up the evaporator, T_e temperature of the evaporator.

Box 9.2. Absorption heat pump.

The working principle of an absorption heat pump is a thermal chemical process depicted in the Figure below. Unlike the compression heat pump which compresses refrigerant vapour, it dissolves the vapour in an absorbent. The principle behind an absorption process is to separate and recombine two fluids (refrigerant and absorbent) to create the cooling effect. Two cycles are used: the NH_3-H_2O cycle or the Lithium bromide cycle. In the NH_3-H_2O cycle, water acts as the absorbent while the ammonia solution acts as the refrigerant. In the LiBr cycle, lithium bromide is the absorbent and water is the refrigerant.

The NH_3-H_2O cycle is applied most frequently and is explained further. The ammonia vapour, which is produced in the evaporator, is transported to the absorber by a vapour pressure gradient resulting from vapour absorption in the absorber. In the absorption liquid heat is released at a higher medium temperature as a result of condensation and binding. The inclusion of liquid refrigerant reduces the uptake of the absorber as ammonia concentration increases. Therefore, liquid from the absorber is pumped to the generator. There, with the help of external heat P_{gen} and pressure, the refrigerant ammonia is evaporated again from the concentrate ammonia solution. This refrigerant vapour is then liquefied again in the condenser. Passing the expansion valve V1, the liquid refrigerant will return to the evaporator to evaporate and take up energy to generate a cooled space. The absorption liquid, that is released from part of the refrigerant, passes from the generator back to the absorber through the regulation valve V2.

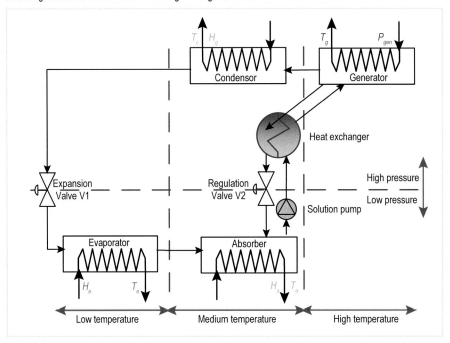

A schematic representation of an absorption heat pump (Drawn after Pocket Book, 2002). P_{gen} is the external energy added to the system by the generator, H_c is the heat flow delivered at the condenser, T_c temperature of the condenser, H_a is the heat delivered at the absorber, T_a temperature of the absorber, H_e is the heat flow (sensible and latent heat) taken up the evaporator, T_e temperature of the evaporator.

In the absorption heat pump, heat is released in both the condenser and the absorber. Heat is taken up in the evaporator. The process is driven by means of heat. An absorption cooling system powered by a concentrating solar collector system will have a coefficient of performance (COP) for cooling of 0.4-0.6 (Sonneveld *et al.*, 2006).

If water is used as a transport medium in the heat pump, then only sensible heat is transported in normally separated circuits. For instance, greenhouse cooling circuit water is cooled at the evaporator and buffer water is heated at the condenser side to store the heat cooled out in a buffer. This process can be reversed (an example is given in Figure 9.6 and Figure 9.7).

9.2.3 Performance of a heat pump – COP

The energy demand of a heat pump is indicated by the coefficient of performance (COP) which is the ratio between useful energy output for cooling (or heating) and energy input to run the heat pump (work) P. A high COP represent a high efficiency. P equals P_{comp} in case of a compression heat pump and generator heat uptake P_{gen} in case of an absorption heat pump. Two COP values are distinguished, one for cooling and one for heating:

$$\text{COP}_c = \frac{H_e}{P} \,;\, \text{COP}_h = \frac{H_e + P}{P} = \frac{H_e}{P} + 1, \text{ so COP}_c = \text{COP}_h - 1 \tag{9.4}$$

COP_c is the coefficient of performance for cooling in which H_e is the heat flow taken up by the heat pump at the evaporator, i.e. the cold side of the heat pump, for both heat pump types (W or W m^{-2}), and P the external energy input (W or W m^{-2}). COP_h is the coefficient of performance for heating in which H_e plus P is the energy usefully released to the system at the condenser, i.e. water or air heated at the warm side of the heat pump. The energy taken up by the compressor is also delivered in the form of work to the refrigerant. In the case of an absorption heat pump, useful energy output can be calculated as $H_e + P_{gen}$, or as the sum of the heat delivered at the condenser and the heat delivered at the absorber $H_c + H_a$ (W or W m^{-2}). Both the COP for cooling and for heating indicate how many units of heat are released by one unit of electricity/heat input. COP_c equals 5 means: for each unit of electricity 5 units of heat are usefully moved within the system. So in that case, you have to invest one fifth of the energy to be cooled from the greenhouse.

The COP of the compression heat pump depends on different factors, the most important being the temperature difference that must be reached between the evaporator (T_e) and the condenser (T_c). The COP value of a compression heat pump decreases with rising temperature difference (Figure 9.5). A condenser temperature of 55 °C is sometimes referred to as the maximum condenser temperature ($T_{c,max}$).

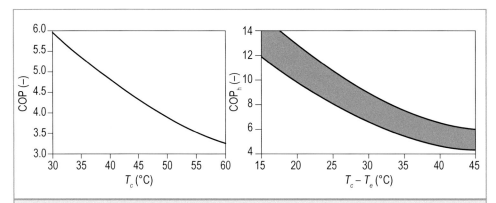

Figure 9.5. Practical coefficient of performance (COP_h) of a compression heat pump with the refrigerant ammonia at a evaporator temperature of 7-8 °C as a function of condenser temperature T_c (left) (De Zwart and Swinkels, 2002). Theoretical COP_h, Carnot Efficiency, at a given temperature difference between condenser and evaporator ($T_c - T_e$) °C (right). The system efficiency is usually 50% to 70% (industrial heat pumps).

Also, the theoretical COP of a heat pump exists, which is called Carnot efficiency (Figure 9.5 right). The Carnot efficiency depends only on the evaporator temperature T_e and condenser temperature T_c both in K:

$$COP_{c,Carnot} = \frac{T_e}{T_c - T_e}; \ COP_{h,Carnot} = \frac{T_c}{T_c - T_e} \tag{9.5}$$

$$COP_h = \eta \cdot COP_{h,Carnot}$$

Where η is the system efficiency, which is usually 50 to 70% as a result of losses and (not further elaborated on) parameters that negatively affect the efficiency.

The compressor is driven by fuel or electricity, the generator by heat. That also comes with an efficiency. To express overall efficiency the primary energy ratio (PER; Box 9.3) is used, which is defined as the ratio between the (yearly) useful energy output and the (yearly) primary energy input required to drive the heat pump from the beginning of the energy chain.

Box 9.3. Primary energy ratio.

In addition to the COP, the primary energy ratio (PER) is also relevant. PER is defined as the ratio of the useful energy output H_c to the primary energy input required to drive the heat pump from the beginning of the energy chain. The difference between the PER and the COP is that PER relates to the total energy system, while the COP refers to the heat pump itself. The PER may be determined as follows:

$$PER = \frac{E_u}{E_{pe}}$$

Where E_u is the useful supplied cold or heat in kWh y^{-1}, E_{pe} is the primary energy consumption in kWh y^{-1}.

9.3 Other cooling and dehumidification methods

9.3.1 Dehumidification by a desiccant

A desiccant is a drying agent, a chemical that picks up water molecules. The use of desiccants for greenhouse dehumidification has often been mentioned and tried (e.g. Milani *et al.* (2014); and the Desicool system by Munters, 2018), however, the main limitation is the high energy use for regeneration of the desiccant. If the regeneration energy is easily available from solar or residual heat, this method is viable. Raaphorst (2013) successfully used a vacuum evaporator to reduce the energy demand for regeneration of the desiccant solution. The dehumidification is integrated with air treatment such as cooling or heating to realise the desired air condition.

An early system for dehumidification was the latent heat converter developed in Israel (Assaf, 1986; Assaf and Zieslin, 2003). The dehumidification allowed for reduced ventilation, heating costs and enhanced effects of CO_2 enrichment at the reduced ventilation rate. It converted vapour into liquid water and heat at a COP_h of 15.4 when recovered sensible heat could all be used for heating the greenhouse. The humid greenhouse air was pulled through a pad wetted with a brine solution. The hygroscopic brine has direct contact with the air to absorb water vapour. In this process brine and air warm up as a result of the release of latent heat. So the system dehumidifies, filters and warms the air returned to the greenhouse. A heat pump is used to separate water from the brine again as the brine dilutes with vapour uptake in the pads alongside. The system is applied in several greenhouses, though not on a large scale. Kempkes *et al.*, 2017b) concluded that with a hygroscopic dehumidification system, sensible and latent heat can be recovered effectively with an expected energy saving of 250 MJ m^{-2} y^{-1} in production greenhouses at northern latitudes.

9.3.2 Dehumidification by air treatment

Whenever natural ventilation is impossible or undesired, such as when thermal screens are closed, it can be replaced by forced ventilation drawing air through the side walls. Usually the fans force external air into perforated tubes that distribute the air in the greenhouse (Campen, 2011). To save energy as well, external air can first be forced through heat exchangers mainly to recover the sensible heat in the outgoing air (energy saving of 130 MJ m^{-2} y^{-1}), or combined with air mixing units to control the humidity of the air by mixing indoor air with outdoor air (Section 4.2.2) (no energy saving, just dehumidification with controlled humidity of the inlet air), or even with a heat pump to dehumidify and retrieve sensible and latent heat (energy saving 225 MJ m^{-2} y^{-1}) (Kempkes *et al.*, 2017b). The energy savings are indications resulting from experiments and simulation of these systems installed in (semi-closed) greenhouses in the Netherlands.

9.3.3 Two-stage or mixed mode cooling and dehumidification

Next to evaporative cooling and heat pump based cooling, there are also two-stage or mixed mode cooling systems (Misra and Ghosh, 2018). Davies (2005) proposed a solar cooling system for greenhouse food production in hot climates. In the proposed cycle, the air is dried in a desiccant pad and heat exchanger prior to entering an evaporative cooler. This lowers the wet-bulb temperature of the air and keeps the desiccant pad cool. The heat exchanger removes heat released as latent heat of condensation and the heat of dilution of the desiccant. The cooling is assisted by using the regenerator to partially shade the greenhouse. Solar energy is used to regenerate the desiccant solution. The case was analysed for the climate around the Gulf based on weather data of Abu Dhabi. Compared to direct evaporative cooling, which lowers average daily maximum temperatures in the greenhouse by about 10 °C, the proposed system lowers maximum summer temperatures by another 5 °C. This will extend the optimum season for lettuce cultivation from 3 to 6 months of the year and, for tomato and cucumber, from 7 months to the whole year. Critical in this system is to generate enough affordable heat to regenerate the desiccant.

9.4 Increasing the solar collector function of a greenhouse

Reducing the ventilation need of a greenhouse has several benefits: it allows carbon fertilisation even under high sunshine (Box 5.4 and Figure 12.5), it increases water use efficiency (Section 13.5) and reduces the chance of pests and their carriers getting in. An additional advantage is that with a heat pump solar energy can be utilised much better. The surplus energy in summer can be harvested and stored in a seasonal buffer for use in winter. As greenhouses even at the latitude of the Netherlands (52°N) have a yearly surplus of energy, this is a way to make them sustainable. This means removing heat and vapour from the greenhouse air during the summer, and using the removed energy as a heat supply during the winter. This is possible through the combination of heat pumps and (underground) storage. It is a proven technology in buildings and factories, and is increasingly being applied in greenhouse horticulture. Box 9.4 illustrates the energy budget of a semi-closed greenhouse.

Aquifers with typical well depths of 20 to 200 m are most feasible for long-term storage (Figure 9.6). In order to operate aquifer systems, the soil in the underground must be suitable for storage. Therefore, the layer has to have a high porosity and must be surrounded by impermeable layers, and there must be (almost) no water flow in this layer. Often a sand layer between two clay layers satisfies these requirements. A storage system consists of two pits: one for the warm water (18-20 °C) and one for the cold water (6-8 °C). During the winter cold is stored by extracting heat from the warm pit and storing the cooled water in the cold pit. The stored water, both warm and cold, is kept separate from the water in the greenhouse systems.

Box 9.4. The benefit of reducing ventilation.

Even at the latitude of the Netherlands (52° N), greenhouses have a yearly surplus of energy. In other words, the energy carried out by ventilation is more than the energy supplied by heating. The problem is the phase shift: there is an excess of energy when there is sunshine and heating is usually required on dark and cold days. If the excess amount of energy in summer is stored such it can be retrieved in the winter, there would be no need for heating. This was the reasoning behind the 'closed greenhouse'. Instead of ventilation, heat is removed from the greenhouse air through cold water circulated in heat exchangers, stored in an underground aquifer, the 'warm well', and then, in the winter, water would be drawn from there, circulated through the heat exchangers and upgraded by a heat pump at the condenser side to warm up the greenhouse air and the warm well water thus cooled at the evaporator side pumped into a 'cold well', which is the source of cold water in the summer.

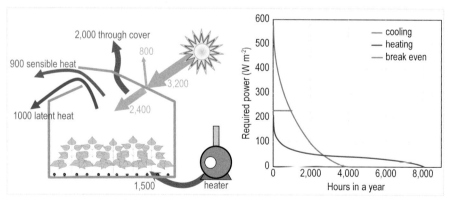

The typical energy budget (MJ m^{-2} y^{-1}) of a reasonably insulated greenhouse in the Netherlands, above left (De Zwart, 2007). Right: yearly distribution of the heating load (red) and cooling requirement (the energy 'ventilated away', blue). The area below the blue and the break-even line (green) is the same as the area below the red line. In other words, a cooling power of some 230 W m^{-2} with adequate storage would make the greenhouse energy neutral at 100% storage efficiency. Ventilation would be required for the hours (some 650 h y^{-1}) in which energy load exceeds 230 W m^{-2} (semi-closed greenhouse) whereas a power of some 550 W m^{-2} would be required to prevent ventilation altogether (closed greenhouse). Normally storage efficiency is about 70% which would bring the break-even line close to 300 W m^{-2}.

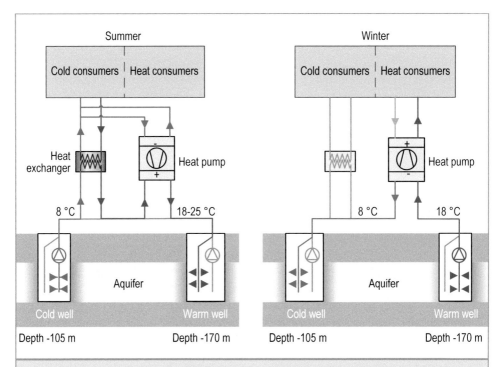

Figure 9.6. Aquifer operation in summer and winter situation (TechWiki). In summer (left) the greenhouse is a cold consumer, meaning the greenhouse is cooled using water from the cold well either directly or deeper cooled to about 5 °C by means of a heat pump which also warms the return water from the greenhouse that is sunk into the warm well. In this example no heat consumer is identified in summer. In winter (right) when the greenhouse is a heat consumer, flow is reversed and water from the warm well is cooled by the heat pump to fill the cold well. The heat pump upgrades the heat removed from the warm well to a medium temperature level at the condenser side to heat the greenhouse. Dehumidification is an optional cold consumer in winter.

Heat exchangers transfer the heat from one water system to the other. The grower can use a heat pump to reach higher, more suitable temperatures in winter and to cool the greenhouse more in summer (Figure 9.6). Another state-of-the-art greenhouse energy system that heats, cools and dehumidifies is shown in Figure 9.7.

This greenhouse uses thermal heat and stores surplus heat in buffers or in the aquifer. In addition, cold water is stored underground. A heat pump brings the water to a higher temperature and also harvests latent heat by dehumidification. The thermal power of a Seasonal Thermal Energy Storage (STES) system (H_{th}) is:

$$H_{th} = \phi_w \cdot \rho c_p \cdot \Delta T \qquad\qquad\qquad\qquad\qquad\qquad\qquad \text{W} \quad (9.6)$$

With ϕ_w the water flow in the system (m³ s⁻¹), ρc_p specific heat of water (J m⁻³ K⁻¹) and ΔT temperature difference between supply and return water (°C or K) (see also Equation 8.12).

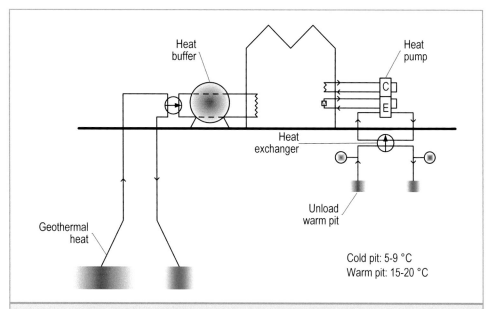

Figure 9.7. A concept for a fossil zero greenhouse with dehumidification and base heating by a heat pump operated with a non-fossil energy source and connected to an aquifer system (right side) and a connection to a geothermal shared heat network in case of a remaining heat demand (left side).

In summer, when the cold pit is unloaded, the heat pump is not needed unless further cooling is required (Figure 9.6), i.e. to extract more energy from the greenhouse air. Either way, the cold water is sent through a heat exchanger system to cool the greenhouse. The water in the pipes warms up in the process and this energy is stored (using another heat exchanger) in the warm pit of the aquifer system. In this way the greenhouse is cooled and solar energy is harvested and stored.

9.5 Concluding remarks

Dehumidification, cooling and heating are closely related processes but targets for dehumidification and temperature management differ. Dehumidification is required all year long and cooling is required only if indoor temperatures exceed the upper temperature limit for optimal production. The main methods of dehumidification are ventilation, cooling and desiccant processes. Several system designs for cooling and dehumidification presented in this chapter allow conservation of (part of) the energy involved and will increasingly reduce the need for ventilation as a mechanism for dehumidification and cooling. This field is still in development but it points to better options for climate management using fewer resources (fuel, CO_2 and water).

Chapter 10
Supplementary lighting

Photo: Paolo Battistel, Ceres s.r.l.

In Northern Europe and Canada supplementary lighting is used to increase the total daily light in winter. Supplementary light improves yield and product quality and allows for year-round production (see Section 7.1). Lamps are also used for day length control of flowering. This chapter deals with the engineering aspects of supplementary lighting.

10.1 Introduction

Since light intensity, spectrum and duration affect several plant growth and development processes (Figure 1.2 and Section 7.1) there are several reasons for providing supplementary lighting. The most obvious is to increase assimilation when natural light is scarce, thereby increasing yield and product quality. No less important is a positive effect on the market value of yield. For instance, off-season production, year-round delivery, and the relative decrease in labour costs thanks to the flat pattern of labour requirement. For instance, rose growers in the Netherlands use large amounts of supplementary light (operating hours and intensity) and obtain a rather flat production pattern throughout the year.

Supplementary light is also applied for day-length control in ornamental crops, to induce or prevent flowering in long-day plants and short-day plants, respectively. Additional light is also used in plant propagation to stimulate rooting. Light spectrum management (which has been made easy by recent developments in light emitting diode (LED) technology) is increasingly applied to manipulate crop morphology, particularly in ornamentals.

There are several types of lamps used in greenhouse horticulture, the choice for one or the other depending on several factors. High intensity is not necessary for day length control, therefore compact fluorescent lamps, fluorescent tubes or LEDs with relatively low light intensity are used for this purpose. Assimilation light, on the other hand, needs to have a relatively high intensity and the most commonly used in greenhouses are high-intensity discharge lamps, particularly high pressure sodium (HPS) lamps. The latter, however, are unsuitable for multi-layer growing systems (such as climate rooms, growth chambers or vertical farms), as the heat and thermal radiation they produce require a distance of at least 1.2 m from the crop. Fluorescent tubes were the lamps of choice for this application, before the LED era.

The development of high power LEDs (watts rather than milliwatts) has also made it possible to use LEDs for assimilation, which is increasingly done (Box 10.1), in addition to all applications that would require a specific combination of wavelengths (light spectrum). LED systems usually consist of red LEDs (which are the most efficient), with a small fraction of blue LEDs. For instance, whereas 12% blue was the norm in 2012 (Dueck *et al.*, 2012) present systems usually have a much smaller blue component. Commercial applications of Far red ($\lambda \approx 720$ nm) as well as white LEDs are being investigated (see Chapter 7, p141).

Whenever spectrum is not the relevant aspect, there may be more than one type of lamp available for a specific task. Then the factors to be considered are as follows:
- investment costs, of which lifetime of the lamp is one aspect;
- the efficiency: how much electricity for a given output of photosynthetic active radiation (PAR) light. This determines:
  - the running costs;
  - the amount of possibly undesired heat that is pushed into the greenhouse in addition to the energy carried by the light;
- how much natural light is cut off by the lamps;
- spatial light distribution.

10.2 Light measurement units and variables used

Light can be measured in several unit systems (Thimijan and Heins, 1983): photometric units, radiometric units and quanta or Einstein units. Photometric units are used for characterisation of light in offices and homes, and thus account for the sensitivity of the human eye (Box 3.2)

Box 10.1. Advantages of LED over HPS lamps.

Morrow (2008) mentions several advantages of LED lighting over the currently used HPS lamps in horticulture:

- More efficient in converting electricity into light: Persoon and Hogewoning (2014) reported that LEDs (95% red: 5% blue) have reached an efficiency of 2.3 µmol J^{-1} electricity, whereas for the best HPS lamps this is 1.85. Efficiencies reported by LED manufacturers vary between 1.4 and 3.0 µmol J^{-1} electricity. Hence the best LEDs would be 60% more efficient than HPS lamps.

- Adjustable spectrum (light quality) which allows optimisation for photosynthesis and plant form and function.

- Low radiative heat emission so can be placed close to the plants (e.g. interlighting).

- Very long life (>50,000 hours).

- Less light pollution: human eye light sensitivity (Box 3.2) is highest in yellow, which is the peak emission in HPS (Figure 10.4). Yellow can be avoided with LED.

- Good safety characteristics.

---

to different wavelengths. Therefore such units are not relevant for crop applications and should not be used in horticulture. Einstein units are the units needed when we discuss light in terms of relevance for photosynthesis and crop growth, since they account for photon numbers, rather than the energy they carry. When the energy aspect of light is the relevant issue, radiometric units are used (Table 10.1; see also Box 3.1). The units of the 3 systems can be converted into each other when the spectrum of the lamp is known, so conversion factors (Box 10.3) are lamp-specific.

The Einstein units account for light as photons. The photosynthetic photon flux (usually indicated by PPF, µmol s^{-1}) indicates the flux of photons generated by the light source. When the direction is relevant (i.e. the light beam is not evenly distributed), then we indicate the intensity (µmol s^{-1} sr^{-1}) per unit solid angle (steradian, sr; Figure 10.1). For crop applications, the relevant unit is the flux density the crop receives (photosynthetic photon flux density (PPFD) in µmol m^{-2} s^{-1}). The photon quantity (µmol) in the Einstein system is comparable with radiant energy (J) in the radiometric system (Box 3.1).

Table 10.1. Three systems (photons, energy, visible light) for measuring and processing light in physical units. Each row indicates the analogy between these systems. Photometric units (Box 10.2) are added for understanding the analogy. It is not advised to use photometric units for crop growth. However, they may be relevant for workers passing through the crop for crop operations.

Einstein units	Radiometric units	Photometric units (not in horticulture)
photon quantity P_q Einstein (µmol)	radiant energy Q_r (J, Wh)	luminance l (Candela cd = 1 lm sr^{-1})
photosynthetic photon flux PPF (µmol s^{-1})	radiant flow q_r (W)	luminous flux φ (lumen, lm)
		light quantity Q (lm s)
PPFD: PPF-density (µmol m^{-2} s^{-1})	irradiation E_r in (W m^{-2})	illumination E (lm m^{-2}, lux)
efficiency electrical input (µmol J^{-1})	radiant efficiency (mW W^{-1})	specific light flow, luminous efficacy (lm W^{-1})

Box 10.2. Photometric units – to be used for conversion purposes only.

Photometric units are in some cases needed for manual calculation of light distribution since many manufacturers publish light distribution diagrams for photometric units only.

The light emitted by the light source is called luminance l in candle cd (1 cd = 1 lm sr^{-1}). This is the amount of lumen sent out by the light source per steradian. This results in a light flow φ in lm in a given direction. Integrated over time this leads to a light quantity Q in lm s. At the receiver side we speak about illumination E in lm m^{-2}, lux which is the light flow intercepted per unit area. To calculate the electricity demand or efficiency of a lamp in terms of light production we use the specific light flow, also called luminous efficacy, in lm W^{-1}.

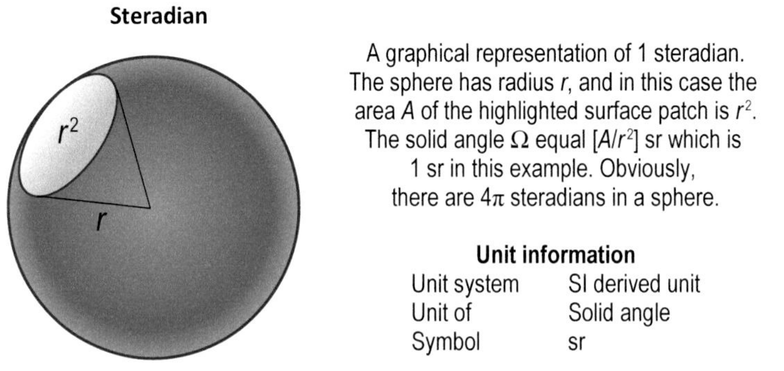

Figure 10.1. Definition of solid angle (Wikipedia).

From an engineering point of view the most relevant question is to design lamps and fittings for uniform light distribution at the required incident quantum or radiation flux at crop level. Incident light is defined by $PPFD = PPF / A$; $E_r = q_r / A$; $[E = \varphi / A]$, with A the area of the receiving surface.

10.3 The amount of light at a certain point

Within a light beam of one steradian, 1 m^2 is illuminated at a 1 metre distance from the light source. At 2 m this is 4 m^2. The area covered by the light beam is proportional to the square of the distance to the light source and the intensity of measured light is inversely proportional:

$$PPFD = \frac{PPF}{A} \text{ (μmol m}^{-2}\text{ s}^{-1}\text{);} \quad E_r = \frac{q_r}{A} \text{ (W m}^{-2}\text{)}$$

A calculation of light intensity at plant level is given in Box 10.4. A lamp doesn't send out a uniform beam in all directions. The secondary optics (reflectors and, possibly, lenses) are designed to distribute the light as uniformly as possible on a horizontal surface at a pre-set distance from the lamp, therefore the angle of incidence of the beam is variable for a horizontal 'receiver'. When the beam is at an angle α from perpendicular, the irradiation measured is $E_{r,\alpha} = E_{r,perpendicular} \times \cos(\alpha)$ in W m^{-2}.

Each point in the crop usually receives light from (many) more than one lamp so the contributions of each one, accounting both for distance and angle of incidence, have to be bundled together, to calculate irradiation. As an example, Figure 10.2 shows two light sources. The illumination at the elementary area A is the sum of the contributions of the lamps L_1 and L_2, positioned at a different angle and distance.

Box 10.3. Conversion factors.

The photometric, radiometric and Einstein units can be converted into each other. The energy content of radiation allows the radiation emitted by a lamp in both Einstein and radiometric units to be quantified, once the spectral distribution is known. However, what we measure is the radiation perceived per unit area of the receiver, so an additional complication is the spectral sensitivity of the receiver, which is not necessarily flat (see Box 3.3). This applies even more to photometric units, which account for the spectral sensitivity of our eye, which is definitely not flat (see Box 3.2).

So, what people do is to use overall conversion factors, which have been determined through measurements by standard sensors, within the PAR waveband (see Thimijan and Heins, 1983). It is important to be aware that, given the possible spectral variability among lamps of the same type (think of reflectors, for instance) and the possible variations in spectral sensitivity among sensors, these are approximate values.

Below are the conversion factors between Einstein, radiometric and photometric measurements, all within the waveband 400 to 700 nm.

Radiation source	Multiply by indicated value					
	PPFD to E_r (µmol to J)	E_r to PPFD (J to µmol)	PPFD to E (µmol to lm s)	E to PPFD (lm s to µmol)	E_r to E (J to lm s)	E to E_r (lm s to J)
Sunlight	0.219	4.57	54	0.019	0.249	4.02
Cool white fluorescent	0.218	4.59	74	0.014	0.341	2.93
Plant growth fluorescent	0.208	4.80	33	0.030	0.158	6.34
High-pressure sodium	0.201	4.98	82	0.012	0.408	2.45
High-pressure metal halide	0.218	4.59	71	0.014	0.328	3.05
Low-pressure sodium	0.203	4.92	106	0.009	0.521	1.92
Incandescent 100 W tungsten halogen	0.200	5.00	50	0.020	0.251	3.99

The conversion number given here for sunlight is more than double the number calculated in Chapter 3 (see Box 3.1). This follows from the fact that more than half the sun's radiation is outside the PAR waveband and the number here converts measurements within that waveband. In addition, as the spectral distribution of solar radiation varies with atmospheric conditions (Hogewoning et al., 2010a), conversion numbers for sunlight are even more approximate than the others.

Box 10.4. Calculation of light intensity at plant level.

A 600 W (1,070 µmol s^{-1}) HPS lamp has an electric efficiency of 35%, while the efficiency of the reflector is 98%. If the grower wants 80 µmol m^{-2} s^{-1} PAR, which area can be illuminated by this lamp assuming an even light distribution? What is the required electricity to provide 12 h of supplementary light?

Answer: 1,070 µmol s^{-1} from 1 lamp with 98% efficiency of the reflector. Grower wants 80 µmol m^{-2} s^{-1} hence 1,070 × 0.98 / 80 = 13.1 m^2 per HPS lamp on average.

The total light sum per day depends on lighting hours and required intensity of lighting. In case of 12 h of lighting, the light sum will be 12 × 3,600 × 80 / 10^6 = 3.46 mol m^{-2} per day. To provide this light sum, electricity need is 600 / 13.1 × 12 × 3,600 / 10^6 = 1.98 MJ m^{-2} per day = 0.55 kWh m^{-2} d^{-1}.

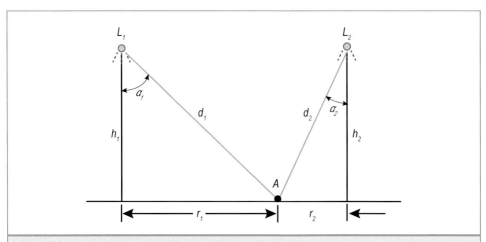

Figure 10.2. Point A is illuminated by two lamps L1 and L2.

From lamp L_1 light has to travel a distance d_1 to illuminate A under a zenith angle α_1 and light is not perpendicular to the elementary area A but with elevation angle 90-α_1. I_{α_1} is the light intensity sent from L_1 in the direction of A. In photometric units the illumination in point A is:

$$E = \frac{I_{\alpha_1} \cdot \cos \alpha_1}{d_1^2}, \text{ with } d_1 = \frac{h_1}{\cos \alpha_1} \tag{10.1}$$

Combining these equations and transforming to Einstein units or radiometric units results in:

$$PPFD = PPF_{\alpha_1} \cdot \frac{\cos^3 \alpha_1}{h_1^2} \text{ or } E_r = q_{r,\alpha_1} \cdot \frac{\cos^3 \alpha_1}{h_1^2} \qquad \text{µmol m}^{-2}\text{ s}^{-1} \text{ or W m}^{-2} \tag{10.2}$$

The same has to be done for L_2. If there are more lamps, n in total, the contribution of each lamp to the total illuminance of the spot has to be calculated and then added to the total as shown in Equation 10.3 and Figure 10.3 for both Einstein and radiometric units.

$$PPFD = \sum_{i=1}^{n} PPF_{\alpha_i} \cdot \frac{\cos^3 \alpha_i}{h_i^2} \text{ or } E_r = \sum_{i=1}^{n} q_{r,\alpha_i} \cdot \frac{\cos^3 \alpha_i}{h_i^2} \qquad \text{µmol m}^{-2}\text{ s}^{-1} \text{ or W m}^{-2} \tag{10.3}$$

As a result of lamp and secondary optics, the intensity of the PPFD or radiant flow changes with the angle to the light source. This light intensity sent out under different angles is mostly expressed in luminous intensity distribution diagrams, also called light distribution diagrams which are drawn in a polar diagram (Box 10.5).

As light intensity and distribution have a large effect on yield and its uniformity, they need to be measured very carefully. Nederhoff *et al.* (2009) and Dueck and Pot (2010) have proposed a standard protocol for taking measurements in research greenhouse compartments.

The curves in a light distribution diagram denote (Box 10.5) the relative or absolute intensity of the light source in different directions. Such diagrams are a tool for determining PPFD at

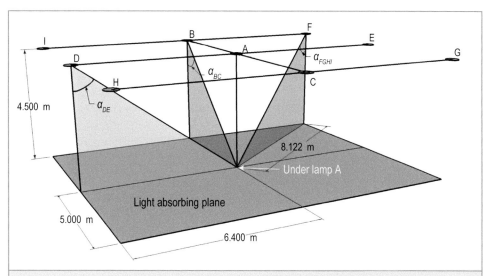

Figure 10.3. Nine light sources A to I contributing to photosynthetic photon flux density in a given point of the reference plane (n=9).

Box 10.5. An example of light distribution in a greenhouse.

The low-blue version of a LED system has a light output of 1,562 µmol s^{-1}. The grower requires an average PPFD of 75 µmol s^{-1} m^{-2}. On average this translates to 20.83 m^2 per luminaire. If the distance between trellis girders is 5 m, the one between the luminaires along the trellis girder should be 20.83 / 5= 4.16 m. The figure shows the light distribution diagram for this LED with a wide radiant fitting (120°) as well as the light distribution with uniformity coefficient of variation of 0.096.

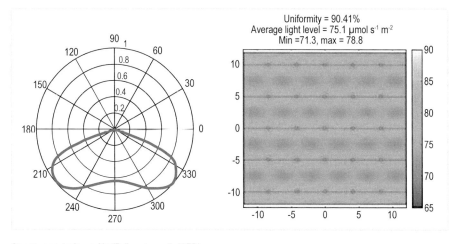

(source: groentenieuws, HortiDaily water cooled LED)

a certain point. The concentric rings indicate equal luminous intensity lines at angle α. Such distribution is more the effect of secondary optics than the type of lamps.

Nevertheless, such diagrams are most useful for providing a quick insight into the characteristics of the beam, since with many light sources (and with LEDs there are very many) they become unworkable: computer programs with ray-tracing algorithms have replaced them.

10.4 Major lamp types and their characteristics

10.4.1 HPS lamps

The HPS lamp was first developed in 1964 and has been used ever since to enhance photosynthesis and therefore crop growth. In greenhouses, 400 W, 600 W and 1000 W lamps are used; the latter are inherently more efficient and are preferred, since fewer fixtures are needed for a given light intensity and thus least block natural light. On the other hand, less fixtures makes it more difficult to obtain an uniform light distribution when low light intensities are desired.

HPS lamps have a maximum efficiency of about 1.85 μmol J^{-1} (Box 10.6). A 1000 W HPS lamp (bulb), has a power uptake of about 1,060 W, converts about 37% of the electric energy provided into PAR; about 39% into heat; some 5% is used in the electrode and 18% is non-PAR radiation (mainly thermal infrared (TIR)). This large production of radiative heat can be a relevant factor for greenhouse crops. All high intensity discharge lamps (also the whitish metal halide) are based on the same type of process and thus have similar energy distribution, the main difference being the colour distribution in the visible range. The spectral distribution of a commercial HPS lamp is shown in Figure 10.4.

HPS lamps (like all high-intensity discharge lamps) have to be fitted into a well-shaped reflector, to ensure that most light is directed and well distributed towards the crop. The effect of a reflector on light distribution is shown in Figure 10.5. Of course even the best reflectors have some dispersion, so the efficiency of the reflector must be taken into account. A much overlooked aspect is that reflectors age very much: humidity and dirt do take a toll. Therefore reflectors should be changed (or at least get a surface processing) every couple of years.

10.4.2 Light emitting diodes

As its name suggests, a LED is a p-n (positive/negative) junction that emits light when a current flows through it. The light emitted is nearly monochromatic, the wavelength depending on the materials used for the junction and the dope. For instance, red LEDs (the first in the visible range) were introduced in the early 1960s whereas reasonably bright blue LEDs only in the early 1990s. Together with green LEDs, this made it possible to generate white light, although of a somewhat unnatural colour. This has been obviated more recently by white LEDs that are,

Box 10.6. Efficiency of current lighting systems.

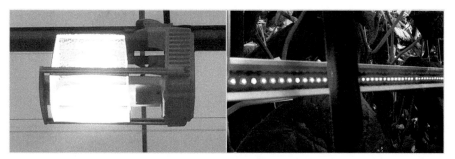

The vast majority of lighting systems in greenhouses use HPS (high pressure sodium) lamps (left), but the market share of LEDs (right) is increasing. There has not been much evolution for the past 20 years in HPS systems, whose maximum efficiency is about 1.85 μmol PAR per J, whereas a state of the art LED system can attain 2.7 μmol J^{-1}, and systems giving 3 μmol J^{-1} are already available.

The efficiency of a lamp depends in the first place on the wavelength of the light, as one can infer from the energy of radiation (Box 3.1). The longer the wavelength the lower the energy content of radiation, so that a μmol of red photons (λ=660 nm) carries 0.181 J and a blue one (λ=450 nm) 0.267 J. The efficiency of the process, converting electricity into photons within the lamp, also determines overall efficiency. This process can be improved (and is being improved) by technological progress; unfortunately the same cannot be said of the laws of physics.

The emission spectrum of HPS is well-known; that of LED can vary. The efficiency of LEDs given above refers to the typical combination of 94% red and 6% blue used in greenhouses. Light with this composition carries 0.186 J μmol^{-1}, so a process generating it with an efficiency of 1 would give 5.37 μmol J^{-1}.

The efficiency must be converted to kWh (1 kWh = 3.6 MJ), since electricity is usually priced in kWh. The HPS lamps thus give 6.7 mol kWh^{-1} and a state of the art LED system 9.7 mol kWh^{-1}.

(photos: Frank Kempkes (left), Cecilia Stanghellini (right), Greenhouse Horticulture, Wageningen University & Research)

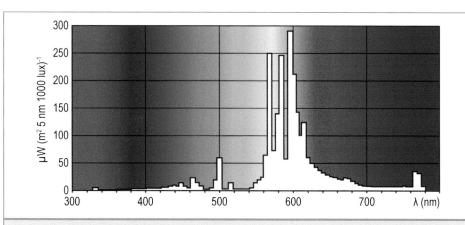

Figure 10.4. The spectral power distribution of the SON-T 1000 W E40 1SL/4 (http://www.lighting.philips.com/main/products/horticulture). For each 5 nm the radiation in μJ per 1000 lm was determined.

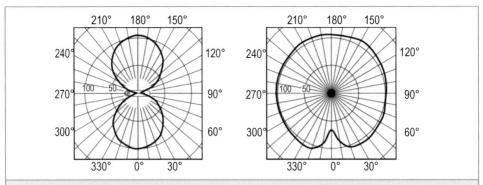

Figure 10.5. Light distribution diagrams with relative radiation pattern of MASTER GreenPower & Agro 250 W – 600 W of lamp (left) and lamp plus fitting (right) (Philips lighting). The concentric circles indicate the relative intensity of the light source in each direction.

in fact, blue LEDs within a fluorescent case that spreads blue light to a more or less 'natural' distribution of longer wavelengths (Figure 10.6).

The energy carried by a photon is inversely proportional to its wavelength, so that more energy is required to create a blue photon than a red one. No less important, however, is the efficiency of the process that produces the photon inside the lamp, which is not related to the wavelength. For instance, in a review of the LEDs it was producing then, Philips reported in

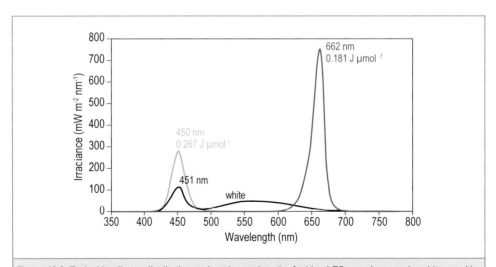

Figure 10.6. Typical irradiance distribution and peak wavelength of a blue LED, a red one and a white one (the black line). As the peak in the blue wavelength of the latter reveals, white light is created by blue light generating fluorescence of dye(s) in the encasing material. The figure also reports the energy (J) carried by a µmol in the blue and red wavelength, respectively, as explained above.

2012 a lower efficiency for deep red (650-670 nm) than for blue (440-460 nm). This is where technological developments are bringing progress.

Yet, as it has been calculated in Box 10.6, LED systems currently in use still have an efficiency of around 50%. The excess energy is mainly given back as convective heat, since the temperature is too low to have a significant fraction of TIR. Most LED systems are encased in a conductive and conveniently shaped support, to disperse heat efficiently, which is necessary since the efficiency of the lamp decreases with temperature. Some companies produce water-cooled LEDs, which give the additional opportunity of buffering the excess heat for later use or use elsewhere.

In addition, in LEDs it is necessary to convey the light where it is needed, so they come enclosed in 'secondary optics', which is basically a reflector on the back and a lens in front that collimates the light to a desired angle, for instance a beam of 80 or 150° (Figure 10.7). Obviously also the efficiency (and aging) of such 'secondary optics' have to be taken into account when determining the efficiency of the system.

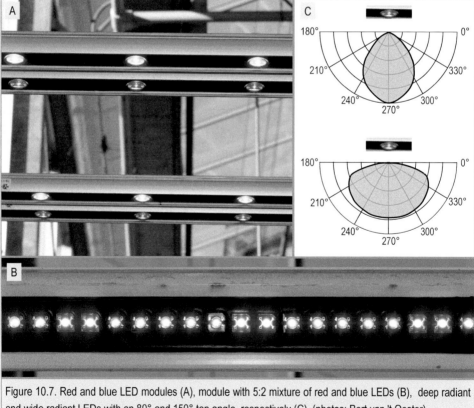

Figure 10.7. Red and blue LED modules (A), module with 5:2 mixture of red and blue LEDs (B), deep radiant and wide radiant LEDs with an 80° and 150° top angle, respectively (C). (photos: Bert van 't Ooster)

The variety of LED-wavelengths marketed today offers the opportunity to control plant growth processes to an extent that was unthinkable only a few years ago (Chapter 7, p141).

In principle, since for photosynthesis 'a photon is a photon', the narrow spectrum of a LED should not negatively affect production (at a given PAR intensity). Nevertheless, it soon became clear that red light alone produces malformed plants, which is the reason why the blue component is commonly used today. Since that time, experiments and first experiences have shown that productivity under (at least bi-chromatic) LEDs is comparable to HPS (Gómez and Mitchell, 2014; Singh *et al.*, 2015), provided the PAR intensity is the same. In heated greenhouses, however, the lack of thermal radiation causes lower crop temperature (and thus productivity) under LED than HPS, something that needs to be obviated by raising the air temperature (Dueck *et al.*, 2012; Dieleman *et al.*, 2016).

10.4.3 Fluorescent light

Fluorescent tubes are not used in greenhouses. The main reason is that a high density of tubes would be needed for a reasonable intensity, which would block too much natural light. Fluorescent tubes are used in climate rooms (vertical farming, germination rooms, tissue culture rooms), although they are gradually being replaced with LEDs. One advantage of fluorescent tubes is that they do not produce a lot of heat, even when very close to the crop. The radiant angle is very wide (160°), which makes it possible to position them very close to the crop and to keep the space between layers small. The white appearance of fluorescent light is created by the peaks in blue, red and yellow (Figure 10.8).

The requirements for lamps for day length extension or night breaks are not strict, as long as they produce the right colour. They may have low power, e.g. 20 W. Compact fluorescent lamps or small fluorescent tubes are used, and increasingly LED. Compared to incandescent

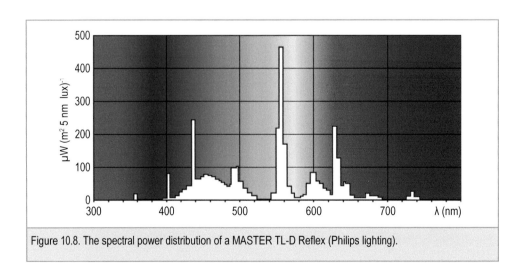

Figure 10.8. The spectral power distribution of a MASTER TL-D Reflex (Philips lighting).

light, fluorescent lamps or tubes produce hardly any far-red light which make them less suitable for day length control.

10.5 Sources of electricity

The electricity required for the lighting system can be provided by the public network or by a co-generator. Electricity from the grid does not require a large investment, however, as in most countries the price of electricity has two components: the actual use and the capacity of the connection , the high capacity that is required, may raise the cost of electricity. Running costs are usually (much) higher than with an internal supply.

The advantage of generating electricity with a co-generator is the high total energy efficiency: the waste heat and carbon dioxide may be used in a greenhouse. As this has an environmental benefit with respect to electricity generated in power plants fired by fossil fuels, in several countries there are/have been subsidies for the installation of such machines at greenhouse enterprises. In addition to the (remaining) investment, drawbacks are possible heat surplus (low heat demand and/or inadequate heat buffer) and the fluctuating price of electricity.

One key element in the choice between electric power supply through the grid or an internal co-generator is the difference between the price received for electricity produced by a (co-) generator and the cost of the fuel/natural gas needed to produce that electricity, the spark spread.

It has to be said that the current (worldwide) trend to reduce use of fossil fuels, both for heating and for generating electricity, is already bringing changes in what was 'accepted wisdom' until today. Electricity that is generated in a sustainable way is a better way to supply energy to a greenhouse than burning gas.

10.6 Concluding remarks

In this chapter an overview of the most relevant properties of lamps and lighting systems has been given. The most important aspects to consider when applying supplementary light have been pointed out. This information will facilitate the selection and management of lighting systems. For an economic evaluation we refer to Section 12.4.1.

Photo: Luca Incrocci, University of Pisa

Enrichment of the greenhouse atmosphere with carbon dioxide (also called CO_2 fertilisation) is one of the most important reasons for the high productivity of high-tech greenhouses. In this section we discuss the possible sources of CO_2 and their features, and how CO_2 is distributed and managed in the greenhouse.

11.1 Introduction

Photosynthesis is the basic process underlying plant production. Light provides the energy for this process and, in the so-called dark reaction, CO_2 and water are combined to produce carbohydrates and oxygen. Therefore CO_2 can be seen as the substrate for photosynthesis and thus the CO_2 concentration in the greenhouse air is a factor greatly influencing plant photosynthesis. This has been shown and discussed in Section 7.3.

The current chapter explains how we can obtain CO_2 for the greenhouse (CO_2 sources), the criteria applied for enriching the greenhouse air with CO_2, what technical installations are used for distribution of CO_2 in the greenhouse and which sensors can deliver feedback on the CO_2 concentration obtained (monitoring).

11.2 Units

The unit used consistently whenever concentration is related to photosynthesis is $\mu mol\ mol^{-1}$, which is the same as vpm (volume parts per million, also often indicated by ppm) thanks to equal molar volume of gases at equal temperature and pressure, as formulated in the ideal gas law. 1 $cm^3\ m^{-3}$ is obviously also 1 vpm. Furthermore, 1 m^3 of pure carbon dioxide equals 1.83 kg of CO_2 (at 20 °C and 101.3 kPa atmospheric pressure). The present atmospheric concentration of 400 ppm of CO_2 therefore equals 732 $mg\ m^{-3}$. Commonly used units for CO_2 concentration and unit conversion are given in Table 11.1.

Table 11.1. Commonly used units for CO_2 concentration and unit conversion.[1]

Concentration CO_2 in air	Units
Volume based	1 vpm = 1 µl l^{-1} = 1 cm^3 m^{-3} = 10^{-4}%
Molecular	1 ppm = 1 µmol mol^{-1}
Mass based	1 mg kg^{-1}= 10^{-4}%
Mass per unit of volume	1 mg m^{-3}= 1 µg l^{-1}
Moles per unit of volume	1 µmol m^{-3}
Partial pressure	1 Pa= 0.01 mbar= 10^{-5} bar

[1] 1 vpm = 1 ppm = (at 20 °C and 101.3 kPa) 1.53 mg kg^{-1} = 1.53×10^{-6} kg kg^{-1} = 1.83 mg m^{-3} = 1.83×10^{-6} kg m^{-3} = 41.6 µmol m^{-3} = 41.6×10^{-9} kmol m^{-3} = 0.1013 Pa.

11.3 CO_2 sources

CO_2 enrichment from the combustion of fossil fuels is the most common method, although not all fossil fuels are equally suitable. This can be combustion of fossil fuel with air heaters inside the greenhouse, or combustion of fuels with a central boiler, often in combination with a heat storage tank. It is important that the combustion is complete, and that the flue gases do not contain any noxious gases like oxides of nitrogen (NO_x), ethene (also known as ethylene), CO or SO_2. When using low-sulphur natural gas and a properly functioning burner, exhaust gases only contain CO_2 and H_2O. Other appropriate fuels for direct CO_2 enrichment are low-sulphur oils, premium kerosene (paraffin) and propane.

However, when using other fuels like heavy oil, the exhaust gases need to be cleaned before they can be used for CO_2 enrichment in the greenhouse. The same is true when using combined heat and power (CHP) with a stroke engine, even when burning natural gas, as the burning process is not constant and the temperature in the combustion chamber (cylinder) is not uniform. In all these cases special catalytic (mainly urea-based) filters are used to remove NO_x from the flues before injection into the greenhouses (Figure 11.1). Ethene is usually monitored but not filtered. When a ceiling concentration is exceeded, the flue gas is not used for carbon dioxide enrichment. Dueck and Van Dijk (2007) measured the period after which a given concentration of NO_x and ethene caused visible damage to the crop (Table 11.2). Though acceptable concentration and duration differ per crop and even variety (flowers of pepper and orchids are very sensitive), the data in Table 11.2 have an acceptable validity as they are based on exposure experiments with several crops. Flue gases of biomass cannot be used, because of the high concentration of SO_2, which is highly negative for plant growth (López *et al.*, 2008).

In all cases, the exhaust gas is cooled down by a heat exchanger to a temperature of approximately 55 °C to remove the water vapour before it is supplied to the greenhouse for CO_2 enrichment. This recovers the latent heat and prevents condensation in the supply lines, which could result in water locks preventing proper CO_2 distribution in the greenhouse.

Figure 11.1. Example of technology required to make exhaust gas of a combined heat and power (CHP) motor suitable for CO_2 enrichment in a greenhouse (see also Figure 8.10). The motor is behind the sound-insulating white wall and liquid urea is sprayed into the exhaust which has a temperature around 120 °C. What we see here is the cooler: the thick insulated duct comes from the motor (up, left, in the back) enters the cooler (centre, left) and is sent to the greenhouse (up right), after that water is condensed and recovered in the pit (centre, bottom) (photo: Paolo Battistel, Ceres s.r.l.).

Table 11.2. Maximum acceptable concentration (parts per billion) of some noxious gases in the greenhouse air, and maximal exposure for damage to plants (Dueck and Van Dijk, 2007).

Gas	Concentration (ppb)	Max duration
NO_x	40	24 hours
	16	1 year
Ethene	11	8 hours
	5	4 weeks

Different fuel sources provide different amounts of CO_2. Burning 1 m^3 natural gas, 1 litre of kerosene or 1 litre of propane provides 1.8 kg, 2.4 kg and 5.2 kg of CO_2, respectively (Portree, 1996, Engineering Toolbox). As mentioned in Section 8.5.1, one W m^{-2} heating demand that lasts for an hour requires about one m^3 gas per ha per h when delivered by a central boiler, which equals 1.8 kg CO_2 per ha per h or 0.18 g m^{-2} h^{-1} (see footnote Table 11.1).

When using natural gas for heating, CO_2 is available for free whenever there is heat demand from the greenhouse. However, there is an obvious phase mismatch: CO_2 enrichment is needed during the day, and heating mostly at night. So, one of the two (heat or carbon dioxide) has to be stored for later use. It is cheaper to store warm water than a gas, so greenhouses equipped with hot water pipe heating from a central boiler usually have a warm water storage tank, the 'heat buffer' (Figure 11.2).

During the day the burner runs to produce CO_2 that is distributed in the greenhouse, and the resulting hot water is stored in the heat buffer, for use when heating is needed. If daytime heat demand is zero, then the storage capacity is calculated, accounting for the energy content of the fuel, the efficiency of the storage, and the temperature rise in the buffer. For instance, with

---

Box 11.1. CO_2 concentration in flue gas: how much air do we need to displace for CO_2 supply?

Let's take a natural gas with the composition indicated below, the rest being almost exclusively nitrogen (please note that there is quite some variation in the composition of gases from different fields). For such a gas, n_p (the number of moles per m^3 at standard pressure) = 44.6.

The reaction equations for burning each component are:

$CH_4 + 2\,O_2 \rightarrow CO_2 + 2\,H_2O$	81.3% methane
$2\,C_2H_6 + 7\,O_2 \rightarrow 4\,CO_2 + 6\,H_2O$	2.9% ethane
$C_3H_8 + 5\,O_2 \rightarrow 3\,CO_2 + 4\,H_2O$	0.4% propane
$2\,C_4H_{10} + 13\,O_2 \rightarrow 8\,CO_2 + 10\,H_2O$	0.2% butane

From these, we can derive that 1 mol natural gas requires 1.76 mol O_2 and produces 0.9 mol CO_2 and 1.74 mol H_2O. With 21% oxygen in air, we need a ratio R_{air} of 1.76/0.21= 8.4 mol air per mol gas. In order to burn 1 m^3 natural gas we need n_{da} moles of dry air: $n_{da} = n_F \times R_{air}$ = 44.6 × 8.4 = 375 mol of air. The total amount of flue gas, n_{fg} in mol, equals 375+44.6= 419.6 mol. At a normal flue gas temperature of 55 °C, the molar volume is 26.9 l mol^{-1}, which adds up to 11.3 m^3 flue gas of 55 °C per m^3 gas. With a surplus air factor (A) of 1.2, used to secure complete burning, this is 13.3 $m^3\ m^{-3}$. CO_2 production n_{CO_2} is 0.9 × 44.6 = 40.2 mol, which results in a maximum flue gas concentration $c_{CO_2,fg}$ of 8 vol-% (10 vol-% with aeration factor 1), with $c_{CO_2,fg} = n_{CO_2} / (A \times n_{da} + n_F) \times 100$.

For optimal distribution of the carbon dioxide in the greenhouse a minimum static pressure inside the PVC injection lines of 7 mm H_2O is needed as well as a pressure difference between the ends of the injection lines of less than 20%. At low CO_2 injection rates dilution of the flue gas may support even distribution. For instance, when 50 kg $h^{-1}\ ha^{-1}$ is required and the required minimum amount of injection gas for a given distribution system is 1,800 $m^3\ h^{-1}$ per ha, what is the dilution factor D needed?

Solution: $D = MF_{min} / MF_{fg}$; $MF_{fg} = l_{CO_2} / m_{CO_2,fg}$; $m_{CO_2,fg} = n_{CO_2} \times M_{CO_2} / V_{s,fg}$, with index fg referring to flue gas, index min to minimum gas flow, D is the dilution factor, m is mass, n is moles, M is molecular mass. The mass carbon dioxide per m^3 flue gas $m_{CO2,fg}$ = 40.2 × 44 / 13.3 = 133 g m^{-3}; MF_{fg} = 50 × 10^3/133 = 376 $m^3\ ha^{-1}\ h^{-1}$; The dilution factor D = 1,800 / 376 = 4.8.

Figure 11.2. Heat storage tank, which is a well-insulated water tank (the largest cylinder in the picture), towering over the technical space housing the co-generator of a large greenhouse complex in the Netherlands (photo: Paolo Battistel, Ceres s.r.l.).

natural gas of 35.2 MJ m^{-3} energy content, storage efficiency of 90% and 50 °C temperature rise:

$$1 \text{ kg CO}_2 \Rightarrow 0.9 \frac{35.2 \text{ MJ m}^{-3}_{\text{gas}}}{1.8 \text{ kg m}^{-3}_{\text{gas}} \cdot 4.19 \text{ MJ m}^{-3}_{\text{water}} \text{ K}^{-1} \cdot 50 \text{ K}} = 0.084 \text{ m}^3_{\text{water}} \text{ kg}^{-1} = 84 \text{ l kg}^{-1}$$

So, for a day with 10 h of CO$_2$ supply at a rate of 100 kg ha^{-1} h^{-1} this adds up to a required capacity of the heat buffer of 84 m^3 ha^{-1} (Box 11.2).

However, the present trend (worldwide) towards reduction of carbon dioxide emissions is decreasing the amount of CO$_2$ available as a by-product of greenhouse heating through two processes: reduction in heating requirement (thanks to insulation and new insights into climate management) and a gradual shift towards carbon-free energy sources. We are still a long way from carbon-free electricity production (see alsSection 14.4) but the trend is there and accelerating. Moreover, there are plenty of conditions/places where CO$_2$ enrichment with exhaust from burners or CHP engines is not an option, such as unheated greenhouses, or heating fuels with unsuitable flue gas composition (such as biomass). On the other hand, the potential market for CO$_2$ enrichment is extending from the typically high-tech regions to the mid-tech greenhouses of the mild-winter regions (Sánchez-Guerrero *et al.*, 2010). In other words, CO$_2$ enrichment will increase and will have to rely on external CO$_2$ sources.

Box 11.2. What can a 100 m³ ha⁻¹ buffer do in terms of separation of CO_2 and heat production?

Assume a 9-hour light period with negligible heat demand. How much CO_2 on average can be delivered to the greenhouse when the buffer is completely filled during this 9-hour period?

$$E_{max} \text{ (buffer)} = V \cdot \rho c_{p,w} \cdot \Delta T = 100 \cdot 4.186 \cdot 50 = 20{,}930 \text{ MJ ha}^{-1}$$

This equation assumes an empty 100 m³ ha⁻¹ buffer tank that heats up by 50 °C. The greenhouse has no heat demand and we assume a thermal efficiency of the central boiler of 90% water sided for the high temperature circuit in which the buffer is integrated. This efficiency is relative to the gross calorific value[1] of natural gas. The hourly averaged mass flow CO_2 MF_{CO_2} available for injection is in that case:

$$MF_{CO_2} = \frac{E_{max}}{t \cdot E_{fuel} \cdot \eta_w} \cdot P_{CO_2} = \frac{20{,}930}{9 \cdot 35.2 \cdot 0.9} \cdot 1.8 = 132 \text{ kg ha}^{-1} \text{ h}^{-1}$$

This is 13.2 g m⁻² h⁻¹. The first ratio in the equation above represents the average hourly natural gas demand to fill the thermal buffer. P_{CO_2} is the specific CO_2 production per m³ natural gas (kg).

---

[1] For definition of the gross and net calorific value of a fuel see Wikipedia (https://en.wikipedia.org/wiki/Heat_of_combustion).

Obviously, there are more than enough potential sources, as most industrial processes generate some CO_2 and the legal framework stimulates/compels its removal from the flues. Indeed, growers in the Westland area of the Netherlands are connected to a piping system carrying waste CO_2 from industrial plants, mainly in the area of the port of Rotterdam (www.ocap.nl). Thanks to the success of the scheme, the number of CO_2 sources is increasing and seasonal storage is being considered (there is more need for CO_2 in summer than winter, whereas sources are relatively flat) in the now empty gas fields under the North Sea. Unfortunately, this system is one of a kind (for now) and additional CO_2 for most growers is bottled and delivered by lorry.

To give an idea how much CO_2 may be necessary for a greenhouse of 1 ha, we have seen in Box 2.6 that 1 kg assimilated CO_2 yields, for instance, 5 kg fresh tomatoes. We have also seen (Box 5.4) that the amount flowing out of the vents can be several times over the amount assimilated. A typical Dutch grower harvests some 600 t ha⁻¹ y⁻¹, for which 120 t ha⁻¹ y⁻¹ have been assimilated. With the current Dutch supply strategy a total of 250 t ha⁻¹ y⁻¹ would be regarded as restrained, so there is room for improvement, through better supply strategies.

11.4 CO_2 supply strategy

In full sunshine a fully grown, healthy, C3 crop assimilates about 5 g m⁻² h⁻¹ (Figure 2.12). In a perfectly airtight greenhouse, a CO_2 supply capacity of 50 kg ha⁻¹ h⁻¹ would then be sufficient to cover the maximum hourly assimilation requirement. Calculations on the amount of CO_2 needed to increase greenhouse air concentration are given in Box 11.3. The CO_2 concentration

---

Box 11.3. How much CO$_2$ is needed to increase greenhouse air concentration?

Assume an empty 6 m high greenhouse. How much CO$_2$ is needed to bring the inside concentration from 400 to 1000 ppm?

   1000 ppm CO$_2$ means that the greenhouse air contains 1000 µl CO$_2$ per litre of air, i.e. 1 l CO$_2$ per m^3 of air. Molecular weight of CO$_2$ is 44 g mol^{-1} (molar volume is 24 litre at 20 °C). So, 1 l CO$_2$ equals 44/24 = 1.83 g (see also Table 11.1). 1000 ppm therefore equals 1.83 g m^{-3}.

   400 ppm CO$_2$ means 1.83 × 400/1000 = 0.73 g m^{-3}.

   Average greenhouse height of 6 m means that each m^2 greenhouse represents 6 m^3 of air. To get from 400 to 1000 ppm CO$_2$, we need (1.83-0.73=) 1.1 g m^{-3} CO$_2$ × 6 m = 6.6 g m^{-2}. Hence, for 1 ha 66 kg CO$_2$ has to be supplied just to raise the concentration when the greenhouse is airtight and no assimilation occurs during injection.

How much CO$_2$ should we supply to maintain 1000 ppm in an empty greenhouse (no crop)? Assume leakage ventilation of 2 m^3 m^{-2} h^{-1} and outside air concentration at 400 ppm.

   To maintain 1000 ppm we need to supply 2 × (1.83-0.73) = 2.2 g m^{-2} h^{-1}, which equals 22 kg ha^{-1} h^{-1}.

If the required supply rate in the same greenhouse with crop were 60 kg ha^{-1} h^{-1}, what would be the (net) assimilation rate of the crop?

   Carbon dioxide balance: I= V+A (see Chapter 5) in kg ha^{-1} h^{-1}

   A = I – V = 60 – 22 = 38 kg ha^{-1} h^{-1} or 3.8 g m^{-2} h^{-1}.

---

at sunrise in such an airtight greenhouse would be high thanks to dark respiration (it easily reaches 700 vpm in real greenhouses as well) and the capacity would be enough to maintain the concentration. The desired CO$_2$ concentration, the CO$_2$ set-point, is between 700 and 1000 vpm for most commercial crops (Figure 2.11).

Unfortunately, as we have seen, sun radiation and ventilation requirement are strongly correlated (in the absence of active cooling), so when the crop would make the most of a high CO$_2$ concentration (at high radiation levels), most carbon dioxide is lost through the vents as soon as the concentration inside the greenhouse is higher than outside (see also Box 5.4).

The CO$_2$ supply rate needed to maintain a certain CO$_2$ concentration (Figure 11.3) depends on several factors, of which ventilation rate and photosynthesis (crop species, leaf area index, light intensity) are the most important. This means that increasing the ventilation rate significantly decreases the efficiency of supply: in other words, an ever smaller fraction of the CO$_2$ supplied ends up in the crop.

An obviously safe strategy would be to try and maintain the outside concentration (and no more) whenever the vents are open, because then there would be no CO$_2$ exchange between inside and outside, and all CO$_2$ introduced in the greenhouse would end up in the crop. A 'null balance CO$_2$ enrichment' controller of this kind was developed in Japan (Kozai *et al.*, 2015) and made commercially available.

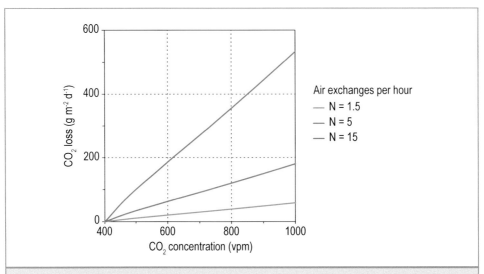

Figure 11.3. Calculated loss of CO_2 through the vents, in relation to the desired CO_2 concentration inside, assuming a CO_2 enrichment period of 8 h d^{-1} and three ventilation rates as indicated. Outside concentration was assumed to be 380 vpm (Incrocci *et al.*, 2008).

Nevertheless, the value of assimilated carbon dioxide (Box 2.6) is usually several times its price, so it may make economic sense to supply it even with a low efficiency, i.e. also with vents open (Stanghellini *et al.*, 2009). Most climate controllers allow for a linear decrease in the CO_2 set-point with increasing vents opening. When the price of CO_2 is 'negligible' the strategy usually applied is a constantly high CO_2 set-point. Thereafter, the capacity of the system is left to limit the concentration that is really attained. The supply capacity is typically between 100 and 200 kg ha^{-1} h^{-1}, which is enough to raise the concentration slightly, even at ventilation rates up to 20 h^{-1}. However, around 70% of the CO_2 supplied is then lost (see also Box 5.4).

A more advanced strategy (very recently launched commercially, Tarnavas *et al.*, 2020) is to let the climate computer determine online the most economical supply rate to be maintained, as shown in Chapter 12.

11.5 CO_2 distribution

11.5.1 Direct release from hot air burners

In the absence of a heating system, heating may be provided by small burners located inside the greenhouse, either suspended or free-standing, which release the flue gas directly into the greenhouse. Such simple systems have an on/off control, which causes dramatically fluctuating concentrations and unequal distribution. Additional problems include excessive CO_2 levels resulting from the heating requirement; possible undesired trace elements in the flue gas;

and heat production coupled to carbon enrichment, which is undesirable most of the time carbon dioxide supply is required. Usually the flues also contain water vapour, which may be undesirable as well. Finally, maintenance and monitoring for incomplete combustion is more difficult with several small burners than with one single central burner.

11.5.2 Flue gas CO_2 distribution in the greenhouse

Because of the disadvantages of direct release of CO_2 from hot air burners listed above, and particularly the synergy to be attained by separating heat and CO_2 production (Box 11.2), the CO_2 is usually brought into the greenhouse either from a central burner or from a liquid storage tank under pressure. In both cases the transport system must be properly designed and dimensioned and the distribution pipes are laid low in the greenhouse. This ensures a relatively high concentration in the crop volume, taking advantage of the fact that CO_2 is heavier than air.

The flue gas with CO_2 is transported to the greenhouse through a PVC duct. The amount (Box 11.1) of flue gases to be transported depends on the desired CO_2 supply and on the minimum amount of air needed for even distribution (pressure and flow in the lines). If the available amount of flue gas to be distributed is less than the minimum gas flow required in the system, the flue gas is automatically diluted with outside air. For this purpose a T-shaped connection pipe is mounted in the flue gas from the stack. The main duct in the greenhouse tapers off towards the end, to maintain an equal gas pressure. From the main duct a net of perforated polyethylene (PE) film lay-flat ducts of 32-50 mm diameter with four 0.8 to 1 mm diameter perforations per 20 to 120 cm, at a distance of 1.6 m or 3.2 m at 70 Pa overpressure, is used as a good and cheap distribution system (Figure 11.4). In practice one per one or two crop rows is normally applied.

Figure 11.4. The distribution system for CO_2 in the greenhouse. Left: the polyethylene ducts for distribution of flue gases (green), placed below a crop gutter. The two irrigation lines (black) can also be seen, and the 'spaghetti' for the drippers. Right: mass flow controller for pure CO_2. (photos: Frank Kempkes, Greenhouse Horticulture, Wageningen University & Research)

11.5.3 Distribution of liquid CO_2

At a temperature around 25 °C and pressure between 1 and 2 MPa, CO_2 is a liquid, with a density similar to that of water ($\approx$1,100 kg m^{-3}). In other words, a reasonable pressure (approx. 10 to 20 times atmospheric pressure) decreases the volume more than 500 times. So, CO_2 is transported and stored in liquid form.

A pressurised tank is usually leased from the company charged with the regular delivery of the CO_2. Incrocci *et al.* (2008) estimated that, in view of the then current prices, the leasing and installation could be economical in the Mediterranean area for greenhouses larger than 1 ha.

The liquid CO_2 changes phase to gas form through expansion and is then heated up to ambient temperature, to prevent condensation along the distribution lines (Figure 11.5). The control systems drive the valves, based on measured and set-point values of CO_2 concentration. The CO_2 can be injected directly into the greenhouse through a ventilator, or through a distribution system of small pipes.

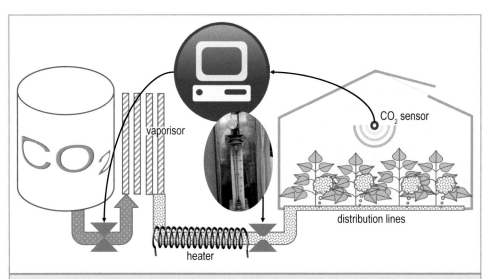

Figure 11.5. Schematic representation of the storage and delivery system of liquid CO_2 (see also the photograph at the beginning of this chapter). The opening/closing of the valve depends on the measured concentration and the set-point. A flow meter (photo inset) is used to set the flow (manually, the blue knob) allowed through the gauge when the valve is open.

11.6 CO_2 sensors

To monitor the carbon dioxide concentration in the greenhouse, several sensors are commercially available, all so-called non-dispersive infrared detectors (NDIR). The working principle is to have an (infrared) source beaming through a chamber filled with the greenhouse air. As the absorption of CO_2 in this waveband is known, the measured absorption is used to determine the quantity of CO_2 in the chamber.

There are also manual sensors based on chemical reactions. By transporting a given amount of air through the sample tube the reactor will change colour and show the concentration on a scale on the tube. However, these sensors obviously cannot be used in the automated system necessary for CO_2 enrichment.

11.7 Concluding remarks

Enrichment of the greenhouse air with carbon dioxide is one of the most important reasons for high productivity in [high-tech] greenhouses. Whereas it was developed as (and still largely is) a by-product of heating through fossil fuels, the trend is to decouple it from the heating requirement and to develop profitable systems for unheated greenhouses. Furthermore, abandoning fossil fuel as energy source, for example in an all-electric greenhouse, will also stop this source of carbon dioxide. There is a huge potential for the use of waste carbon dioxide, recovered from industrial processes, in greenhouses. An adequate regulatory framework and pricing should be in place to reap the obvious environmental benefit. Concentrating CO_2 from the air (e.g. Marshall, 2017) has an even larger potential, and will be the only possibility in a fossil fuel free future. However, before that becomes a feasible option for greenhouse CO_2 enrichment, improvements in the technology should reduce both investment costs and energy requirement. In all cases, the distribution and control system should ensure homogeneity and sensible application, in view of ventilation losses. Algorithms are available for the optimised injection on an economical basis (Section 12.4.2). Their application can help minimise CO_2 emissions in greenhouse production.

Chapter 12
Managing the shoot environment

Photo: Paolo Battistel, Ceres s.r.l.

Insights in plant production and greenhouse physics alone don't make a good grower. Managing the greenhouse is about making choices. Almost anything is possible, whether it relates to climate or to production. But a lot of the possible choices would not be economically justified. The grower can only make the right decisions if there is sufficient information and tools are available that make it possible to quantify both the economic and cultivation consequences of possible actions.

12.1 Introduction

The ability to modify the shoot environment is what makes protected horticulture different from open field agriculture. In the previous chapters we have built up knowledge about the factors that determine crop productivity; how these factors can be modified by technology and which physical processes determine the outcome (and the un-intended consequences) of possible actions. The purpose of this chapter is to show how informed choices at all levels can be based on what we have learned up to now. In the first place how one can best adapt to conditions and thereafter how knowledge of the processes can be used to get more benefit from what one has. Finally we will see how application of technology (one factor at a time) can be optimised based on economics.

The amount of decisions a grower has to take expands with the amount of 'actuators' that are available. For instance, the decisions that have to be taken in the typical 'Canarian' greenhouses of Morocco (Figure 12.1) are the dates of application and removal of whitewash, its 'thickness', the ventilation openings can be made smaller of larger (albeit not daily), besides all decisions about crop management, growing medium, irrigation and fertilisation. Fertigation computers are becoming rather common in relatively low-tech greenhouses. For instance, 50% of the greenhouses in Almeria had one in 2009 (Cuadrado Gómez, 2009: p64) and the present trend is towards a (very) limited amount of technology and simple steering programs.

The amount of actuators (and thus of possible choices) available to growers operating high-tech greenhouses is much more, with almost unlimited chances of synergy or interferences among the possible actions. Indeed, the (short term) choices for such greenhouses are handled by computer programs (Box 12.1), on the basis of set-points and factors influencing them. Even

Figure 12.1. Left: 'Canarian' greenhouses in Morocco. The cover are two iron nets enclosing plastic (usually polyethylene) sheets separated by a 'ventilation' gap, fitted with a net against insects. Side wall ventilation is ensured by pulling down the lateral sheets, the opening fitted as well by insect screens. All gaps can be manually adjusted. Right is an inside view of such a greenhouse, with a melon crop (photos: Paolo Battistel, Ceres s.r.l.).

then, knowing 'how things work' ensures the best use of all options within such programs. Indeed, it is a well-known fact that the majority of growers are not aware of (or even get confused by) the huge amount of options offered by advanced climate computers, and never bother to check whether the 'default' inputs provided by the manufacturer to obviate this problem are the most suitable to their particular conditions.

12.2 The limiting factor

The concept of the limiting factor is a very important insight to make the right management decisions. If all climate, nutrition and crop factors are perfect, but one is not, this one factor ensures that production is reduced. It is like the weakest link. For instance, we are well aware that scarcity of irrigation water can be the limiting factor for vegetable production. That other factors may be limiting is less intuitive, an example is shown in Figure 12.2, which combines the response to light and $[CO_2]$ shown in Section 2.2.7.

Of course we know that other factors (not shown here) may be limiting: for instance temperature (Figure 7.13 and 7.14). Both coldness and heat affect development of the crop. Another well-known limiting factor is nutrition: if just one of the essential elements is not sufficiently available, the crop will show deficiency symptoms and production falls. Good management starts with improving the limiting factor. There is no point in improving the factors that are not limiting.

Obviously, good management must also weigh the cost of improving the limiting factor(s) against the expected increase in income. For instance this is what is implicitly done by

Box 12.1. The basics of a standard climate control program.

Climate control programs require the grower to input set-point values for a number of target variables, such as air temperature and (relative) humidity, carbon dioxide concentration or light.

Within the program, a number of factors can 'influence' these target values: think for instance of different day/night temperature set-points (or even multiple periods during day or night), but also of allowing higher temperature when sun radiation is above a given level. The number of 'influences' that may be defined, changes with the complexity of climate control programs.

Thereafter a proportional control (the corrective measure proportional to the deviation from the target) is applied to ensure rather steady values of the target variable, which means the grower has to input for each set-point also the width of the proportional P-band. Think for instance how many degrees is the air temperature allowed to fall short of the heating set-point before the heating system is pushed to full capacity.

A range between a control action and its opposite ensures steady climate: the set-point for opening a screen, for instance, is higher than the set-point for closing it and the set-point of ventilation is higher than the set-point of heating. This difference is called the 'deadband' (the range where no control action is required) and may be input into the program.

Target	Influences	Actuators	Over-rule all
temperature	day/night sun radiation humidity	heating vents opening shadow screen energy screen mist or fog or pad&fan roof spray air conditioning	rain wind speed wind direction frost pest treatments light pollution regulations
humidity	day/night	vents opening heating mist or fog or pad&fan air circulator air conditioning	
carbon dioxide	day/night light sun/ supplemental vents opening	carbon dioxide supply	
light	sun radiation	shadow screen supplemental light	

So far for what the grower is required to do. Life is made easier by the fact that most climate control programs have default values wherever no input is given. In this way, however, a large fraction of control options is given up.

In reality, none of these targets can be modified directly: what can be controlled for instance, is the opening of a valve in the heating system; the (fraction of largest possible) opening of the vents; or the switch of the motor operating a screen or a pump. A climate control program transforms the relevant target values into instructions for the actuators, at each controller cycle.

Some actuators appear more than once in the table above (the vents are actuators for temperature and humidity; shadow screens for light and temperature, etc.). In this case a hierarchy is set within the program: most often, for instance, humidity control is allowed only within the deadband of the temperature control, or a shadow screen is considered only as an actuator for light and not temperature.

Finally, there are conditions that overrule a specific action: for instance, no attempt to open the vents is allowed when it is freezing, or darkening screens must not be open whatever the temperature.

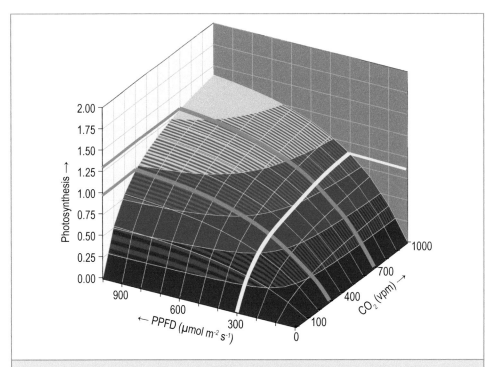

Figure 12.2. Schematic trend of assimilation as function of photosynthetic photon flux density (PPFD), and CO_2. The increase of assimilation for a given increase of light and/or CO_2 gets smaller and smaller with increasing light and/or CO_2, until a 'plateau' is reached: the 'saturated' level. However, the saturated level with respect to one variable, depends on the value of the other variable, which may thus be 'limiting'. So, even at a very high PPFD, the assimilation rate may be limited by a sub-optimal CO_2 concentration (Stanghellini *et al.*, 2008).

selecting the 'growing season' in each place of the world. This is explained in Figure 12.3. For instance, tomato requires at least about 15 °C mean temperature, which means that heating would be necessary several months a year, both in the Netherlands and in Beijing. Indeed, all Dutch tomato growers have heating. Similarly, a rule of thumb for tomato (and several other horticultural crops) is that below some 6-8 MJ m^{-2} d^{-1} sun radiation there is hardly any net assimilation (respiration almost balances photosynthesis), so supplementary light would be needed for the winter months in Holland. Prior to the introduction of supplementary light, growers had just to adapt to this, but cogeneration has made electricity and thus supplementary lighting cheaper. So now quite many growers choose to use lights to compensate the winter darkness. On the other hand, what does limit (or even prevent) growth in Southern Spain is the high summer temperature (compounded by the poor ventilation of the typical greenhouses there), too high for fruit set.

As cooling is not economical, this determines the growing cycle in Almeria and in all 'mild winter' (and hot summer) regions of the world (Figure 1.8). Vegetable growers in such regions transplant at the end of the hot period and harvest starts early autumn and continues until

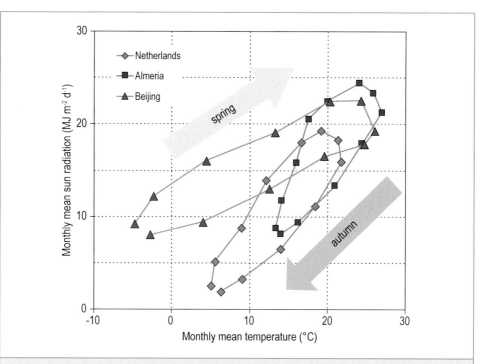

Figure 12.3. Mean monthly sun radiation (MJ m^{-2} d^{-1}) available at three sites (De Bilt, Netherlands, 52.1°N; Almeria, Southern Spain, 36.9°N and Beijing, China, 39.9°N), vs the mean monthly temperature. Each point represents a month, starting from January (in all cases the coldest). Such a plot is useful in pin-pointing the periods in which either light or temperature (or both) may be limiting for growing a given crop in a place, without modification of the microclimate.

late spring. They also remove the growing tip of the plants, like their dark winter colleagues (Box 12.2), but that happens somewhere in early spring. In Almeria most growers don't have heating, but some do. Indeed, Vanthoor *et al.* (2012) have shown that expected net financial benefit (NFB) from tomato in a well-ventilated multi-tunnel in Almeria was not improved by adding heating. They did show that NFB in a unheated greenhouse was most sensitive to variations in weather, whereas in a heated greenhouse it was to variations in prices. So, probably the few growers in Almeria who do have heating, manage by relatively long-term sales and prices agreements.

In spite of being about the same latitude as Almeria, Beijing is much colder in winter, even colder than the Netherlands. This is a climate in which it is very expensive to grow year round. Both in winter and in summer temperature is the limiting factor. The traditional, cheap approach to these circumstances is a greenhouse exploiting to the maximum heat storage (Box 3.8) and night-time insulation, to dampen the daily and, somewhat, the yearly cycle of temperature.

Box 12.2. The typical cropping cycle of 'dark winter' regions.

In the regions with limiting light in winter, growers who do not have supplementary light usually transplant in early winter. The clean white mulch ensures that each small plant gets as much light as possible. The amount of sun radiation increases rapidly with time, and growers usually allow a few more stems (Figure 2.8) to grow from late winter, and another few from early spring. In early autumn the growing tip of the plant is removed, so that no assimilates are wasted into stem, new leaves and fruits that would not reach maturity. Since light levels go down in autumn, in this way the lower assimilates source is balanced with a lower sink (less growing trusses; see also Section 7.1.1).

A daytime (transparent) energy screen was deployed when the photo was taken.

12.3 Managing climate: source/sink balance

Aereal management is about balancing temperature, light (and carbon dioxide), which is about balancing source and sink (see also Section 7.1.1). Source means the total production of sugars. Sink is the total demand for these sugars from the growing organs, like young leaves, roots, flowers and fruits.

In very young plants the source is bigger than the sink, but in adult plants usually the sink is much bigger than the source (Figure 7.3), but there may be exceptions. Plants grown under low light and high temperature are source limited. The temperature stimulates production of organs but due to the low light there are not enough assimilates for these organs. As a result they grow small, thin and long. Under high light and low temperature plants develop thick and short leaves and stay short. Such plants are sink limited. The produced assimilates have to go somewhere but due to the low temperature not enough new organs are formed, so the assimilates go to existing organs, which become bigger.

Temperature in a greenhouse with automatic climate control (Box 12.1), is maintained in principle between a heating set-point (minimum temperature) and a ventilation set-point (maximum temperature), which both usually have a daytime and a lower night-time value and are pre-determined for the crop that is grown. In the 'deadband' in between (usually a few degrees), either heating or ventilation (or both; see also Section 6.3.2) may be activated to 'control' humidity. An additional control on each set-point is the P-band (proportional band) that determines the extent of reaction when a set-point is exceeded. For instance, it makes a big difference whether the vents are 100% open when the temperature is 1 °C above the

Box 12.3. Old and new heating and ventilation management in Dutch glasshouses.

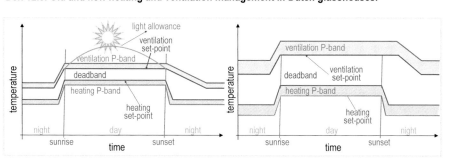

As in most Dutch greenhouses heating is produced by burning natural gas, the need to cut back on that has brought new insights into the way climate is controlled (Geelen *et al.*, 2019). Whereas (left) the aim was to have a strict control on temperature (small bands), only allowing for an increase linked to sun radiation, now (right) the natural behaviour of the greenhouse as a solar collector is exploited, by increasing the deadband between heating and ventilation (and the P-bands on each). Sunny days will naturally cause higher temperatures than cloudy days. A relatively low heating set-point will compensate for the higher daytime temperature, so that the desired long-term mean temperature is maintained.

ventilation set-point or when it is 10 °C above (Box 12.3). A large P-band allows for a smooth control but also for some deviation from the 'ideal'. Obviously, the same applies to the P-band on the heating set-point: it determines how cold the greenhouse may be before the full power of the heating system is applied.

Good management means balance between light and temperature. If light is poor, the crop cannot be pushed by a high temperature. The optimum temperature increases with increasing light intensity and CO_2 concentration (see Figure 7.13 and 7.14). which is another factor that increases source, of course. So good management does not mean maintaining pre-selected set-points as good as possible. It means flexible set-points: lower temperature on a cloudy day, higher temperature on a sunny day. The night temperature must ensure enough transport of assimilates out of the leaves, so it should be higher after a sunny day. Ripening is almost exclusively determined by temperature, so a high overall temperature fastens ripening (but not photosynthesis), and results in smaller fruits.

Since the pioneering work of De Koning (1990), a lot of evidence has been collected that growth and development of most greenhouse crops is determined by long-term (several days) mean of temperature, with little or no effect of short-term (few days) deviations from the mean (see also Section 7.2.2). Indeed, a very strict control of temperature may cause more (energy) costs than it brings in additional yield, and 'temperature integration' is adopted (in a way or the other) in modern climate computers. See for instance: Körner and Challa (2003).

12.4 Optimizing a single climate factor

Good management is weighing costs and benefits. The reality of agricultural production is that very often the benefits are difficult to quantify (what is the effect of a single action on final yield?) and there may be an additional uncertainty on the benefit, that is price of the yield. In addition, very often there is interaction (or synergy) among the various factors that determine production, so that usually one has to rely on ideal models of the crop, taking as many factors as possible into account. Nevertheless, there are simple decisions, affecting a single growth factor, that we may attempt analysing. What is needed is to know the response of the crop to that single factor.

Let's take for instance, crop assimilation as shown in Figure 12.2. At a given light (or CO_2) level, the response to CO_2 (or light) has the shape shown in Figure 12.4. The cost of supplying CO_2 (or light) increases with the amount supplied, after the investment has been done for a CO_2 supply (or lighting) system.

The optimal level of supply of factor F is the level that maximizes NFB, that is the difference between income, Y and expenses, E:

$$NFB \equiv Y - E = [\max] \rightarrow \frac{\partial(Y-E)}{\partial F} = 0 \rightarrow \frac{\partial Y}{\partial F} = \frac{\partial E}{\partial F} \tag{12.1}$$

In Equation 12.1 we have made use of the fact that in the maximum of a function the slope is horizontal, which is like saying that the derivative is 0. There may be more points in a function where the slope is horizontal (also in a minimum it is), so some common sense is

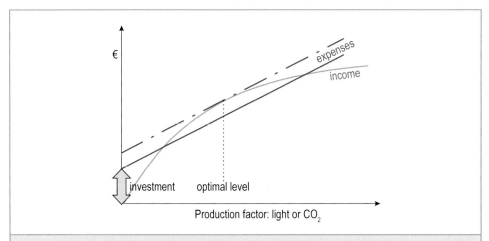

Figure 12.4. Schematic representation of the optimization of a single production factor whose effect on income is known. Optimal management is to maximize net financial benefit (income – expenses). That is at the level where the tangent to the income line is parallel to the expense line (marginal income equal to marginal expense).

needed. When in doubt, one should check that the second derivative is negative or calculate the value of the function at points slightly before and after the 'maximum', and check that they are, indeed, lower.

In most cases, the expenses are proportional to the supply (after the initial investment), so the derivative of the expenses is the slope of the straight line in Figure 12.4. Equation 12.1 is more general: it would hold also for non-linear expenses (for instance: electricity or water prices that increase or decrease with consumption).

12.4.1 Optimal management of supplementary light

Additional light increases assimilation ($\rightarrow$ production) through a higher light intensity or a longer day or both, more commonly (see Section 7.1). In the Netherlands light is the limiting factor during several months. A lot of growers use an additional light intensity of 150 μmol m^{-2} s^{-1}. Twelve hours of such an intensity cumulate to a number comparable to the natural radiation available within a greenhouse in a day mid-February or mid-October. Some growers use higher intensities, for instance, 200 μmol m^{-2} s^{-1}. Most mentioned reasons for one choice or the other are: the crop and whether or not a grower has a co-generation system. As an example, we show here that we can explain the choice, and identify the factors that determine it, extending the method first applied by Challa and Heuvelink (1993).

Let's assume a grower wants to find out the optimal intensity of additional light. The NFB (revenue minus the cost) is:

$$NFB = R_{mg} A(I_{tot}, [CO_2]) - C_{\mu mol} (\eta_{light}, P_{kWh}) I_{sup} \qquad \text{€ m}^{-2}\text{ s}^{-1} \quad (12.2)$$

where:

R_{mg}	€ mg^{-1}	is the value of 1 mg assimilated CO_2;
A	mg m^{-2} s^{-1}	is net assimilation, a function of light and carbon dioxide concentration;
$[CO_2]$	μmol mol^{-1}	is the carbon dioxide concentration;
I_{tot}	μmol m^{-2} s^{-1}	is the total light intensity (natural and supplementary);
I_{sup}	μmol m^{-2} s^{-1}	is the light intensity from the supplementary lighting system;
$C_{\mu mol}$	€ μmol^{-1}	is the cost of one μmol of additional light, which is given by the ratio between price of electricity and the efficiency of the lighting system;
η_{light}	μmol kWh^{-1}	how efficient is the lighting system in converting electricity into micromoles PAR;
P_{kWh}	€ kWh^{-1}	the price of electricity.

Let's assume that the assimilation can be represented by the product of two independent functions, one of light and one of carbon dioxide concentration. Then:

$$A(I, [CO_2]) \equiv A_{MAX} h(I) m([CO_2]) \qquad \text{mg m}^{-2}\text{ s}^{-1} \quad (12.3)$$

where A_{MAX} is the 'unlimited' photosynthesis (mg m^{-2} s^{-1}) and h and m are dimensionless functions of light intensity and carbon dioxide concentration, respectively, smaller than 1, accounting for the presence of limiting factors. Indeed, such a function was used to draw Figure 12.2.

The optimal supplementary light level is the one that maximizes NFB, that is, the derivative with respect to supplementary light must be zero:

$$R_{mg} A_{MAX} m ([CO_2]) \frac{dh(I_{tot})}{dI} = \frac{P_{kWh}}{\eta_{light}} \qquad\qquad \text{€ µmol}^{-1} \quad (12.4)$$

which gives:

$$\frac{dh(I_{tot})}{dI} = \frac{P_{kWh}}{\eta_{light} R_{mg} A_{MAX} m ([CO_2])} \qquad\qquad \text{m}^2 \text{ s µmol}^{-1} \quad (12.5)$$

that is, the optimal light intensity level (identified by the slope of the assimilation response curve to light, Figure 12.4) is determined by the two factors mentioned above: price of electricity and value of yield; but also by the efficiency of the lighting system and the carbon dioxide concentration one is able to maintain. A way of reading Equation 12.5 is to say that a high price of electricity makes for a high slope, whereas a high value of yield, a high efficiency of the lighting system and a high carbon dioxide concentration all make the slope smaller. Please have a look at Figure 12.4: the slope decreases with increasing intensity. So a high slope is at a small light intensity and a low slope at higher intensities.

In Chapter 2 (Box 2.6) we have seen how the amount of yield generated by assimilation depends on harvest yield, *HI,* and dry matter content of harvest, *dm,* then the value of 1 mg assimilated CO_2 is:

$$R_{mg} \approx P_{mg} \left(0.5 \frac{HI}{dm} \right) \qquad\qquad\qquad (12.6)$$

with P_{mg}, the selling price of 1 mg harvest.

A possibly less intuitive observation is that at given values of P_{kWh}, R_{mg}, η_{light} and $[CO_2]$ there is an optimum light intensity. Whenever this is achieved naturally, it does not make economic sense to use additional light. Indeed, most climate control computers allow growers to input the intensity (Box 12.1) of sun radiation at which the lighting system is switched off. The calculations we have done here allow to quantify better the switch-off criterion. To do it on-line would have the advantage of accounting for variable prices of electricity.

In principle, when sun radiation does not reach the intensity for 'optimal' switch off, the best value for money is given by supplementary light only up to the optimal intensity. In practice this may be difficult because most lighting systems (such as high pressure sodium (HPS) lamps) are not tuneable. The grower can turn them on or off: all, of perhaps half of them, but

Box 12.4. Optimal lighting in 'discrete' production systems.

(photo: Leo Ammerlaan)

Some crops are not sold by weight, but as numbers of a pre-defined size. Think for instance about seedlings or pot-plants. In such case the yield response curve to light is 'discontinuous', in the sense that adding a few grams dry weight to a pot-plant does not increase its value, until a higher 'quality' value is reached. The value of additional light can then be better measured in the time that it saves to reach a given plant size, rather than in additional dry matter. In places with marked seasonal cycle of radiation, the length of growing cycle is not constant, longer in low-radiation periods and shorter in summer. This reflects the fact that a plant is ready when a given amount of dry matter $\rightarrow$ assimilation $\rightarrow$ photosynthetic active radiation (PAR) light has been cumulated (assuming a given CO_2 level). With information about yearly trend of sun radiation, the greenhouse transmissivity (see Equation 3.2) and length of the growing cycles, it is possible to calculate the amount of PAR radiation cumulated in each growing cycle, which should be rather constant (Box 7.2). With this figure it is possible to calculate the benefits of additional light, assuming the light use efficiency (LUE) is constant at all times and unaffected by the light spectrum. Then one is able to quantify the time gained in each cycle, by a given intensity and duration of supplemental light. In this way it is possible to quantify the time saved (or the number of cycles that can be added) in one year (Box 7.4).

nothing in between. A light emitting diode (LED) system may be tuneable. However, this requires an additional investment and the efficiency decreases much while reducing power.

Obviously, where there is much sun radiation the 'switch off' conditions happen often. This does not mean that a lighting system cannot make economic sense in places with a high intensity of sun radiation. Whenever sun radiation is below the optimal light intensity for a reasonable fraction of the time, it may make economic sense to have supplemental light. Indeed, the optimal light intensity can be very high for very high-value crops (Box 12.5), and rather high also for reasonably priced crops, with an efficient lighting system and cheap electricity.

Box 12.5. Optimal light intensity.

A few examples can be useful to understand what determines optimal light intensity and how. We take a typical example (good value tomato 1 € kg^{-1} fresh, HPS lamps, cheap electricity: 0.05 € kWh^{-1}), and then we change one factor at a time. In all cases we assume that CO_2 concentration is non-limiting and that A_{MAX} is 2 mg m^{-2} s^{-1}, which is reasonable for a fruit crop, for instance.

	P_{kWh} € kWh^{-1}	η_{light} mol kWh^{-1}	R_{mg} € kg^{-1}	dh/dl (m^2 s µmol^{-1})
Reference	0.05	6.5	5	7.69 10^{-4}
Good LED	0.05	10	5	5 10^{-4}
Expensive kWh	0.1	6.5	5	1.54 10^{-3}
Very high value crop ('cannabis')	0.05	6.5	40	9.62 10^{-5}

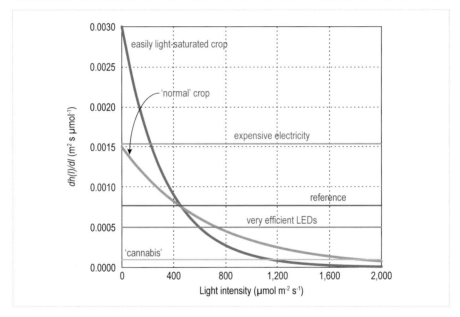

Above the two thick blue lines are the slope of the crop response to light curve (Figure 12.4) in the case of a crop saturated at relatively low intensity (dark blue) and a crop less easily saturated (light blue). The horizontal lines show the 'optimal' slope, calculated with Equation 12.5, the values given explicitly in the rightmost column of the table. The optimal light intensity is the x-axis value of the intersection point between the horizontal line corresponding to a given case and the relevant slope line for the crop.

One can see that a relatively high price of electricity makes supplemental light easily uneconomic, whereas the 'sky is the limit' for the light intensity that can be given to a (very) high-value crop.

To read the x-axis in terms of sun radiation, please remember that for the sun 1 W m^{-2} ≈ 2 µmol m^{-2} s^{-1} (Box 3.1) and that a greenhouse cover only transmits about 2/3 of it. So, saying that supplemental light should be switched off at a natural light intensity of, say, 800 µmol m^{-2} s^{-1} is like saying that it should be switched off when sun radiation exceeds 600 W m^{-2}.

Even when the intensity of the system and the level of natural radiation at which it has to be switched off have been selected, there remains the issue of when and how long to use it. In principle, the most value for money with light is reached in the night, because then the effect of each µmol light is strongest (Figure 12.2). Of course specific crop characteristics have to be respected: many crops do need some hours of darkness (see Section 7.1.1). In addition, the balance of sources and sinks makes it likely that the 'value' of each additional mol light in a day decreases, that is: the light use efficiency decreases with light integral during a day.

Using such a model and a method similar to the one applied here above, Van 't Ooster *et al.* (2008) calculated the break-even point of the lighting time interval in relation to the daily integral of natural radiation, for a HPS system of 180 µmol m^{-2} s^{-1} on tomato, and economic conditions typical of the Netherlands. They showed that the optimal interval decreased linearly from 20 hours at a sun radiation of 4 MJ m^{-2} d^{-1} to 6 hours at 6.5 MJ m^{-2} d^{-1}. As Figure 12.3 shows, this means that in the months November through at least January one should give light as long as the crop can stand it.

12.4.2 Optimal supply of carbon dioxide

Carbon fertilization – made possible by the direct application of heating fumes – is one of the factors leading to the high productivity in heated greenhouses. Energy saving and application of renewable energies ensure that there are less fumes around, a slump gradually made up by piped or bottled CO_2. Bottled CO_2 is increasingly sold at competitive prices also for the unheated greenhouses of the Mediterranean region.

As the crop inside the greenhouse is a 'sink' for CO_2, the CO_2 for assimilation must come from outside, in a greenhouse without carbon dioxide supply. That means that $[CO_2]$ inside must be lower than outside (Box 5.4) and the ventilation rate is the limiting factor. This is particularly true in the relatively cold months (when the product prices are high) and hardly any ventilation is needed for temperature control. On the other hand, under sunshine (when a high CO_2 concentration would be most desired; Figure 12.2) supplying CO_2 in order to maintain a concentration higher than external would obviously result in a low efficiency of carbon fertilization, since some CO_2 would flow through the vents. Altogether, there is a need for optimal management of supply, in order to ensure the maximum net benefit from cost of carbon dioxide supplied and increase in yield (Stanghellini *et al.*, 2012b).

We can use the same approach as shown for additional light, but we have to take into account that part of the supplied CO_2 will vanish through the open vents and will not contribute to production:

$$\text{Net Financial Benefit (NFB)} = R_{CO_2}\, A(I, [CO_2]) - P_{CO_2}\, S \qquad \text{€ m}^{-2}\text{ s}^{-1} \quad (12.7)$$

where:

R_{CO_2}	€ kg^{-1}	is the value of 1 kg assimilated CO_2;
A	kg m^{-2} s^{-1}	is net assimilation, a function of light and carbon dioxide concentration;
I	µmol m^{-2} s^{-1}	is the light intensity;
$[CO_2]$	µmol mol^{-1}	is the carbon dioxide concentration;
P_{CO_2}	€ kg^{-1}	is the cost of one kg of CO_2;
S	kg m^{-2} s^{-1}	is the CO_2 supply rate, which must balance assimilation and CO_2 loss for ventilation.

$$S = A + V = A_{MAX}\, h(I)\, m\, ([CO_2]_{in}) + g_V([CO_2]_{in} - [CO_2]_{out}) \qquad \text{kg m}^{-2}\text{ s}^{-1} \quad (12.8)$$

with:

g_V	m^3 m^{-2} s^{-1} = m s^{-1}	the specific ventilation rate.

Here Equation 12.3 and Equation 5.5 have been used, and the suffix *in* and *out* to the carbon dioxide concentration mean inside and outside, respectively. To obtain the $[CO_2]$ whereby the NFB is maximum, the derivative of NFB with respect to $[CO_2]$ must be zero:

$$\frac{\partial}{\partial [CO_2]_{in}} \left\{ (R-P)_{CO_2}\, A_{MAX}\, h(I)\, m([CO_2]_{in}) - P_{CO_2}\, g_V([CO_2]_{in} - [CO_2]_{out}) \right\} = 0$$

$$(R-P)_{CO_2}\, A_{MAX}\, h(I)\frac{dm([CO_2]_{in})}{d[CO_2]_{in}} - P_{CO_2}\, g_V = 0$$

$$\frac{dm([CO_2]_{in})}{d[CO_2]_{in}} = \frac{P_{CO_2}\, g_V}{(R-P)_{CO_2}\, A_{MAX}\, h(I)} = \frac{g_V}{\left(\left.\frac{R}{P}\right|_{CO_2} - 1 \right) A_{MAX}\, h(I)} \qquad \text{m}^3\text{ mg}^{-1} \quad (12.9)$$

Please refer again to Figure 12.4 and to our previous observation that the slope decreases with increasing concentration: a high slope is at low concentrations and low slope at higher concentrations. With this in mind, Equation 12.9 tells that the optimal carbon dioxide concentration increases (not surprisingly) in the measure that assimilated CO_2 is more worth than it costs; it also increases with available light and it decreases with the ventilation rate. On the other hand, it may be more surprising that the optimal CO_2 concentration does not depend on the concentration outside. However, please be aware that acceptable solutions of Equation 12.9 must satisfy the condition $CO_{2,in} > CO_2$,out since we have given a price to the carbon dioxide transferred by ventilation. When looking for a strategy that does not 'throw CO_2 out of the windows', it is easy to see that optimal management is to maintain $CO_{2,in} = CO_{2,out}$ (see Section 11.4), that is, to maximise assimilation under such constraints.

Equation 12.8 and 12.9 can be combined to calculate the economically optimal supply rate once the ventilation rate is known (or can be calculated; see Box 5.5), and the prices and values are given. An algorithm based on this concept has been recently implemented in a commercial climate control system (Tarnavas *et al.*, 2020).

12.5 Concluding remarks

In this chapter we have seen how knowledge of the crop response can be used to manage technology in the most economic way. Of course the very first decision to be taken is about the amount (and type) of technology that is sensible to have in given economic (and climate) conditions (e.g. Box 12.6). The issue is that, whereas the investment costs are well-known and often high (or, at least, perceived as such), the benefits are very difficult to quantify. Indeed, a given technology has a bearing usually on more than one factor affecting production, during various periods and conditions, so that only rather complex crop-greenhouse climate models can help to develop long-term crop response functions to particular technologies. It was through such a method, for instance, that Stanghellini and De Zwart (2015) concluded that a good energy screen could be a worthwhile investment even for a low-tech greenhouse as shown in Figure 12.1.

What is very wise to do is to share information. Dutch growers do it more often than growers elsewhere, and it is a common view that everybody benefits.

Box 12.6. Management options.

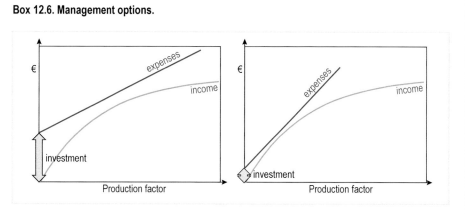

It may happen that the expense line does not cross at all the income line. That is, there is no way that the action one is considering might make economic sense. Then it is worthwhile analysing the reason: for instance in the example on the left, it is the high initial investment, whereas in the example right it is the high price of the resource (carbon dioxide, electricity, etc.). In one case it may be smart to find ways to reduce the initial investment (for instance sharing a CO_2 storage system or a co-generator among a number of growers), whereas in the second the issue would be to look for cheaper alternatives (other providers of CO_2; producing own electricity with a co-generator, etc.).

Figure 12.5 shows that CO_2 supply may be economic, even with relatively expensive carbon dioxide, such as bottled CO_2 in the Mediterranean region. Looking at the left panels in Figure 12.5 (optimal injection rate), one can see that increasing the value of the crop (from top to bottom) justifies ever larger injection capacities, to cope with ventilation. The right panels, however, show that the most benefit is to be got at very low ventilation rates, which only a cooling system can make compatible with high sun radiation. This is, indeed, the rationale behind the closed/semi-closed greenhouse concept.

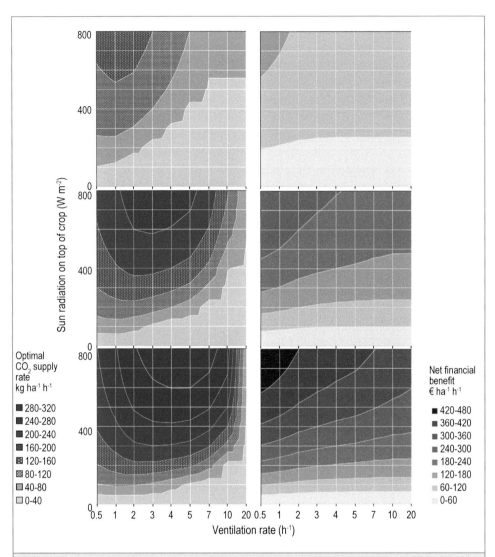

Figure 12.5. Optimal carbon supply rate (kg ha^{-1} h^{-1}, left) calculated combining Equation 12.8 and 12.9 and expected net financial benefit (€ ha^{-1} h^{-1}, right), for various combinations of sun radiation (y-axis) and ventilation rates (x-axis, mean greenhouse height 4 m). Price of bottled carbon dioxide is assumed to be 0.20 € kg^{-1} throughout (no investment costs are considered) and value of produce is 0.5, 1.0 and 1.5 € kg^{-1}, respectively, from top to bottom (Stanghellini *et al.*, 2009).

Indeed, conditions are never the same, one greenhouse or one year or one crop is not the other and there are more factors affecting production than one is able to consider. Exactly for this reason, plots such as the one in Figure 12.6 can be helpful, since they compile much information in an easy to grasp property of the growing system: the efficiency of use of solar radiation.

This should make it more easy to quantify the benefit of technology, which follows from two factors: the slope of the line and its length (the radiation cumulated in a growing season). For instance, improved ventilation management and heating in Sicily, result in a higher productivity (the slope of the line) but also allow a longer growing season (much more sun radiation to be cumulated). More data would make it possible to split the effect of ventilation from the effect of heating. It is good to observe that the productivity of a high-tech greenhouse seems to be unaffected by external climate, as far as this very limited compilation of growers data can be solid.

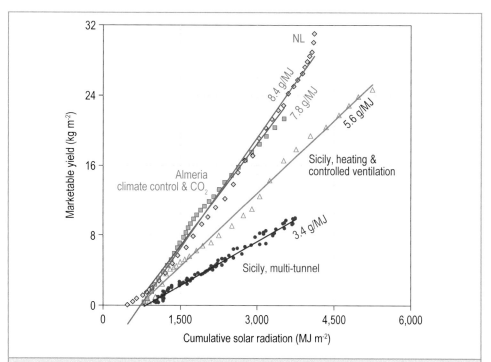

Figure 12.6. Efficiency of use of solar radiation of cherry tomato (more than one variety) in three regions (the Netherlands, Almeria and Sicily) and three growing systems (high-tech, mid-tech and low-tech). Each colour is one crop, except violet that is a crop in three similar greenhouses, in the same year. In the y-axis is cumulated (marketable, fresh) yield and x-axis cumulated solar radiation outside the greenhouse, and each point is one harvest event. The slope of the best-fit lines is a 'radiation use efficiency' of the growing system.

Chapter 13
Root zone management: how to limit emissions

Photo: Paolo Battistel, Ceres s.r.l.

A significant amount of the 'environmental footprint' of agriculture is linked to irrigation and fertilisation, and greenhouse horticulture is no exception. The difference is that, more often than in open field crops, greenhouse growers have the skills, the means and the tools to manage the root zone environment in such a way that water wastage and chemical emissions can be minimised.

This chapter deals with the tools for water and fertiliser supply that can be used in a greenhouse, and their management to minimise environmental impact. There will be no information on soil fertility or structure, nor on specific substrates, and we will assume that the fertiliser requirement for the crop is known and that the requisite nutrients can be supplied with irrigation water (fertigation). For nutrition recipes and the function of specific nutrients the reader is referred to specialist books, such as Plant Nutrition of Greenhouse Crops (Sonneveld and Voogt, 2009). Plant Physiology in Greenhouses (Heuvelink and Kierkels, 2015) gives an easy-to-read summary of the function(s) of each nutrient. A review of the technologies that make fertigation possible has been published in the framework of a EU-H2020 project (Thomson *et al.*, 2018) and can be downloaded free of charge at http://www.fertinnowa.com/the-fertigation-bible.

There are only two ways to ensure minimal emissions (and thus maximal resource use efficiency): precision irrigation, i.e. an irrigation strategy aimed at supplying the exact amount of water and nutrients that plants require; and (endless) re-collection and re-use of drainage water. As we will see later, neither can totally prevent emissions if the irrigation water is of poor quality. Since the knowledge required and the technology available for the two are very different, we will first deal with them separately, and thereafter with the consequences of the quality of irrigation water.

13.1 Precision irrigation

Even without explicit knowledge of crop response to drought (see Section 7.5.1), all growers know that there are few costs associated with over-irrigation, whereas there is always a risk of yield loss with insufficient irrigation, not only due to drought but also possible accumulation of salts in the root zone. Excess irrigation, which is often called leaching in agronomic literature, is therefore the norm, in spite of the waste and pollution it causes. There are three preconditions for precision irrigation: (1) knowledge of the crop's water and nutrient requirements; (2) technologies and methods to deliver them; and (3) water of good quality, something we will deal with later.

13.1.1 Crop water requirements

Much has been written on this topic, that we do not need to repeat here. A very good review of factors affecting crop water requirement and current irrigation practices, particularly in unheated greenhouses in 'mild winter' regions is the dedicated chapter (Gallardo *et al.*, 2013) in the FAO handbook on Good Agricultural Practices for greenhouse vegetable crops.

Approximately 90% of the water taken up by the roots is transpired; only about 10% is stored in the fresh biomass, so crop transpiration (see Section 6.3) is a good indicator of crop water requirements. As we have seen in Equation 6.9, crop transpiration depends on available (net) radiation, the vapour saturation deficit of the air and (to a lesser extent) on temperature, in addition to characteristics of the crop (such as stomatal resistance, leaf area and light interception), which may be variable. It will be no surprise, then, that simpler equations are usually applied for the purpose of estimating crop water requirements (see, for instance, Allen *et al.*, 1998).

As solar radiation is generally the largest source of energy available for transpiration, linear empirical equations for solar radiation are usually quite good at predicting total transpiration of a given crop, at a time interval of at least one day. As a glance at Equation 6.9 will show, such empirical equations merge the plant variables and the effect of saturation deficit (and its correlation with solar radiation) into the linear coefficients. Therefore, one cannot apply such an empirical equation outside the crop(s) and the conditions for which it was determined. Voogt *et al.* (2000) extended such a model to account for the energy released by the heating system in Dutch glasshouses and Voogt *et al.* (2012) explained at length how the parameters could be modified to account for various crops and their development stages, while also taking into account the greenhouse transmission and the use of screens and supplemental lighting.

A more general approach is a multilinear equation for both sun radiation and saturation deficit. As variations of temperature within a greenhouse are rather limited, the regression coefficients largely depend only on the crop and its development. Such empirical equations have been published for several greenhouse crops: a review with a useful compilation of the regression coefficients was published by Carmassi *et al.* (2013).

Other methods applied to field crops (e.g. Doorenbos and Pruitt, 1997) are based on applying an empirical 'crop coefficient' to a reference evaporation, either from a standard pan of water or a standard crop (e.g. Fernández *et al.*, 2010). However, this method is only reliable with crop coefficients determined in conditions and for crop stages similar to the ones where it has to be applied, for instance the irrigation program PrHo developed for the greenhouse crops of the Spanish region of Almeria (Fernández *et al.*, 2009).

13.1.2 Nutrient requirement and uptake

Growing plants require minerals. As minerals are a rather constant percentage of dry matter content, nutrients are required (and absorbed) as far as there is growth (dry matter production).

In greenhouses, nutrients are most commonly supplied dissolved in water. The resulting electrical conductivity of water (EC, measured in dS m^{-1}; Box 13.1), which can be easily and cheaply measured, is used as an indicator of their concentration, although it does not provide information about the concentration of individual nutrients. There are standard nutrient solutions for crops, with adaptations to systems (i.e. growing medium) or growing stages (Sonneveld and Voogt, 2009). These can be converted to fertiliser solutions following standard rules (applied, for instance, by the free-download calculation method of Incrocci, 2012). The proportion in which minerals have to be supplied (Table 13.1) roughly reflects the mineral composition of a plant (Box 13.2). In total, the EC of a nutrient solution is typically 1.5-2 dS m^{-1}.

Box 13.1. Electrical conductivity and salt concentration.

Conductivity is a measure of water's ability to transmit electricity, which is directly related to the concentration of ions in the water. When salts dissolve in water, they split into positively charged (cation) and negatively charged (anion) particles. This means that, even though the conductivity of water increases with dissolved salts, it remains electrically neutral, as anyone who swims in the sea can attest. Electrical conductivity (EC), usually expressed in dS m^{-1}, is therefore a good indicator of the total amount of salts dissolved in water, although it does not say anything about their nature or composition.

Each (fertiliser-)salt is built up of different ions, each with a specific molecular weight. When dissolved, it will generate ions of different weight and electric charge. Each fertiliser therefore has a specific EC value, defined as the EC increase when 1 g of this salt is added to 1 litre of water (for example, the EC of KNO_3 is 1.35 dS m^{-1}). The final EC of a mixed solution can be calculated by adding the totals of all individual specific EC values. Alternatively, the EC can be estimated from the molar concentration of either cations or anions. For the combination of salts typically present in a nutrient solution, Sonneveld *et al.* (1999) proposed a simplified formula:

EC ≈ 0.1C^+ (dS m^{-1})

where $C^+ = C_{NH_4} + 2C_{Ca} + 2C_{Mg}$ with C_i the molar concentration (mmol l^{-1}) of the ion *i*. The molar concentration of cations and anions must be the same, so that $C^+ = C^- = C_{NO_3} + 2C_{SO_4} + C_{H_2PO_4}$.

Obviously, if NaCl is present in the irrigation water, the concentration of Na$^+$ (or Cl$^-$) should also be accounted for. It should also be noted that the simplified formula as well as the specific EC of salts is only linear for dissolved quantities up to concentrations around 3.5 dS m^{-1}. For higher levels, the relationship becomes logarithmic and can only be calculated using a complicated mathematical formula (Mc Neal *et al.*, 1970).

Table 13.1. Recommended concentration (mmol l⁻¹) of macronutrients in the nutrient solution for some crops. In order to get similar root-zone concentrations, slightly different concentrations are recommended for free-drain or closed irrigation systems (from Sonneveld and Voogt, 2009: Appendix C, where many more crops can be found). The recommended concentration can vary somewhat with crop stage.

		NO_3	NH_4	H_2PO_4	K	SO_4	Ca	Mg
open	tomato	15.5	1.5	1.5	7.75	3.5	5.5	2.5
closed	tomato	12.5	1.1	1.2	7	2.5	4.5	1.5
	sweet pepper	13.5	1	1	6.75	1.5	3.5	1.25
	cucumber	13	1	1.15	7.5	1.25	2.75	1
	rose	7.5	0.5	0.75	3.5	1	3	1

Box 13.2. Mineral composition of the 'average' plant.

There is usually a large difference in dry matter content of fruits, leaves, stems and roots even within a species, let alone between species. Therefore, please be aware that the following numbers are only rough estimates.

Of one kg fresh biomass, some 130 g are dry matter and of that, 110 g are carbohydrates, which leaves some 20 g (2% of weight) of minerals. These are several elements whose relative presence (in weight) is represented in the pie chart (below left). Only a tiny percentage (not even 0.3% of mineral content) is the sum of the so-called microelements, listed below right.

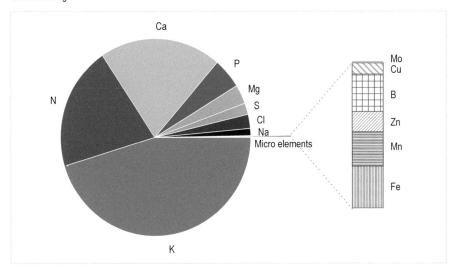

Note that the Na and Cl in this figure are not essential components but are present in the water, the growing medium or the fertilisers used.

Nutrients are absorbed via water uptake. So, if water uptake were proportional to photosynthesis, the 'uptake concentration' would be constant. Of course, the main driving factor of both transpiration and photosynthesis is radiation, so there is a correlation between them. However, we know that the trends do not match: transpiration of a well-watered crop continues to increase indefinitely with radiation, whereas photosynthesis does not. This means that we should expect uptake concentration to decrease with amount of water uptake: i.e. at high transpiration rates plants absorb relatively more water than nutrients (Figure 13.1, left), with a notable exception, calcium, whose uptake is proportional to water uptake (Figure 13.1 right).

The consequence for nutrient concentration is that the EC of the nutrient solution supplied should decrease with increasing water uptake and, indeed, some advanced fertigation programmes do decrease EC of the irrigation with increasing solar radiation, which is taken as an indicator of transpiration, hence of water uptake. However, how this really affects the EC in the root zone environment depends, obviously, on the amount of water in the root zone (Box 13.3) and the amount of excess irrigation (usually indicated as the 'drain fraction').

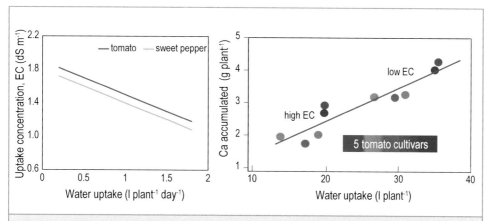

Figure 13.1. Left: overall uptake concentration in various experiments manipulating water and nutrient supply (sweet pepper) and transpiration to assimilation ratio (tomato). The lines shown are the best fit on a large number of points, with large variations (re-drawn from the data of Stanghellini *et al.*, 2003a and other unpublished data). Right: Ca uptake is proportional to water uptake, an experiment with young plants of five tomato cultivars, subjected to two salinity levels (redrawn from Ho *et al.*, 1995).

Box 13.3. How much water is in the root system?

To estimate the amount of water in a given growing system, we start with an estimate of the maximum volume of water that could be there, accounting for the volume of the substrate (or the depth of the root system in a soil), and for the volume that is occupied by the substrate/soil itself. This is shown in the first column. If multiplied by the typical volumetric water content (%) at field capacity (the second column), we get an estimate of the total volume of water in the system. This is added, in the third column, to an estimate of the amount of additional water in the storage tanks and in the piping, to quantify the volume of water in each system. The two bottom rows refer to soils, whereby the depth of the root zone is assumed to be 35 cm and values are indicative.

	Water volume in the root system at saturation ($l \, m^{-2}$)	Volumetric water content at field capacity (%)	Typical volume of water in the system ($l \, m^{-2}$)
NFT (nutrient film technique)	2	100	3
DFT (deep flow technique)	100	100	100
Ebb and flow	50	90	45
Stone wool	12	80	10
Perlite	15	65	10
Pumice	15	40	6
Coir	15	67	10
Fine peat	15	80	15
Coarse peat	15	55	10
Coarse sandy soil	165	18	54
Clay soil	180	25	75

13.1.3 Irrigation strategies

Even with perfect knowledge of water requirements, it is still necessary to define the amount and timing of irrigation events. A useful review of sensors and strategies for precision irrigation has been given by Pardossi *et al.* (2009). Indeed, any soil or substrate has a limited retention capacity (Box 13.4) so that excess water will percolate into the sub-soil or run-off, either way being lost. The irrigation system that can be best controlled is drip irrigation, whereby water is brought in small amounts to each plant (Figure 13.2). As this is the most common system, also in low-tech greenhouses (99.9% of the protected cultivation area in Almeria, Cuadrado-Gomez, 2009: p104) we will not discuss the others. However, there is also a lower limit to the amount that can be supplied at any time, as small amounts would not reach all drippers. Pressure-compensated drippers do improve distribution, but even then there will be heterogeneity in water supply among drippers (even in brand-new drip irrigation systems); there will be a spatial difference in the water-holding capacity of the soil/substrate and on top of that there is spatial variation in transpiration.

Therefore, as far as the amount of water is concerned, it is wise to give as little as is compatible with the (expected) inhomogeneity in the water distribution system and water demand during

---

Box 13.4. Water-holding capacity and available water.

No soil or substrate is a solid; it is an aggregate of particles of various sizes, interspaced by gaps (pores). Gravity will 'flush out' water from large pores that are above the soil water table, but capillarity will work against it in small pores, due to the cohesion force between water and the pore wall. The size (diameter) of the pores determines the force: the smaller the pore, the more forcefully it will retain water. So, the amount of water that a soil (or substrate) can retain depends on the size distribution of its pores, which depends on the size, form and distribution of its particles. So, clay soils, with aggregates of various sizes and shapes, retain water more easily than sand, with a more even size and shape of particles. The amount of water contained by a soil (or substrate) that has been saturated (all pores filled with water) and then left to drain is called field capacity. Since gravity works against the capillary force, there is a relationship between height above the soil water table and water content in the soil (see drawing below left), unless the connection to the water table is 'broken'.

It is easy to understand that the amount of suction (negative pressure) needed to extract water from the soil increases with the height above the soil water table, and that the speed at which it increases with height depends on the relationship between water content and height, which depends on the pore size distribution. The water retention curve (pF curve, below, right) quantifies for each soil/substrate the relationship between the volumetric water content and the suction (actually its logarithm) required to extract water from it. The three colours represent three different soils or substrates.

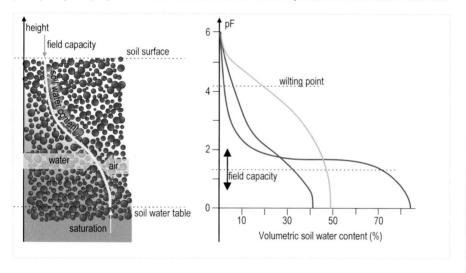

The pF at field capacity can change among media. The wilting point is the suction level at which it becomes impossible for plants to extract water from the root medium, usually around a pF=4.2. So, the water that roots can easily extract from a unit volume of soil (the 'available' water) is the difference between water content at field capacity and at wilting point.

---

each irrigation event. Since we know that transpiration (thus water uptake) is highly variable during a one-day period, the scheduling of the irrigation events will have to change in relation to the water uptake, either measured (Box 6.5) or estimated. Advanced irrigation programmes often use cumulative radiation as an indicator of crop transpiration, which makes sense, as we have seen that the (radiation) energy available is the main driver for plant transpiration. Such programmes start a new irrigation event whenever a given, preselected amount of measured

Figure 13.2. Drip irrigation in a commercial net house near Johannesburg (South Africa, left) and in a greenhouse near Riyadh, Kingdom of Saudi Arabia. (photo: Cecilia Stanghellini (left) and Jouke Campen (right), Greenhouse Horticulture, Wageningen University & Research)

(solar) radiation has been accumulated. The main issue here is the link between the amount of energy and the water actually transpired, which depends on many factors, as discussed above. In practice these programmes are calibrated, whenever possible, with observed drainage. However, this is only applicable for (soilless) culture systems where the drainage rate can be measured on a short-term basis. Drainage is not measurable for most soil-grown crops and, even if there is a drainage system, fluctuations in the groundwater table may interfere with readings. The only possibility is to install a lysimeter, but then calibration is only possible over longer periods due to the large buffer of the soil.

Another option is to use a measurement of soil water content (Box 13.5) as a proxy for crop water requirement. In this case, an irrigation event is started whenever the soil water content drops below a pre-selected value. Here as well, the issue is the calibration of the threshold value of soil water content. Nevertheless, when done well (and with good-quality water), this system does ensure precision irrigation and nutrition (Table 13.2). Incrocci *et al.* (2014) demonstrated the feasibility of such a method when applied to container crops in a plant nursery, but also its limits whenever there is a variety of crops within an irrigation sector.

On the other hand, Tüzel *et al.* (2009) demonstrated that an irrigation strategy allowing depletion of only 20% of available water (Box 13.4), resulted in 40% more irrigation events but only 5% more water use than the control, yet 19% more yield of a greenhouse cucumber crop in Turkey.

Box 13.5. Sensors for steering irrigation and nutrition.

The development of new, affordable, types of root zone sensors has enabled the widespread application of 'smart' (precision) irrigation and fertilisation. These sensors are based on the effect water and salts have on the electrical properties of soils, and are replacing the sensors (tensiometers, right) that are based on measuring the pressure in a porous cavity in hydraulic equilibrium with the soil.

Dielectric sensors apply an alternating field and determine the dielectric characteristics of the medium, which relate in a known way to water content and to

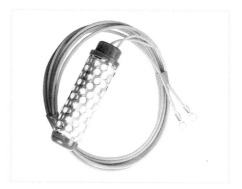

soil/substrate properties. They therefore need to be calibrated for a given soil/substrate, see e.g. Incrocci *et al.* (2009).

Several sensors are available, depending on the signal they use and its processing. Time Domain Reflectometry was the first method developed. The advantage of frequency domain (reflectometry and capacitance) is that it requires less expensive equipment (below, left). As the reading is affected by temperature and salt concentration in pore water, with some additional wiring, EC of pore water can be determined as well (below, right).

Table 13.2. Total use of water and nitrogen for a lettuce crop and resulting yield. Treatment A was irrigated as usual; irrigation in the others was soil-sensor driven, to maintain a pre-set soil water content, under pre-selected fertilisers' supply rates (Balendonck *et al.*, 2010).

Treatment	Water use (mm)	N Fertiliser (kg ha^{-1})	Mean head weight (g)	Class 1 (%)
A (ref)	186	100	516	98.6
B	70	100	528	98.8
C	70	83	592	97.2
D	70	58	595	98.4

13.1.4 Irrigation and nutrition strategy

When reliable sensors of root zone (pore) EC are available (Box 13.5), it is even possible to fine-tune nutrition, as demonstrated by Stanghellini *et al.* (2003a). By drawing irrigation of a greenhouse sweet pepper crop from either a normal or low-concentration nutrient solution tank, depending on the EC measured in the stone wool substrate, and striving to maintain volumetric soil water concentration at field capacity to prevent drainage, they were able to halve water use, without any effect on the uptake of either water and nutrients or on crop yield (Table 13.3). Long-term application of such a method on multi-annual, soil-grown crops may require periodic 'reset' irrigation events, to offset emerging local differences in EC and water content (e.g. Voogt, 2014).

13.2 Closed irrigation systems

The percentage of greenhouse crops being cultivated in soil-less systems, i.e. systems with a confined root environment (Box 13.6), is also increasing in regions where greenhouses are typically rather low-investment. This is because the increased amount of control that such systems allow usually results in higher production, with a higher uniformity and quality, so that revenues justify the larger investment required. Caballero and De Miguel (2002), for instance, made a compelling comparison of net earnings of soil-less sweet pepper vs traditional cultivation in the region of Almeria. In addition, soil-less growing is one of the possible

Table 13.3. Resulting values for a stone-wool-grown sweet pepper crop, whose irrigation was controlled on the basis of an on-line sensor of volumetric water content and electrical conductivity (EC) of pore water in the root zone. Irrigation was performed according to standard Dutch procedure (control) or aiming for a volumetric water content of 50%. In this latter case the irrigation was either with the control nutrient solution or with a nutrient solution of 1 dS m^{-1} whenever EC in the root zone exceeded 3 dS m^{-1}, and switched back to normal whenever it dipped below 2.8 dS m^{-1} (Stanghellini *et al.*, 2003a).

	Units	Control	No-drainage	% of control
Mean EC irrigation	dS m^{-1}	2.37	1.51	64
Mean EC slab and drainage	dS m^{-1}	2.9	2.9	100
Volumetric water content	%	63	49	78
Water supply	l plant^{-1}	345	157	46
Water uptake	l plant^{-1}	143	139	97
Transpiration	l plant^{-1}	137	132	97
Mean leaf area index	–	4.9	4.7	96
Biomass (no root)	kg plant^{-1}	6.5	6.4	98
Yield	kg plant^{-1}	4.2	4.2	100

Box 13.6. What is soil-less?

We define a soil-less growing system as a system where the roots are in a confined volume, disconnected from the soil. As the table below shows, there is a large variety of soil-less systems. In addition to price, availability and technology, the following factors may affect the choice for one or the other: water retention properties and, increasingly, environmental impact of production and/or recycling.

			Used in ...
hydroponics		NFT (nutrient film technique)	Leafy vegetables, herbs, strawberries
		DFT (deep flow technique)	Leafy vegetables, herbs
growing media	inorganic	Stone wool	All fruit vegetables, some cut flowers
		Perlite	All fruit vegetables, some cut flowers
		Pumice	All crops, when available
		Sand	All crops, mixed with organic manure, often used as artificial soil in low-tech greenhouses
	organic	Ebb and flow	Potted plants on concrete floor or tablets
		Coir	All fruit vegetables, berries, some cut flowers
		Peat	Orchids and berries, often used in combination with coir
		Compost	Mainly used as amendment

---

answers to the worldwide ban on methyl bromide formerly used as a soil disinfectant. Raviv and Leith (2008), published a comprehensive review/handbook about soil-less systems.

Soil-less systems make it possible to recollect and re-use any excess irrigation and nutrition that may have been given. It is important to be aware that a soil-less system is not the same as a closed irrigation system: indeed, in most low-tech greenhouses fitted with soil-less systems, drainage water is not recollected. Recollection and, possibly, disinfection systems require an additional investment, on top of the investment in the root zone system. Thanks to the resulting savings in fertilisers, the additional investment can be recovered in a reasonable amount of time. For instance, Pardossi *et al.* (2011) estimated from the results in Table 13.4 that the investment could be recovered in two years, a period that would become longer if any disinfection system were to be installed, which is advisable (see for a review, e.g. Van Os and Blok, 2016).

Such 'closed' growing systems are seen as the solution both to the environmental problems caused by leaching of fertilisers and the scarcity of water in many regions.

Table 13.4. Total use of water and fertilisers for tomatoes grown in stone wool, either with or without re-use of drain water (closed-loop irrigation). Leaching from the open system was estimated as the difference between supply and uptake, which was independently determined. Commercial yield and its quality (°Brix) were the same in the two treatments (Pardossi *et al.*, 2011).

	Supply – open system	Supply – closed system	Saving	Leaching
Water (m^3 ha^{-1})	8,632	6,831	21%	1,682
N (kg ha^{-1})	1,591	1,032	35%	266
P (kg ha^{-1})	306	244	20%	25
K (kg ha^{-1})	2,422	2,000	17%	343

13.3 Saline irrigation water

However, when the irrigation water contains (or the fertilisers used carry) non-nutrient salts (sodium chloride, NaCl, is the most common, but there may be others), such salts will accumulate in the closed loop. Since a high salt concentration is known to cause yield loss (see also Section 7.5.1), at some point the salts will have to be leached out, by draining the loop, at least in part, Figure 13.3.

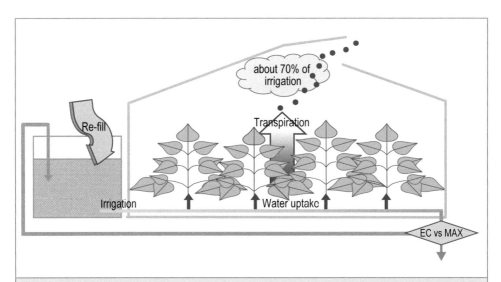

Figure 13.3. Schematic representation of a closed irrigation system fed by water containing non-nutrient salts. The water that leaves the system (transpiration and biomass) does not contain salts yet is substituted with water that does. The salts will accumulate in the system and the concentration (EC) will increase. At some point it is necessary to discharge the water that has exceeded a 'maximal' concentration, EC MAX.

As NaCl is the most common 'disturbing' salt, we will limit our discussion to this. There may be (quality) issues for growers to prevent the accumulation of other elements (such as boron), other than the yield loss described by the 'Maas-Hoffman' response (e.g. Beerling *et al.*, 2014). Nevertheless, the discussion and the methods applied in the following hold for any non-nutrient element, taken up in limited measure by the roots, and for whose allowed concentration there is an upper boundary.

A concentration deemed 'unacceptable' obviously depends on the crop and on the preferences and perceptions of the grower. Figure 13.4 shows the trend of salt concentration in such a system, Table 13.5 gives an indication for Na for the most common crops.

13.3.1 Salt concentration in the root zone

The variable steepness of the growing branches of Figure 13.4 implicitly shows that, although salinity will increase with time, it is not time that determines what happens. The following example makes it clear. Let's assume a grower has irrigation water with an Na concentration of 2 mmol l^{-1}. That will obviously be the concentration in the slabs at the start of the crop. Each litre refill brings in another 2 mmol Na. When the crop has taken up the equivalent of the whole volume of water in the system (and all Na has been left behind), each litre has been

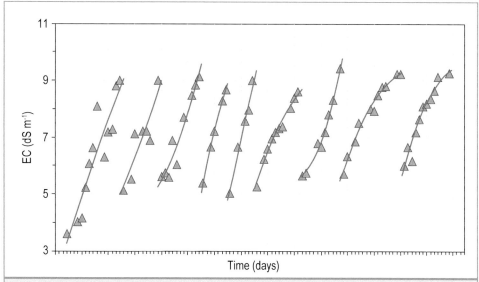

Figure 13.4. Evolution of daily readings of electrical conductivity (EC) in the stone wool slabs of a tomato crop, during an experiment simulating a water source containing 12 mmol l^{-1} of sodium, used to refill a basic nutrient solution of 1.6 dS m^{-1}. Whenever EC in the slabs exceeded 9 dS m^{-1}, the mixing tank was emptied, and the slabs were left to drain for a couple of hours (Stanghellini *et al.*, 2003a).

Table 13.5. Indicators of salinity response of typical greenhouse crops. For each crop indicated in the first column, the second column gives the sodium concentration that had to be reached in a closed loop, before leaching was allowed according to now replaced Dutch regulations. The third column gives the range for the slope of the Maas-Hoffman response (see 7.5.1), the Salinity Yield Decrease (after Sonneveld, 2000: p 39, 53; Kaya *et al.*, 2007 for strawberry and Xu and Mou, 2015, for lettuce).

Crop	Max Na concentration (mmol l^{-1})	SYD (% per dS m^{-1})
Tomato	8	2-7
Sweet pepper; cucumber; eggplant; melon; squash (zucchini); beans	6	5-10
Lettuce	5	4-20
Strawberry	3	>10
Rose; gerbera	4	5-10
Anthurium, lilium, bouvardia, iris	3	5-15
Carnation, amaryllis	4	2-4
Orchid	0	
Others	5	

replaced, and the concentration has doubled to 4 mmol l^{-1}, independent of the time required for a given volume to be taken up.

In more general terms, let's say that accumulated water uptake of the crop at time t is U_t [l] and that amount of water is re-filled with water with concentration C_{in} [kg (or mmol) l^{-1}] of a given salt. Therefore, $U_t C_{in}$ [kg or mmol] are brought into the system and, if V [l] is the volume of water in the system, the concentration increases by $U_t C_{in} / V$ [kg (or mmol) l^{-1}] and, at time t, the concentration C_t in the system is:

$$C_t = C_{in} \frac{U_t}{V} + C_{in} \Rightarrow \frac{C_t}{C_{in}} = 1 + \frac{U_t}{V} \qquad \text{kg (or mmol) l}^{-1} \quad (13.1)$$

The uptake at which the discharge concentration C_{MAX} is reached, is calculated by inverting Equation 13.1 to get:

$$U_{C_{MAX}} = \left(\frac{C_{MAX}}{C_{in}} - 1 \right) V \qquad \qquad \text{l} \quad (13.2)$$

Assuming that the whole volume can be discharged (possible, for instance, in nutrient film technique (NFT) or deep flow technique (DFT)), the total water use, W, i.e. uptake and leaching, L, is:

$$W = U + L = U_{C_{MAX}} + V = \frac{C_{MAX}}{C_{in}} V \qquad \qquad \text{l} \quad (13.3)$$

and the ratio of total water use and leaching to crop water uptake is respectively:

$$\frac{W}{U} = \frac{C_{MAX}}{C_{MAX} - C_{in}} \Rightarrow \frac{L}{U} = \frac{C_{in}}{C_{MAX} - C_{in}} \tag{13.4}$$

As the slope of the concentration vs uptake is constant, it is easy to understand that Equation 13.4 also holds, even if it is not possible to discharge the whole system every time. In this case the concentration in the system after discharging is higher than the concentration of the irrigation water (see Figure 13.4, for instance), and it will require a lower water uptake to reach the ceiling value C_{MAX} again. Equation 13.4 is nicely dimensionless, so it applies whatever units are used for the concentrations, mmol l^{-1} as in Table 13.5, or in terms of EC, dS m^{-1}.

When the crop takes up some of the accumulating salts Equation 13.4 gives an overestimate of total water use since the concentration increases less steeply with water uptake (Box 13.7). Voogt and Van Os (2012) have shown examples of the required discharge from the system, showing that the rate of discharge (and thus the water use efficiency (WUE)) greatly depends on the maximum concentration allowed for Na, as this is the bottle-neck element, since its uptake concentration is nearly always lower than the uptake concentration of Cl (Box 13.7).

The whole reasoning above would also apply to a soil crop under precision irrigation, since its purpose is to replace exactly the water that is taken up by the plants. It is true that the distribution of salts in a soil layer under drip irrigation is probably less uniform than in most soil-less systems. However, as plants seem to be able to cope with inhomogeneity in the root zone (Box 7.5), we may assume that Equation 13.4 is a reasonable estimate of minimal water use in any system fed with poor-quality water.

In practice, water use (and connected emissions) may be much higher: a survey of 53 growers in Almeria (Thompson *et al.*, 2006) concluded that excess irrigation was within 50% of the dose locally recommended in 68% of the cases, the rest of the growers irrigating more. Excess irrigation in soil-less systems with open-loop irrigation may well be at levels comparable to soil-grown systems.

13.3.2 Emissions

The consequence of poor quality irrigation water is pollution (Box 13.8). Therefore, incentives to use low-quality water may well reduce the pressure on good water resources but may significantly increase pollution caused by leaching.

Indeed, the Dutch regulations underlying Table 13.5 were soon replaced by rules simply compelling growers to use good-quality irrigation water by whatever means possible. As a result, nearly all Dutch growers have a rain collection basin. In this respect, the Netherlands has the obvious advantage of an equilibrium between rainfall and potential evaporation, and

Box 13.7. Crops absorbing non-nutrient salts.

Although uptake of sodium chloride is negligible for most crops, hardly any crop is able to fully exclude sodium (or chlorine) from water uptake. It is easy to imagine that the amount of salts absorbed with each unit of water (the uptake concentration) will increase with the concentration of the salt in the root environment. The table here gives the uptake concentration of sodium and chlorine, p, as a percentage of the root zone concentration of each, for a number of greenhouse crops (after Sonneveld, 2000, p. 41 and 52). If pC is the uptake concentration of a given salt when the concentration of the salt is C, the variation in the time interval dt of salt mass, M, in a system of volume V is:

	p (%)	
	Na	**Cl**
Sweet pepper	2.5	5.5
Tomato	6.2	7.5
Cucumber	9.8	13.5
Rose	~0	~0
Carnation	2.9	4.2
Gerbera	7.4	18.9
Lily	10.3	9.2
Aster	18.2	19.5

$$\frac{dM}{dt} = \left(C_{in} - p\frac{M}{V} \right)\frac{dU}{dt}$$

dividing by dU/dt we get:

$$\frac{dM}{dU} = C_{in} - p\frac{M}{V}$$

which has a known solution:

$$M_U = \frac{C_U}{V} = \frac{C_{in}V}{p} - \left(\frac{C_{in}V}{p} - C_{in}V \right)e^{-p\frac{U}{V}}$$

dividing by $C_{in} V$ (the initial mass in the system):

$$\frac{C_U}{C_{in}} = \frac{1}{p} - \left(\frac{1}{p} - 1 \right)e^{-p\frac{U}{V}}$$

The equilibrium concentration is: $\dfrac{C_{U_\infty}}{C_{in}} = \dfrac{1}{p}$

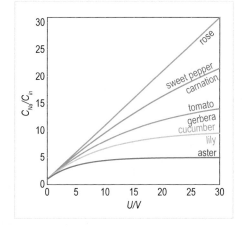

The water uptake needed to reach the discharge concentration and the consequent leaching can thus be calculated:

$$\frac{U_{C_{MAX}}}{V} = \frac{1}{p}\ln\frac{(1-p)C_{in}}{C_{in} - pC_{MAX}} \quad \Rightarrow \quad \frac{L}{U} = \frac{p}{\ln\dfrac{(1-p)C_{in}}{C_{in} - pC_{MAX}}}$$

a rather even distribution of rainfall (Figure 13.5), so that a rain collection basin of 1000 m³ ha⁻¹, for instance, covers 65 to 80% of crop water requirements (KWIN, 2017, Table 44), the rest being covered either by well or tap water or desalinated (surface) water. However, the shallowness of the groundwater table makes it impossible to dig deep basins, which, in turn, increases occupation of expensive and potentially productive land. Indeed, calculations of the integral costs of rain basins (KWIN, 2017, Table 47) show that the cost of storage per cubic metre may exceed the cost of desalination in some cases.

Box 13.8. How to calculate emissions.

Let's take, for instance, a Dutch tomato grower with irrigation water containing 1 mmol l^{-1} Na, discharging whenever a concentration of 8 mmol l^{-1} is attained. Water uptake for the typical Dutch cropping cycle (descibed in Box 12.2) is about 850 l m^{-2} y^{-1} (KWIN, 2017, Table 38) and water volume in a stone wool growing system is about 10 l m^{-2} (Box 13.3). According to Equation 13.4, 1/7 of water uptake, i.e. 120 l m^{-2} y^{-1}, is discharged. Assuming that the water that is discharged has the composition of the nutrient solution, Table 13.1, it is easy to calculate the amount of salts that are deposited in the environment. Let's take N (atomic weight 14), for instance: 12.5 mmol l^{-1} = 175 mg l^{-1} × 120 l m^{-2} y^{-1} = 21 g m^{-2} y^{-1} = 210 kg N ha^{-1} y^{-1}.

Even accounting for the uptake of Na by tomato, and then using the correct equation (Box 13.7), the water uptake needed to reach the discharge concentration would be about 10 times the volume, instead of 7, as calculated above. The water discharge would still be about 85 l m^{-2} y^{-1} and the N discharge 149 kg N ha^{-1} y^{-1}. Note that 1 mmol Na l^{-1} would be considered irrigation water of very good quality in nearly the whole Mediterranean basin, for instance.

A mitigating strategy was successfully applied by Massa *et al.* (2010), halting nutrient supply (yet re-filling water) whenever the discharge concentration for Na was reached, and letting the tomato crop deplete the nutrients from the solution (down to a N concentration of 1 mmol l^{-1}) prior to discharge. In this way N emission could be reduced by some 90%, without yield loss.

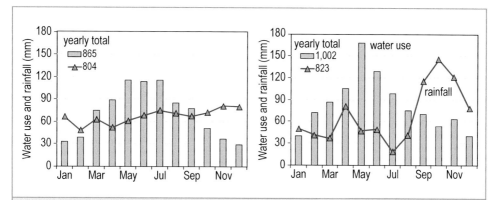

Figure 13.5. Monthly rainfall (line) and water use of greenhouse roses (bars) in the Netherlands (left), and the Versilia region of Italy (right). The numbers near the labels are the annual totals (mm). Rose greenhouses are whitewashed in Versilia from June to August, which explains the relatively low water use in the summer (Stanghellini *et al.*, 2005).

Incrocci *et al.* (2010, 2019) have applied a sensor-based multiple-sources irrigation strategy such as described in Table 13.3 to (open air) nursery container crops, whose irrigation water was either good quality, scarce groundwater or reclaimed waste water, with a sodium content of 10 mmol l^{-1}. Various strategies to schedule irrigation (by a model or by root zone sensors) and to switch between the water sources, were evaluated with respect to water use, leaching and production of timer-controlled either 100% groundwater or 100% reclaimed water. They demonstrated that smart scheduling of irrigation could reduce water use by 24 to 46% and emissions by 17 to 84%, without having an effect on biomass production (which was much reduced in the wastewater control treatment). Sensor-based source selection could save fresh water (51 to 73%) in a salt tolerant species and limit (but not prevent) salinity damage in a salt sensitive species.

13.4 Optimal management of irrigation

Since a mildly raised EC (above the concentration of the nutrient solution) decreases fresh weight but not necessarily dry matter production (for instance, tomato: Li *et al.*, 2001, 2004), for a few fruit crops it may make economic sense to cultivate with a slightly raised EC, if the increased quality is valued in the final market. This is indeed what most Dutch greenhouse tomato growers do. The salinity of irrigation water in other regions may not leave growers with much of a choice, but then, indeed, they may exploit the competitive advantage of producing high-quality (cherry) tomato, for instance, as is done in the southernmost tip of Sicily.

The use of low-quality irrigation water, even at no cost, is seldom (if ever) the economically wise choice if better water is available. Stanghellini *et al.* (2007) quantified the intuitive observation that, whenever the potential income loss with salinity is larger than the potential irrigation costs (the cost of using good quality water), the best option is to use good water. Potential irrigation costs are seldom more than $1 \in m^{-2} y^{-1}$, since cost of desalination is around $1 \in m^{-3}$ and typical uptake of greenhouse crops usually less than $1 m^3 m^{-2} y^{-1}$ (Figure 13.5 and KWIN, 2017, Table 38). On the other hand, the value of greenhouse crops can be anything between 10 and $100 \in m^{-2} y^{-1}$, and yield decrease with salinity is at least a few percent of that (Table 13.5), so that saving on irrigation costs would be a reasonable choice only in crops of very little value that were very tolerant of salinity (and then only with relatively good brackish water at a very cheap price).

The truth is that there are plenty of regions in the world where good-quality water is simply unavailable or regions (such as the European Union) where regulations aim to limit emissions. In both cases a grower has to work out the optimal strategy when faced with such constraints.

Let's assume a crop in optimal conditions generates a gross revenue $Y (\in m^{-2} y^{-1})$ that decreases by a fraction *SYD*, for each dS m^{-1} that the salinity *EC* in the root zone environment increases (Table 13.5), and that yield loss depends on the average salinity in the root zone, with no effect of the amplitude of its fluctuations. Then:

Gross revenue $= Y (1 - SYD \cdot EC_{root})$ $\hspace{2cm} \in m^{-2} y^{-1}$ (13.5)

The potential yield, $Y [\in m^{-2} y^{-1}]$ is typical of the crop, and *SYD* [Table 13.5, m dS^{-1}] is a measure of the crop sensitivity to salinity, here expressed as a fraction instead of a percentage (the higher *SYD*, the faster yield drops with increasing salinity). Now, let's assume that there is irrigation water with $EC = EC_{in}$, whose price is P [€ m^{-3}], and that the potential water uptake of the crop is U [m^3 m^{-2} y^{-1}]. A grower managing a closed irrigation system can choose the EC_{MAX} to discharge the system, which has a bearing on water use, Equation 13.4, so that net profit is maximal. Similarly, a grower managing an open system could choose the leaching fraction, which would be the same.

Assuming that all other production costs are independent of the quality of irrigation water, net financial benefit (NFB), i.e. the balance of gross revenue and irrigation costs is:

$$NFB = Y\left(1 - SYD\,\frac{EC_{in} + EC_{MAX}}{2}\right) - \frac{EC_{MAX}}{EC_{MAX} - EC_{in}}\, U \cdot P \qquad \text{€ m}^{-2}\text{ y}^{-1} \quad (13.6)$$

Water of poor quality also carries a hidden cost, i.e. the cost of fertilisers that are expelled with leaching, Equation 13.4. Taking this into account would simply result in a higher price of water, so that P can be regarded as incorporating both water and fertiliser costs. And the maximal profit is given by:

$$\frac{\partial NFB}{\partial EC_{MAX}} = -\frac{Y \cdot SYD}{2} + \frac{U \cdot P \cdot EC_{in}}{(EC_{MAX} - EC_{in})^2} = 0 \qquad \text{€ m}^{-2}\text{ y}^{-1}\text{ m dS}^{-1} \quad (13.7)$$

Whose only possible solution is:

$$EC_{MAX} = EC_{in} + \sqrt{\frac{2U \cdot P \cdot EC_{in}}{Y \cdot SYD}} \qquad \text{dS m}^{-1} \quad (13.8)$$

Maintaining our previous estimate of $U \cdot P \sim 1$ € m^{-2} y^{-1} at most, it is possible to estimate how much the concentration should be allowed to increase by a grower of cherry tomatoes in the Mediterranean region ($Y \sim 20$ € m^{-2} y^{-1} and SYD $\sim 1.5\%$) and by a rose grower in the Netherlands ($Y \sim 100$ € m^{-2} y^{-1} and SYD $\sim 5\%$): about 2.6 $\sqrt{EC_{in}}$ and 0.6 $\sqrt{EC_{in}}$, respectively.

But then leaching, which can be calculated from Equation 13.4:

$$\frac{L}{U} = EC_{in}\sqrt{\frac{Y \cdot SYD}{2U \cdot P \cdot EC_{in}}} = \sqrt{\frac{Y \cdot SYD \cdot EC_{in}}{2U \cdot P}} \qquad (13.9)$$

could reach 160% of uptake (Dutch rose grower with $EC_{in} \approx 1$ dS m^{-1}) or worse (at higher EC_{in}), which would not comply with leaching regulations. Indeed, in the presence of limits to leaching, the freedom to choose EC_{MAX} is limited by the composition of the discharge, for which a depletion strategy can be applied. Whenever the composition of the discharge is given, a limit to leaching forces the grower to select an adequate quality of irrigation water, through Equation 13.4.

As an example, let's determine the maximum possible Na concentration of the irrigation water for a Dutch rose grower willing to comply with a leaching norm for N of 80 kg ha^{-1} y^{-1} maximum, yet not willing to allow the Na concentration in the irrigation loop to exceed 4 mmol l^{-1} (Table 13.5).

Assuming that the discharge has the same N concentration as the nutrient solution, which contains 8 mmol l^{-1} (Table 13.1), which is 112 mg l^{-1}, the maximum amount that can be discharged is 71.5 l m^{-2} y^{-1}. Water uptake for a rose crop in the Netherlands is about 1000 l m^{-2} y^{-1} (Figure 13.5 and KWIN, 2017, Table 38). Substituting these values into Equation 13.4:

$$\frac{71.5}{1000} = \frac{C_{Na}}{4 - C_{Na}} \Rightarrow 286 = 1,071.5 C_{Na} \Rightarrow C_{Na} \approx 0.27 \qquad\qquad mmol_{Na} \ l^{-1} \quad (13.10)$$

Only rain water is that good, which explains the abundance of rain basins in a country otherwise not known for water scarcity.

13.5 Water use efficiency of growing systems

By comparing water productivity WP (kg produced with 1 m^3 irrigation water) of various growing systems for tomato (from field production to high-tech greenhouse production) Stanghellini *et al.* (2003b), showed that there was easily a factor 10 between the lowest and the highest WP. The main reason by far is that the productivity per unit surface increases significantly with the control of the production process, which is not the case for crop transpiration, since the main driver for transpiration is solar radiation (e.g. Fernández *et al.*, 2010; Bacci *et al.*, 2013; Gallardo *et al.*, 2013).

Nevertheless, there are other additional contributing factors. Stanghellini (2014) proved that the greenhouse environment, being more humid than the field, leads to a higher transpiration efficiency (the amount of dry matter produced per unit water transpired) and Katsoulas *et al.*, (2015) demonstrated that WP is inversely related to the air exchange rate, which is in any case lower in a greenhouse than in the field. Table 13.6 shows that the water productivity of tomato in Spain is about four times higher in a greenhouse environment than in the field. It also shows that 'good management', which was the subject of this section, delivers another factor 2.5, both in greenhouses and in the field.

Table 13.6. Water productivity of tomatoes (kg produced per m^3 irrigation water) in different growing systems in Spain. The effect of farm management was evaluated using the method described by Hsiao *et al.*, 2007 (E. Fereres, personal communication, 2014).

Management	Environment	
	Outdoors	Greenhouse
Average	1.4	8.2
Excellent	4.8	18.9

However, as Figure 13.2 makes clear, most irrigation water is transpired by the plants and then carried away by ventilation. Cooling is increasingly applied in semi-closed, closed greenhouses and plant factories, to dispose of (or reduce) the need for ventilation (Section 5.5 and Box 9.4). This development even offers the exciting perspective of recollecting and re-using the water that the crop has transpired for irrigation. Of course, cooling is also de-humidification, and the water condensed at the cooling elements is transpired water that can be re-used for irrigation, thus closing the transpiration loop.

Indeed, Van Kooten *et al.* (2008) have reported that a closed greenhouse in the Netherlands could produce 1 kg tomato with 4 l water, which is four times better than the performance of the best closed-irrigation-loop greenhouses in the Netherlands. Product water use (WPU in l kg^{-1}) in a perfectly closed greenhouse with a perfectly closed irrigation loop (where all water ends up in biomass) should be the reverse of the harvest index, which would be some 1.5 l water for each kg tomato. Therefore, the 'closed' greenhouse was not perfectly closed after all, which makes economic sense.

However, it has been proven that, in particular market conditions and in the case of extreme scarcity of water, it would be possible in a high-tech greenhouse or a plant factory to grow vegetables with unimaginably low levels of water use (Campen, 2012).

13.6 Concluding remarks

We have seen how knowledge of the processes, coupled to the means for precision application of water and nutrients, may greatly help reduce waste of water and emission of fertilisers. Regulations to limit emissions are equivalent to requiring growers to use good-quality irrigation water. In fact, even the economically optimal management of most closed-loop crops in greenhouses requires irrigation water of good quality. Desalination is a viable option only when the value of the crop warrants it.

A greenhouse environment naturally leads to higher water productivity than in the field; greenhouse growers usually have better tools and skills for water management than open field growers and farmers; and, last but not least, the high economic productivity of water in greenhouses warrants the necessary investments in infrastructure (e.g. precision irrigation, desalinisation).

A high-value crop is then the surest means to save water and prevent emissions, since there is no way that low-value crops using poor irrigation water would still be profitable under stricter environmental rules.

Photo: Luuk Graamans, Wageningen University & Research and Delft University of Technology

The development of high power light emitting diodes (LEDs) has spawned the growth of plant factories, that is enclosed places, where crops are grown under non-natural light and perfectly controlled climate and root zone conditions, to deliver constantly high yields of nutritious and pesticide-free products. In this section we examine how a plant factory works, and then how production in a plant factory compares with production in a [high-tech] greenhouse.

14.1 Introduction

Local scarcity of resources determines the suitability of production systems. Greenhouses provide different levels of control over the environmental conditions, depending on the available equipment. As we have seen, in high-tech greenhouses, temperature is controlled by heating and cooling, humidification and de-humidification are possible, CO_2 can be supplied and light levels can be controlled using supplementary lighting or shading screens. However, seasonal differences caused by the influence of sun radiation and outside temperature, will still exist to some extent, and climate factors cannot be controlled independently from each other, unless very large capacities are installed. For example, opening the vents to reduce temperature, automatically will also reduce CO_2 level, whenever a concentration higher than external is maintained. The only way to prevent it is to have a cooling capacity equal to the peak radiation and/or a CO_2 supply system able to throw away carbon dioxide at the speed that would be required.

Recently, there is a lot of attention for so-called vertical farms or plant factories with artificial lighting (PFAL; Kozai *et al.*, 2015) which allow for season-independent, full control of environmental factors. Vertical farm refers to a plant production facility with a thermally insulated and nearly airtight warehouse-like structure, where multiple culture layers with lamps on each shelf are vertically stacked. Undoubted advantages of a plant factory over a conventional, high-tech greenhouse system are:

- constant, high quantity and quality of product is possible, every day of the year;
- no pesticide use, very high land, water and nutrient use efficiency (closed system);
- it can be built anywhere because neither solar light nor [suitable] soil is needed;
- makes short logistic chains possible: freshness and wider choice of varieties (since shelf life and resistance to transport become less relevant);
- less food waste due to uniform quality and absence of dirt, high harvest index, and shorter distance to consumer.

Of course when a plant factory is compared to field production, the list of advantages becomes much longer (such as reduced risk of drought or flooding, e.g. Cicekli, and Barlas, 2014). However, the gains that could be attained as well by (high-tech) greenhouse (rather than field) production, should not appear among the pros of a plant factory, without a discussion of which of the two systems may be the most suitable for a particular case. Plant factories require an investment 4 to 10 times higher than a very well equipped high-tech greenhouse (Rabobank, 2018b), and require electricity to provide light, whereas greenhouses rely (at least in part) on sunlight.

In principle, anything can be grown in a plant factory. In fact, only a limited number of crops are suitable for commercial production in such an expensive growing system. Their necessary characteristics are:

› high-value, which can be intrinsic, such as seedlings and medicinal plants and/or high market value attained thanks to the growing system (pesticide free; freshness; cleanliness; taste, etc.);
› small volume: short in height (30 cm or less) and high plant density;
› not requiring insect pollination;
› high harvest index;
› growing well under relatively low light intensity.

A very similar list applies to crops suitable for future space missions (Meinen *et al.*, 2018), in view of even worse constraints on resources such as space and electricity.

Indeed, it will not be surprising that mostly it is medicinal plants, leafy vegetables and micro-greens that are grown in plant factories. Seedlings (lettuce and tomato) as well as ornamentals (Wollaeger and Runkle, 2014) and strawberries are also successfully produced in vertical farms. Total global number of vertical farming projects is estimated between 200 and 300, with a total growing area probably less than 30 ha (Rabobank, 2018b).

14.2 Light and energy

Light, artificial or natural, carries energy (Box 3.1). We have seen (Box 10.6) that a light mix typical of LED growing systems (94% red, λ=660 nm and 6% blue, λ=450 nm) carries an energy of 0.186 J/μmol, ($\rightarrow$ 5.4 μmol J^{-1}) that is the energy that would be required to generate light with this composition, with an efficiency of 1. Current systems give 2.7 μmol J^{-1} and the best LEDs coming now on the market give 3 μmol J^{-1}. In other words, significantly more energy is supplied to the system than comes back in the form of light, nearly all the difference being heat released in the process. After all, then, LEDs are not so 'cool', although they are cooler than all other lighting systems.

This means that, since plant factories are usually well-insulated, they need to be cooled while lights are on. In addition, leaves have the habit of transpiring when energy is available (and there is plenty of it in a plant factory), which means that de-humidification is necessary to manage relative humidity. Kozai *et al.* (2015), estimated that light accounts for 70 to 80% of electricity use of a plant factory, the rest being made up by air conditioning (cooling and de-humidification), a distribution confirmed by the model calculations of Graamans *et al.* (2017).

14.3 Resource requirement

It is a fact that a plant factory disposes of the one resource for crop production that is freely available, that is sun light. Therefore, economical production in a vertical farm can be possible only if its advantages outweigh the cost of producing the light that would carry no cost otherwise.

It is a fact that the transparency required of greenhouse covers (to take advantage of sunlight) goes with a relatively high heat transfer coefficient, which does have drawbacks: at high latitudes in winter there is little gain from sun radiation and a lot of heat loss to a cold environment. At low latitudes the reverse is true: there is no (natural) way to dissipate the excess energy collected by the greenhouse.

This reasoning prompted the desk study by Graamans *et al.* (2018) who compared (through model calculations) the resource use efficiency of lettuce production in plant factories and in greenhouses, in three different climate zones and latitudes (24-68°N). To be able to compare year round production in the three places (northern Sweden, the Netherlands and United Arab Emirates) they had to 'fit' the high-latitude greenhouse with supplementary light, and the low-latitude one with cooling (heat pump). In terms of (overall) energy efficiency, plant factories (0.71 g dry weight MJ^{-1}) outperform even the most efficient greenhouse (Sweden with supplementary lighting; 0.59 g dry weight MJ^{-1}). However, all this energy is purchased in a plant factory and it is not in a greenhouse, even a heated, lighted or cooled one. For the processes that involve a heat pump (cooling, de-humidification) it is necessary to account for the coefficient of performance (COP, see also Section 9.2.3) in transforming electricity into useful energy. Transforming in this way all energy requirements into electricity (with a COP between 3.4 and 10 for heating, cooling and de-humidification, depending both on process and site) the production of 1 kg dry weight of lettuce requires an input of 247 kWhe (the 'e' meaning electricity) in a plant factory, compared to 70, 111, 182 and 211 kWhe in greenhouses in respectively the Netherlands, United Arab Emirates and Sweden with and without supplementary lighting (Figure 14.1). On the other hand, plant factories achieve higher productivity for all other resources (water and CO_2, Figure 14.1 and, obviously, land).

Nevertheless, Rabobank (2018b) estimated that operational costs of a vertical farm are between 2.5 and 5 times the operational costs of a Dutch high-tech glasshouse, per square meter crop.

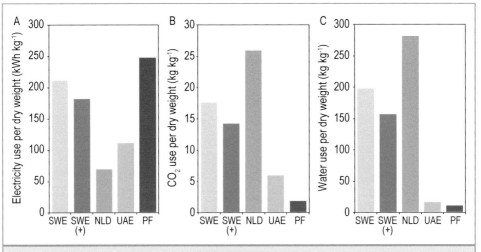

Figure 14.1. Resource use for electricity (A), CO_2 (B) and water (C) of a plant factory (PF) and greenhouses (Sweden without and with supplementary lighting (SWE, SWE(+)), the Netherlands (NLD), and United Arab Emirates (UAE)), normalised for total dry matter production (kg) (reprinted from Graamans *et al.*, 2018).

Altogether (including capital costs) and accounting for the higher productivity, they estimated the cost per unit produce to be about double than that of a good greenhouse. Hence for a plant factory to be profitable there must be a significant marketing gain possible or the value of other [scarce] resources (land, water) must offset the electricity cost required for lighting.

Indeed, rich-world consumers are willing (and can afford) to pay a premium price for residue-free products, and this is probably the single most effective marketing value of vertical farms products. There may be more, obviously: for instance freshness when the logistic chain is shortened. On the other hand, scarcity of water may make a closed system the sensible choice in a desert environment, and scarcity of land can play a role when production in urban environment would deal with food security or messy logistics.

14.4 Environmental footprint

A claim that is often made about plant factories is that they are sustainable or even 'zero footprint'. It is true that when they are well managed there is no need for plant protection products, the water use is minimal, nutrient use efficiency approaches 1, so they may be indeed emission-free. However, Eigenbrod and Gruda (2015) for instance, questioned the concept of 'local at any price' and advocated that sustainability should consider all aspects.

As there is little doubt that input of all resources other than electricity is lower in a plant factory than in a greenhouse, it is worthwhile having a closer look to how the electricity consumption could be sustainable. The usual definition would be that whenever the carbon dioxide assimilated is equal or larger than the carbon dioxide emitted during the production process, the process is sustainable. This overlooks the fact that rapidly decaying (or digested) vegetable material fixes carbon dioxide only for a short time. Real sustainability is reached when no additional carbon dioxide is generated in the process. This is equally true for any production system of fresh vegetables. Since here we deal mainly with comparisons, we will stick with the definition 'assimilation equal or larger than emission'.

Let's assume that a plant factory creates such an optimal environment for growth that light use efficiency (LUE) is 5 g CO_2 assimilated per mol_{PAR} (that is four times the value calculated in Box 2.7 for lettuce) and that the light source is a LED system that gives 10 mol kWh^{-1} (that is about 2.78 µmol J^{-1}). As at least another 25% electricity is needed, on top of lighting, for cooling and dehumidification (Section 14.2), that means 0.125 kWh are used in total to generate 1 mol PAR, that is to fix 5 g carbon dioxide. In other words, zero carbon footprint electricity use in plant factories is possible if the electricity is generated by a process with a CO_2 emission intensity lower than 40 g kWh^{-1}.

Of the 28 EU countries (European Environmental Agency, 2017), in 2014 only Sweden (10.5 g kWh^{-1}) and France (34.8 g kWh^{-1}) were in this range of CO_2 emission intensity, and then thanks to the large fraction of electricity generated in nuclear power plants (78% of production in France and 43% in Sweden). The fraction coming from renewables was 17% and 56% in France and Sweden, respectively. Whenever a significant fraction of the energy requirement is generated by burning fuels (that is the case in all other EU countries, for instance), the numbers are at least 5 times larger than this emission range. The average carbon footprint of electricity generation in 2014 in the EU (28 countries, including France and Sweden) was 276 g kWh^{-1}. Things may change and they are changing, for sure in the EU. However, only an electricity mix that is at least 95% renewable (or nuclear) would support the claim of sustainability (including that of the electricity supply) of plant factories.

The opportunity of a system generating its own electricity in a renewable fashion is obviously very attractive. Placing photovoltaic panels on the roof is an option brought up often. Let's quantify it. A lighting system of 200 µmol m^{-2} s^{-1}, switched on 18 hours a day, provides in one year 4,730 mol m^{-2}, at the rate of 0.125 kWh per mol, that would require 591.5 kWh electricity. How many photovoltaic panels are required depends obviously on the place. For instance, the expected yearly output of a solar panel in the Netherlands is 120 to 150 kWh m^{-2} (Schoenmakers, 2018), 4 to 5 m^2 photovoltaic panels are required per square meter crop. A place with double the amount of solar radiation (for instance, Southern Spain) would require less (but more than half) surface of solar panels. However, as the whole point of a plant factory is stacking the crop layers, the roof (and also the gables) would not go anywhere near making the system self-supporting. Obviously the solar panels can be placed elsewhere, this is just to make clear that there is no free lunch: some additional resource (soil or roof surface, in this case) is needed.

Box 14.1. Is 0 km always better?

A short distance between production and consumers disposes of the concept that long shelf life is an essential quality aspect in variety selection, and does increase the chance that what is on the plate is fresh. An additional advantage, often taken for granted, is that the absence of transport is 'better for the environment'.

This can be quantified. Truck transport has an emission of 62 g CO_2 per ton per km (source: NEN-EN 16258:2012). Let's assume that cooled transport adds 20% to it, to get 74 mg CO_2 kg^{-1} freight per km and that 15% of the weight is packaging and pellets. That adds up to a CO_2 emission of 87.5 g per kg fresh produce per 1000 km road transport.

Compare it, for instance, with the carbon footprint of gas-heated lettuce production in the Netherlands, of about 1 kg per kg lettuce (KWIN, Table V8). In short, transporting lettuce some 10,000 km by truck would still have a lower carbon footprint than producing it in a heated greenhouse, but then quality at the end of the ride would probably not be better.

14.5 Future perspective

Vertical farms are often portrayed as the solution to many of the problems brought about by the ongoing urbanization of the world population. For example, Despommier (2010a; 2010b) describes vertical farming as a 21st century hunger and conservation solution that promises urban renewal, sustainable production of a safe and varied food supply (year-round crop production), and the eventual repair of ecosystems that have been sacrificed for horizontal farming. This can be debated and is debated. For instance, Bugbee (2015) gives a thorough argumentation on 'Why vertical farming won't save the planet'. Indeed, up to now, the one crop that may be grown economically in a plant factory is leaves (green, red, mustard, crispy, micro, nutritious), all of them low-calories, low protein. Not a healthy, fulsome diet for the undernourished population of the (future) megacities. Wheat, rice and potatoes will be grown in the field, relying on solar radiation, for decades to come.

On the other hand, in spite of the high investment and operational costs, various business models for vertical farms are viable, or expected to be in the near future (Rabobank, 2018b), the most obvious being medicinal plants, seedlings and breeding. High-quality, fresh food production for remote communities seems to be a promising option as well.

With respect to the most obvious market (local initiatives with consumers willing to pay high prices), the numbers given above put the premium price required for a plant factory to be viable to some 200% the price of a similar product grown conventionally, in a high-tech greenhouse. Whenever there are consumers willing to pay the premium for the advantages (real or perceived) and/or other resources (as, for instance, agricultural soil in Singapore) are limiting, plant factories may have a serious business case also in this market segment. Indeed, vertical farms are becoming increasingly important for commercial production of leaf vegetables and other micro-greens for local consumption in urban areas in Japan (Kozai, 2013). Nevertheless, it is estimated that only 25% are making a profit (Kozai *et al.*, 2015), the rest still relying on private equity funds. We hope to have made clear here that the business plan for a viable vertical farm must rely on a good knowledge of plant growth processes and a realistic analysis of the marketing perspective.

Chapter 15
The future

Photo: https://upload.wikimedia.org/wikipedia/commons/2/2c/Hotel_Pyramide_Fuerth.jpg

Greenhouses have brought (and continue to bring) unprecedented variety and quality to the diet of billions, at extraordinarily low prices. It is not far-fetched to think that in the near future nearly all vegetables that are consumed fresh will have been grown 'under cover'; after all, this is very nearly the case already in Western Europe. Protected cultivation is bound to expand dramatically under the pressure of demographic, economic and social factors. Nevertheless, the same factors will create new constraints to which greenhouse production will need to adapt.

15.1 The future of greenhouses

According to a recent study (Krishna Bahadur *et al.*, 2018), the world is not growing enough fruit and vegetables for everyone to eat healthily. This research quantified the mismatch between current agricultural production worldwide (FAO, 2016, food balance sheet) and Harvard School of Public Health's 'healthy eating plate' provided, ideally, to the world population. In particular, the production of fruit and vegetables would have to triple to provide a healthy diet for everyone, and the world population will keep growing for some time. Even if only a fraction of the fruit and vegetables additionally required will be grown in greenhouses, the total surface of protected cultivation in the world will continue to increase. It grew approx. 500% in the years from 2009 to 2017 (Hickman, 2018).

Although more and more outdoor crops will be cultivated under cover, it will not be protected horticulture that is going to solve the world food problem: this will be attained by improving the production of commodities such as corn, wheat, rice and pulses. Were such crops to be grown at a profit in greenhouses, they would be unaffordable for the majority of the global population. Nevertheless, greenhouse horticulture is very important for what is called 'nutrition security', i.e. ensuring sufficient vitamins and minerals. Consumer preferences and economy will require an increased variety of crops to be grown in greenhouses: from 'new' (exotic) vegetable crops to special (medicinal) crops for content substances (Bakker, 2015).

Other factors will ensure that an increasing percentage of vegetables consumed fresh worldwide will be grown 'under cover': increasing welfare results in a growing demand for high-quality products. Wealthy markets are 'addicted' to a reliable year-round offer of excellent, yet relatively cheap products. Besides the visual aspects of quality, such as the absence of damage by rain, hail or sunburn, consumers increasingly appreciate the higher food safety resulting from the reduced need for crop protection chemicals in an enclosed environment.

However, consumers are also increasingly aware of the environmental impact and lack of sustainability of present agricultural practices, including those in protected horticulture. Social and regulatory pressure are reducing the number of crop protection products that can be used, are hindering the leaching of nutrients and are limiting the burning of fossil fuels for heating and electricity production. Water scarcity and competition from other economic sectors is reducing the water available for agriculture: the much higher water-use efficiency of protected cultivation is an additional reason for greenhouse areas to expand in arid regions, something that is already happening. However, that very scarcity makes evaporative cooling increasingly unacceptable. In summary: there is a bright future for greenhouse horticulture, provided it can become sustainable in a relatively short time span.

15.2 The greenhouses of the future

The variety of climates and the differences in disposable income around the world will continue to stimulate the diversity among greenhouses that we see today (Chapter 1). Most new protected cultivation area will, as always, be at the cheapest end. On the other hand, taking advantage of the opportunities and meeting the challenges outlined above requires increasing control of the production process. That will be attained by the progressive implementation of technology in the existing greenhouse area: not unlike building a pyramid by adding (ever larger) layers from below and raising the level of the existing ones.

Predicting the future has always been tricky. Nevertheless, present trends in greenhouse development allow for some guesswork concerning the top of the pyramid (the high-tech greenhouses of the near future). As heating by fossil fuels is phased out, the additional energy required for greenhouse production will come from available renewable resources, such as biomass or geothermal heat and increasingly from sustainably produced electricity. This, and the improving performance of light emitting diodes (LEDs), will ensure that supplemental light for intensified and out-of-season production will become widespread. Heat pumps, which are able to cool as well as heat, will be used in preference to burners and will also be installed in regions where greenhouses presently do without heating.

The difference between greenhouses and vertical farms will become more blurred: greenhouses will have more vertical-farm features (lighting, cooling) and vertical farms will adopt greenhouse features (in the properties of the shell) that will make them more economical, in view of the external environment in which they will operate (Graamans *et al.*, 2020). Greenhouse cover materials will be tailored to take the best advantage of solar radiation, in view of the crop inside and the climate outside (Fleischer and Dinar, 2014). 'Active window' materials, with properties that can be modified at will (Casini, 2018) will trickle down from present high-end applications to mainstream use in greenhouse covers (Baeza *et al.*, recent unpublished data).

LEDs will make it possible to exploit the knowledge that is being generated on spectral crop response, to tailor production to specific needs, such as the content of particular metabolites in food/medicinal crops and visual aspect in ornamentals (Dieleman *et al.*, 2016). By exploring the boundaries of the flexibility of crops, greenhouses may help smooth out the mismatches between demand and supply of renewable electricity, which hinder its wholesale application. A 'standing army' of beneficial organisms will ensure that harmful chemicals are no longer needed for crop protection (De Boo, 2018; Messelink, 2013) and 'smart' irrigation practices (Chapter 13) will ensure that there will be no emission of nutrients.

The progressive affordability of a variety of sensors, and information and communication technologies, coupled to a shortage of skilled labour, will ensure that a lot more automation will be present in the greenhouses of the future than is now the case. The autonomous greenhouse challenge (Hemming, 2018) is only the first attempt to apply artificial intelligence to greenhouse management. Thanks to the absence of 'disturbances' caused by the variability of weather, the first commercial applications of artificial intelligence will most likely be in vertical farms, later to be extended to greenhouses. With the development and application of artificial intelligence, machine vision and robotics, the management of crop operations and quality of products will further improve.

Greenhouse horticulture will ensure the sustainable production of high-quality products with the most efficient use of all resources, the ultimate example of sustainable intensification of plant production. In this way, greenhouse horticulture will remain the leading and trend-setting sector for the world's total crop production, thereby providing innovations for possible applications in open field cultivation.

References

Abdel-Ghany, A.M. and T. Kozai, 2006. Cooling efficiency of fogging systems for greenhouses. Biosystems Engineering 94(1): 97-109.

Ahn, S.J., Y.J. Im, G.C. Chung, B.H. Cho and S.R. Suh, 1999. Physiological responses of grafted-cucumber leaves and rootstock roots affected by low root temperature. Scientia Horticulturae 81(4): 397-408. https://doi.org/10.1016/S0304-4238(99)00042-4.

Albright, L.D., 1990. Environment control for animals and plants. American Society of Agricultural Engineers. USDA, Washington, DC, USA.

Alduchov, O.A. and R.E. Eskridge, 1996. Improved Magnus form approximation of saturation vapor pressure. Journal of Applied Meteorology 35(4): 601-609.

Allen, R.G., L.S. Pereira, D. Raes and M. Smith, 1998. Crop evapotranspiration – guidelines for computing crop water requirements. FAO irrigation and drainage paper No. 56. FAO, Rome, Italy. Available at: http://www.fao.org/docrep/X0490E/X0490E00.htm.

American Society of Heating Refrigeration and Air-Conditioning Engineers (ASHRAE), 1989. Psychrometrics. ASHRAE handbook fundamentals. ASHRAE, Atlanta, GA, USA.

American Society of Heating Refrigeration and Air-Conditioning Engineers (ASHRAE), 2017. ASHRAE handbook fundamentals. ASHRAE, Atlanta, GA, USA.

Assaf, G., 1986. LHC – the latent heat converter. Journal of Heat Recovery Systems 6(5): 369-379.

Assaf, G. and N. Zieslin, 2003. Novel means of humidity control and greenhouse heating. Acta Horticulturae 614: 433-437.

Bacci, L., G. Carmassi, L. Incrocci, P. Battista, B. Rapi, P. Marzialetti and A. Pardossi, 2013. Modelling evapotranspiration of greenhouse and nursery crops. Italus Hortus 20(3): 69-78. Available at: http://www.soihs.it/ItalusHortus/Review/Review%2021/06%20Pardossi.pdf.

Baeza, E.J., 2007. Optimización del diseño de los sistemas de ventilación en invernadero tipo parral. Tesis doctoral en Ingeniería Agronómica por la Universidad de Almería, Almería, Spain, 204 pp.

Baeza, E.J., J.I. Montero, J. Pérez-Parra, B.J. Bailey, J.C. López and J.C. Gázquez, 2014. Avances en el estudio de la ventilación natural. Documentos Técnicos n. 7, Cajamar Caja Rural, Almeria, Spain, 60 pp. Available at: http://www.publicacionescajamar.es/pdf/series-tematicas/centros-experimentales-las-palmerillas/avances-en-el-estudio-de-la-ventilacion.pdf.

Baeza, E.J. J., Pérez-Parra, J.C. López, J.C. Gázquez and J.I. Montero, 2008. Numerical analysis of buoyancy driven natural ventilation in multispan type greenhouses. Acta Horticulturae 797: 111-116.

Baeza, E.J., J.J. Pérez-Parra, J.I. Montero, B.J. Bailey, J.C. López and J.C. Gázquez, 2009. Analysis of the role of sidewall vents on buoyancy-driven natural ventilation in parral-type greenhouses with and without insect screens using computational fluid dynamics. Biosystems Engineering 104(1): 86-96.

Bailey, B. and A. Raoueche, 1998. Design and performance aspects of a water producing greenhouse cooled by seawater. Acta Horticulturae 6: 109-121.

Baille, A., 2001. Trends in greenhouse technology for improved climate control in mild winter climates. Acta Horticulturae 559: 161-168. https://doi.org/10.17660/ActaHortic.2001.559.23.

Baille, M., A. Baille and D. Delmon, 1994. Microclimate and transpiration of greenhouse rose crops. Agricultural and Forest Meteorology 71: 83-97. https://doi.org/10.1016/0168-1923(94)90101-5.

Bakker, J.C., 1991. Analysis of humidity effects on growth and production of glasshouse fruit vegetables. PhD-thesis, Wageningen University, 155 pp. Available at: https://edepot.wur.nl/206443.

Bakker, J.C., 2015. De toekomst van de glastuinbouw. Kas techniek 3: 43. Available at: https://tinyurl.com/y929p4hp.

Bakker, J.C. and J.A.M. Van Uffelen, 1988. Effects of diurnal temperature regimes on growth and yield of glasshouse sweet pepper. Netherlands Journal of Agricultural Science: Issued by the Royal Netherlands Society for Agricultural Science 36(3): 201-208. Available at: https://wur.on.worldcat.org/oclc/1054744083.

Bakker, J.C., H.F. De Zwart and J.B. Campen, 2006. Greenhouse cooling and heat recovery using fine wire heat exchangers in a closed pot plant greenhouse: design of an energy producing greenhouse. Acta Horticulturae 719: 263-270.

Balendonck, J., A.A. Sapounas, F. Kempkes, E.A. van Os, R. van der Schoor, B.A.J. van Tuijl, L.C.P. Keizer, 2014. Using a Wireless Sensor Network to determine climate heterogeneity of a greenhouse environment. Acta Horticulturae. 1037: 539-546. https://doi.org/10.17660/ActaHortic.2014.1037.67.

Balendonck, J., A. Pardossi, H. Tuzel, Y. Tuzel, M. Rusan and F. Karam, 2010. FLOW-AID – a deficit irrigation management system using soil sensor activated control: case studies. Transactions of the 3rd International Symposium on Soil Water Measurement using capacitance, impedance and TDT. April 7-9, 2010. Murcia, Spain. Available at: http://edepot.wur.nl/139559.

Bantis, F., S. Smirnakou, T. Ouzounis, A. Koukounaras, N. Ntagkas and K. Radoglou, 2018. Current status and recent achievements in the field of horticulture with the use of Light-Emitting Diodes (LEDs). Scientia Horticulturae 235: 437-451. https://doi.org/10.1016/J.SCIENTA.2018.02.058.

Baptista, F.J., B.J. Bailey, J.M. Randall and J.F. Meneses, 1999. Greenhouse ventilation rate: theory and measurement with tracer gas techniques. Journal of Agricultural Engineering Research 72: 363-374.

Beerling, E.A.M., C. Blok, A.A. Van der Maas and E.A. Van Os, 2014. Closing the water and nutrient cycles in soilless cultivation systems. Acta Horticulturae 1034: 49-55. https://doi.org/10.17660/ActaHortic.2014.1034.4.

Bertin, N. and E. Heuvelink, 1993. Dry-matter production in a tomato crop: comparison of two simulation models. Journal of Horticultural Science 68(6): 995-1011. https://doi.org/10.1080/00221589.1993.11516441.

Bielza, P., V. Quinto, C. Gravalos, E. Fernandez and J.J. Abellan, 2008. Impact of production system on development of insecticide resistance in Frankliniella occidentalis (Thysanoptera: Thripidae). Journal of Economic Entomology 101: 1685-1690.

Bontsema, J., J. Hemming, C. Stanghellini, P.H.B. De Visser, E. Van Henten, J. Budding, T. Rieswijk and S. Nieboer, 2008. On-line estimation of the transpiration in greenhouse horticulture. Proceedings IFAC, Agricontrol 2007, pp. 29-34.

Bottcher, R.W., G.R. Baughman, R.S. Gates and M.B. Timmons, 1991. Characterizing efficiency of misting systems for poultry. Transactions of the American Society of Agricultural Engineers 34(2): 586-590.

Bruggink, G.T., 1992. A comparative analysis of the influence of light on growth of young tomato and carnation plants. Scientia Horticulturae 51(1-2): 71-81. https://doi.org/10.1016/0304-4238(92)90105-L.

Brunt, D., 2011. Physical and dynamical meteorology. Cambridge University Press, Cambridge, UK.

Buck, A.L., 1981. New equations for computing vapor pressure and enhancement factor. Journal of Applied Meteorology 20(12): 1527-1532.

Bugbee, B., 2015. Why vertical farming won't save the planet. Utah State University Extension. Available at: https://www.youtube.com/watch?v=ISAKc9gpGjw.

Buwalda, F., A.A. Rijsdijk, J.V.M. Vogelezang, A. Hattendorf and L.G.G. Batta, 1999. An energy efficient heating strategy for cut rose production based on crop tolerance to temperature fluctuations. Acta Horticulturae 507: 117-126. https://doi.org/10.17660/ActaHortic.1999.507.13.

Caballero, P. and M.D. De Miguel, 2002. Costes e intensificación en la hortofruti-cúltura Mediterránea. In: Garcá, J.M. (ed.) La Agricultura Mediterránea en el Siglo XXI. Instituto Cajamar, Almería, Spain, pp. 222-244. Available at: http://www.publicacionescajamar.es/publicaciones-periodicas/mediterraneo-economico/mediterraneo-economico-2-la-agricultura-mediterranea-en-el-siglo-xxi/.

Caja Rural Intermediterránea, 2005. La economía de la provincia de Almería. Cajamar. Available at: http://www.juntadeandalucia.es/educacion/vscripts/wbi/w/rec/4625.pdf.

Cajamar, 2017. Análisis de la campaña hortofrutícola de Almería. Campaña 2016-2017. Available at: https://tinyurl.com/ycddyp3o.

Campen, J.B., 2005. Greenhouse design applying CFD for Indenesian conditions. Acta Horticulturae 691: 414-424.

Campen, J.B., 2009. Dehumidification of greenhouses. Thesis, Wageningen University, Wageningen, the Netherlands, 117 pp. Available at: http://edepot.wur.nl/12805.

Campen, J.B., 2011. Dehumidification using outside air. Wageningen UR Greenhouse Horticulture: Oral presentation. Available at: http://edepot.wur.nl/178729.

Campen, J.B., 2012. Towards a more sustainable, water efficient protected cultivation in arid regions. Acta Horticulturae 952: 81-85. https://doi.org/10.17660/ActaHortic.2012.952.7.

Carmassi, G., L. Bacci, M. Bronzini, L. Incrocci, R. Maggini, G. Bellocchi, D. Massa and A. Pardossi, 2013. Modelling transpiration of greenhouse gerbera (*Gerbera jamesonii* H. Bolus) grown in substrate with saline water in a Mediterranean climate. Scientia Horticulturae 156: 9-18. https://doi.org/10.1016/j.scienta.2013.03.023.

Carvalho, S.M.P., E. Heuvelink, R. Cascais and O. Van Kooten, 2002. Effect of day and night temperature on internode and stem length in chrysanthemum: is everything explained by DIF? Annals of Botany 90(1): 111-118. https://doi.org/10.1093/AOB/MCF154.

Carvalho, S.M.P., F.R. Van Noort, R. Postma and E. Heuvelink, 2008. Possibilities for producing compact floricultural crops. Wageningen UR Greenhouse Horticulture / Productschap Tuinbouw, Report 173. Available at: http://edepot.wur.nl/11685.

Carvalho, S.M.P., S.E. Wuillai and E. Heuvelink, 2006. Combined effects of light and temperature on product quality of Kalanchoe Blossfeldiana. Acta Horticulturae 711: 121-126. https://doi.org/10.17660/ActaHortic.2006.711.12.

Casini, M., 2018. Active dynamic windows for buildings: a review. Renewable Energy 119: 923-934.

Cavcar, M., 2000. The international standard atmosphere (ISA). Anadolu University, Turkey.

Centraal Bureau voor de Statistiek (CBS), 2018. Landbouw; economische omvang naar omvangsklasse, bedrijfstype. Available at: https://tinyurl.com/y8ymuuhf.

Challa, H. and E. Heuvelink, 1993. Economic evaluation of crop photosynthesis. Acta Horticulturae 328: 219-228. https://doi.org/10.17660/ActaHortic.1993.328.21.

Cheng, C.-H. and F. Nnadi, 2014. Predicting downward longwave radiation for various land use in all-sky condition: Northeast Florida. Advances in Meteorology, Article ID: 525148. https://doi.org/10.1155/2014/525148.

Christiaens, A., M.C. Van Labeke, B. Gobin and J. Van Huylenbroeck, 2015. Rooting of ornamental cuttings affected by spectral light quality. Acta Horticulturae 1104: 219-224. https://doi.org/10.17660/ActaHortic.2015.1104.33.

Cicekli, M. and N.T. Barlas, 2014. Transformation of today greenhouses into high technology vertical framing systems for metropolitan regions. Journal of Environmental Protection and Ecology 15: 1779-1785.

Cockshull, K.E, D. Gray, G.B. Seymour and B. Thomas, 1998. Genetic and environmental manipulation of horticultural crops. Cab International, Wallingford, UK.

Cooper, AJ., 1973. Influence of rooting-medium temperature on growth of Lycopersicon Esculentum. Annals of Applied Biology 74(3): 379-385. https://doi.org/10.1111/j.1744-7348.1973.tb07758.x.

Cuadrado Gómez, I.M. (ed.), 2009. Caracterización de la explotación hortícola protegida de Almería. Fundación para la Investigación Agraria en la Provincia de Almería, 178 pp. Available at: http://www.publicacionescajamar.es/pdf/series-tematicas/centros-experimentales-las-palmerillas/caracterizacion-de-la-explotacion.pdf.

Cuartero, J., M.C. Bolarín, M.J. Asíns and V. Moreno, 2006. Increasing salt tolerance in the tomato. Journal of Experimental Botany 57(5): 1045-1058. https://doi.org/10.1093/jxb/erj102.

Davies, P., 2005. A solar cooling system for greenhouse food production in hot climates. Solar Energy 79(6): 661-668.

De Boo, M., 2018. Back-up troops for a healthy harvest. Wageningen World 2: 16-23. Available at: https://tinyurl.com/ycy2r5ok.

De Gelder, A., E.H. Poot, J.A. Dieleman and H.F. De Zwart, 2012. A concept for reduced energy demand of greenhouses: the next generation greenhouse cultivation in the Netherlands. Acta Horticulturae 952: 539-544. https://doi.org/10.17660/ActaHortic.2012.952.68.

De Koning, A.N.M., 1990. Long-term temperature integration of tomato. Growth and development under alternating temperature regimes. Scientia Horticulturae 45(1-2): 117-127. https://doi.org/10.1016/0304-4238(90)90074-O.

De Koning, A.N.M., 1993. Growth of a tomato crop. Measurements for model validation. Acta Horticulturae 328: 141-146.

De Koning, A.N.M., 1994. Development and dry matter distribution in glasshouse tomato: a quantitative approach. Available at: http://edepot.wur.nl/205947.

De Koning, A.N.M., 2000. The effect of temperature, fruit load and salinity on development rate of tomato fruit. Acta Horticulturae 519: 85-94. https://doi.org/10.17660/ActaHortic.2000.519.7.

De Koning, A.N.M. and I. Tsafaras, 2017. Real-time comparison of measured and simulated crop transpiration in greenhouse process control. Acta Horticulturae 1170: 301-308. https://doi.org/10.17660/ActaHortic.2017.1170.36.

De Ruijter, J.A.F., 2002. Kansen voor lage-temperatuurwarmte in combinatie met warmtepompen en ondergrondse energieopslag bij (bijna) gesloten kassen: deel 1 en 2: haalbaarheid warmtepompgebaseerd koel-/ontvochtigings- en verwarmingssysteem met warmteopslag in aquifers. KEMA, Arnhem, the Netherlands.

De Visser, P.H.B., G.H. Buck-Sorlin and G.W.A.M. Van der Heijden, 2014. Optimizing illumination in the greenhouse using a 3D model of tomato and a ray tracer. Frontiers in Plant Science 5: 48. https://doi.org/10.3389/fpls.2014.00048.

De Zwart, H.F., 1996. Analyzing energy-saving options in greenhouse cultivation using a simulation model. Wageningen University, Farm Technology Group, 236 pp.

De Zwart, H.F., 2004. Praktijkexperiment duurzame energieverzameling door middel van daksproeiers. Wageningen University, Wageningen, the Netherlands. (in Dutch)

De Zwart, H.F., 2007. Overall energy analysis of (semi) closed greenhouses. Acta Horticulturae 801: 811-818.

De Zwart, H.F., 2012. Lesson learned from experiments with semi-closed greenhouses. Acta Horticulturae 952: 583-588.

De Zwart, H.F., 2016. Radiation monitor. GreenTech 2016. Available at: http://www.glastuinbouwmodellen. wur.nl/radiationmonitor/?user=Mard_ENC and http://edepot.wur.nl/388294.

De Zwart, H.F. and G.L.A.M. Swinkels, 2002. De kas als zonne-energie oogster. Wageningen University, Wageningen, the Netherlands.

Demirbas, A., 2009. Hydrogen from mosses and algae via pyrolysis and steam gasification. Energy sources, Part A: Recovery, Utilization, and Environmental Effects 32(2): 172-179.

Despommier, D., 2010a. The vertical farm: feeding the world in the 21st century. St. Martin's Press, New York, NY, USA.

Despommier, D., 2010b. The vertical farm. TEDxWindyCity. Available at: https://www.youtube.com/watch?v=XIdP00u2KRA.

Dieleman, J.A., F.W.A. Verstappen and D. Kuiper, 1998. Root temperature effects on growth and bud break of rosa hybrida in relation to cytokinin concentrations in xylem sap. Scientia Horticulturae 76(3-4): 183-192. https://doi.org/10.1016/S0304-4238(98)00131-9.

Dielcman, J.A., P.H.B. De Visser and P.C.M. Vermeulen, 2016. Reducing the carbon footprint of greenhouse grown crops: re-designing LED-based production systems. Acta Horticulturae 1134: 395-402.

Doorenbos, J. and W.O. Pruitt, 1997. Crop water requirements. FAO irrigation and drainage, Paper 24. Food and Agriculture Organisation, Rome, Italy. Available at: http://www.fao.org/docrep/018/s8376e/s8376e.pdf.

Dueck, T. and S. Pot, 2010. Lichtmeetprotocol: lichtmetingen in onderzoekskassen meet LED en SON-T belichting. Wageningen UR Greenhouse Horticulture/Plant Dynamics and Wageningen University, Wageningen, the Netherlands.

Dueck, T.A. and C. Van Dijk, 2007. Veilig CO_2 blijven doseren uit WKK. Groenten & Fruit, week 31: 16-17.

Dueck, T.A., J. Janse, B.A. Eveleens, F.L.K. Kempkes and L.F.M. Marcelis, 2012. Growth of tomatoes under hybrid LED and HPS lighting. Acta Horticulturae 952: 335-342. https://doi.org/10.17660/ActaHortic.2012.952.42.

Ehret, D.L. and L.C. Ho, 1986. The effects of salinity on dry matter partitioning and fruit growth in tomatoes grown in nutrient film culture. Journal of Horticultural Science 61(3): 361-367. https://doi.org/10.1080/14620316.1986.11515714.

Eigenbrod, C. and N. Gruda, 2015. Urban vegetable for food security in cities: a review. Agronomy and Sustainable Development 35: 483-498. https://doi.org/10.1007/s13593-014-0273-y.

Elings, A, H.F. De Zwart, J. Janse, L.F.M. Marcelis and F. Buwalda, 2006. Multiple-day temperature settings on the basis of the assimilate balance: a simulation study. Acta Horticulturae 718: 227-232. https://doi.org/10.17660/ActaHortic.2006.718.24.

Elings, A., F.L.K. Kempkes, R.C. Kaarsemaker, M.N.A. Ruijs, N.J. Van de Braak and T.A. Dueck, 2005. The energy balance and energy-saving measures in greenhouse tomato cultivation. Acta Horticulturae 691: 67-74.

European Environmental Agency, 2017. Overview of electricity production and use in Europe. Indicator assessment. Prod-ID: IND-353-en, also known as: ENER 038. Available at: https://tinyurl.com/yd249ev6.

European Food Safety Authority (EFSA) Panel on Plant Protection Products and their Residues (PPR), 2010. Scientific Opinion on emissions of plant protection products from greenhouses and crops grown under cover: outline of a new guidance. EFSA Journal 8(4): 1567. https://doi.org/10.2903/j.efsa.2010.1567.

Fanasca, S., A. Martino, E. Heuvelink and C. Stanghellini, 2007. Effect of electrical conductivity, fruit pruning, and truss position on quality in greenhouse tomato fruit. Journal of Horticultural Science and Biotechnology 82(3): 488-494. https://doi.org/10.1080/14620316.2007.11512263.

Fanourakis, D., D. Bouranis, H. Giday, D.R.A. Carvalho, A.R. Nejad and C.-O. Ottosen, 2016. Improving stomatal functioning at elevated growth air humidity: a review. Journal of Plant Physiology 207: 51-60. https://doi.org/10.1016/J.JPLPH.2016.10.003.

Farquhar, G.D., S. Von Caemmerer and J.A. Berry, 2001. Models of photosynthesis. Plant Physiology 125(1): 42-45. https://doi.org/10.1104/PP.125.1.42.

Fernández, M.D., E. Baeza, A. Céspedes, J. Pérez-Parra and J.C. Gázquez, 2009. Validation of On-Farm Crop Water Requirements (PrHo) model for horticultural crops in an unheated plastic greenhouse. Acta Horticulturae 807: 295-300. Available at: https://www.actahort.org/books/807/807_40.htm.

Fernández, M.D., S. Bonachela, F.R. Orgaz, J.C. Thompson, M.R. López, M. Granados, M. Gallardo and E. Fereres, 2010. Measurement and estimation of plastic greenhouse reference evapotranspiration in a Mediterranean climate. Irrigation Science 28: 497. https://doi.org/10.1007/s00271-010-0210-z.

Fleischer, M. and M. Dinar, 2014. How to tailor-make a greenhouse cover. Acta Horticulturae 1015: 259-261.

Food and Agriculture Organization of the United Nations (FAO), 2016. FAOSTAT: food supply – crops primary equivalent. Available at: http://www.fao.org/faostat/en/#data/CC.

Foyer, C.H., J. Neukermans, G. Queval, G. Noctor and J. Harbinson, 2012. Photosynthetic control of electron transport and the regulation of gene expression. Journal of Experimental Botany 63(4): 1637-1661. https://doi.org/10.1093/jxb/ers013.

Fuchs, M. and C. Stanghellini, 2018. The functional dependence of canopy conductance on water vapor pressure deficit revisited. International Journal of Biometeorology 62(7): 1211-1220.

Gallardo, M., R.B. Thompson and M.D. Fernández, 2013. Water requirements and irrigation management in Mediterranean greenhouses: the case of the southeast coast of Spain. FAO Plant Production and Protection Paper 217. FAO, Rome, Italy, pp. 109-136. Available at: http://www.fao.org/3/a-i3284e.pdf.

Garcia Victoria, N., F.L.K. Kempkes, J.C. Löpez Hernández, E.J. Baezo Romero, L. Incrocci and A. Pardossi, 2012. Evaluation trials of potential input reducing developments in 3 test locations. Almeria/Bleiswijk/Pisa. European Commission, Brussels, Belgium. Available at: http://edepot.wur.nl/222835.

García Victoria, N., F.L.K. Kempkes, P. Van Weel, C. Stanghellini, T.A. Dueck and M. Bruins, 2012. Effect of a diffuse glass greenhouse cover on rose production and quality. Acta Horticulturae 952: 241-248. https://doi.org/10.17660/ActaHortic.2012.952.29.

Gatley, D.P., 2004. Psychrometric chart celebrates 100th anniversary. Ashrae Journal 46(11): 16.

Geelen, P.A.M., J.O. Voogt and P.A. Van Weel, 2019. Plant empowerment. LetsGrow.com. Available at: www.plantempowerment.com.

Ghannoum, O., 2018. C3 photosynthesis. In: Atwell, B.J., P.E. Kriedemann and C.G.N. Turnbull (eds.) Plants in action: adaptation in nature, performance in cultivation. Macmillan Education AU, South Yarra, Australia. Available at: http://plantsinaction.science.uq.edu.au/content/21-c3-photosynthesis.

Ghosal, M.K., G.N. Tiwari and N.S.L. Srivastava, 2003. Modeling and experimental validation of a greenhouse with evaporative cooling by moving water film over external shade cloth. Energy and Buildings 35(8): 843-850.

Gieling, T.H., 1998. Sensors and measurement, a review. Acta Horticulturae 421: 19-36. https://doi.org/10.17660/ActaHortic.1998.421.1.

Golovatskaya, I.F. and R.A. Karnachuk, 2015. Role of green light in physiological activity of plants. Russian Journal of Plant Physiology 62(6): 727-740. https://doi.org/10.1134/S1021443715060084.

Gómez, C. and C.A. Mitchell, 2014. Supplemental lighting for greenhouse-grown tomatoes: intracanopy LED towers vs overhead HPS lamps. Acta Horticulturae: 855-862.

Goudriaan, J., 1977. Crop micrometeorology: a simulation study. Wageningen Centre for Agricultural Publishing and Documentation, 249 pp. Available at: http://edepot.wur.nl/166537.

Graamans, L., A. Van den Dobbelsteen, E. Meinen and C. Stanghellini, 2017. Plant factories; crop transpiration and energy balance. Agricultural Systems 153: 138-147.

Graamans, L., E. Baeza, A. Van den Dobbelsteen, I. Tsafaras and C. Stanghellini, 2018. Plant factories versus greenhouses: comparison of resource use efficiency. Agricultural Systems 160: 31-43. https://doi.org/10.1016/j.agsy.2017.11.003.

Graamans, L., M. Tenpierik, A. van den Dobbelsteen, C. Stanghellini, 2020. Plant factories: Reducing energy demand at high internal heat loads through façade design, Applied Energy, 262: 114544. https://doi.org/10.1016/j.apenergy.2020.114544.

Grashoff, C., M.G.M. Raaphorst, J.W.M. Kempen, J. Janse, J.A. Dieleman and L.F.M. Marcelis, 2004. Temperatuurintegratie kleine gewassen. Plant Research International, Nota 307. Available at: https://www.kasalsenergiebron.nl/content/research/Eindrapport_11293.pdf.

Grieve, C.M., S.R. Grattan and E. Van Maas, 2012. Plant salt tolerance. Agricultural Salinity Assessment and Management, 2nd edition. American Society of Civil Engineers Reston, Virginia, USA, pp. 405-459

Hanan, J.J., 1998. Greenhouses: advanced technology for protected horticulture. CRC press, Boca Raton, FL, USA, 708 pp.

Hao, X., J. Zheng, and C. Little, 2015. Dynamic temperature integration with temperature drop to improve early fruit yield and energy efficiency in greenhouse cucumber production. Acta Horticulturae 1107: 127-132. https://doi.org/10.17660/ActaHortic.2015.1107.17.

Harvard School of Public Health (Harvard University), 2011. Healthy eating plate & healthy eating pyramid. Available at: https://tinyurl.com/zn5849w.

Haynes, W.M., 2014. CRC handbook of chemistry and physics. CRC press, Boca Raton, FL, USA.

Hemming, S., 2015. Diffuse light – benefits for crops and growers. Agricultural film conference, Barcelona. Available at: http://edepot.wur.nl/385695.

Hemming, S., 2018. International challenge of self-cultivating greenhouses. Greenhouse Horticulture, Wageningen UR, Wageningen, the Netherlands. Available at: https://tinyurl.com/ycpwm9e9.

Hemming, S., G.L.A.M. Swinkels, A.J. Van Breugel and V. Mohammadkhani, 2016. Evaluation of diffusing properties of greenhouse covering materials. Acta Horticulturae 1134: 309-316. https://doi.org/10.17660/ActaHortic.2016.1134.41.

Hemming, S., J. Balendonck, J.A. Dieleman, A. De Gelder, F.L.K. Kempkes, G.L.A.M. Swinkels, P.H.B. De Visser and H.F. De Zwart, 2017. Innovations in greenhouse systems – energy conservation by system design, sensors and decision support systems. Acta Horticulturae 1170: 1-16. https://doi.org/10.17660/ActaHortic.2017.1170.1.

Hemming, S., T.A. Dueck, N. Marissen, R.E.E. Jongschaap, F.L.K. Kempkes and N.J. Van de Braak, 2005. Diffuus licht: het effect van lichtverstrooiende kasdekmaterialen op kasklimaat, lichtdoordringing en gewasgroei. Rapport 557, Wageningen UR Agrotechnology & Food Innovations, Wageningen, the Netherlands, 97 pp. Available at: http://edepot.wur.nl/295718.

Hemming, S., V. Mohammadkhani and J. Van Ruijven, 2014. Material technology of diffuse greenhouse covering materials – influence on light transmission, light scattering and light spectrum. Acta Horticulturae 1037: 883-895. https://doi.org/10.17660/ActaHortic.2014.1037.

Hernández, R. and C. Kubota, 2016. Physiological responses of cucumber seedlings under different blue and red photon flux ratios using LEDs. Environmental and Experimental Botany 121: 66-74. https://doi.org/10.1016/J.ENVEXPBOT.2015.04.001.

Heuvelink, E., 1989. Influence of day and night temperature on the growth of young tomato plants. Scientia Horticulturae 38(1-2): 11-22. https://doi.org/10.1016/0304-4238(89)90015-0.

Heuvelink, E., 1996. Tomato growth and yield: quantitative analysis and synthesis. PhD-thesis, Wageningen University, Wageningen, the Netherlands, 326 pp. Available at: http://edepot.wur.nl/206832.

Heuvelink, E. 1997. Effect of fruit load on dry matter partitioning in tomato. Scientia Horticulturae 69: 51-59. https://doi.org/10.1016/S0304-4238(96)00993-4

Heuvelink, E. (ed.), 2018. Tomatoes. CABI, Wallingford, UK, 256 pp.

Heuvelink, E. and T. Kierkels, 2015. Plant physiology in greenhouses. Horti-Text BV, Woerden, the Netherlands, 128 pp. Available at: http://www.ingreenhouses.com/books.

Heuvelink, E., J.H. Lee, R.P.M. Buiskool and L. Ortega, 2002. Light on cut Chrysanthemum: measurement and simulation of crop growth and yield. Acta Horticulturae 580: 197-202. https://doi.org/10.17660/ActaHortic.2002.580.25.

Heuvelink, E., M. Bakker and C. Stanghellini, 2003. Salinity effects on fruit yield in vegetable crops: a simulation study. Acta Horticulturae 609 133-140. https://doi.org/10.17660/ActaHortic.2003.609.17.

Heuvelink, E., M.J. Bakker, A. Elings, R.C. Kaarsemaker and L.F.M. Marcelis, 2005. Effect of leaf area on tomato yield. Acta Horticulturae 691: 43-50.

Heuvelink, E., M.J. Bakker, L. Hogendonk, J. Janse and R. Kaarsemaker, 2006. Horticultural lighting in the Netherlands: new developments. Acta Horticulturae 711: 25-34. https://doi.org/10.17660/ActaHortic.2006.711.1.

Hickman, G.W., 2018. International greenhouse vegetable production – statistics. Cuesta Roble Greenhouse Consultants, Mariposa, CA, USA, 170 pp. Available at: http://www.cuestaroble.com/statistics.htm.

Higashide, T. and E. Heuvelink, 2009. Physiological and morphological changes over the past 50 years in yield components in tomato. Journal of the American Society for Horticultural Science 134(4): 460-465.

Ho, L.C. and P.J. White, 2005. A cellular hypothesis for the induction of blossom-end rot in tomato fruit. Annals of Botany 95(4): 571-581. https://doi.org/10.1093/aob/mci065.

Ho, L.C., P. Adams, X.Z. Li, H. Shen, J. Andrews, and Z.H. Xu, 1995. Responses of Ca-efficient and Ca-inefficient tomato cultivars to salinity in plant growth, calcium accumulation and blossom-end rot. Journal of Horticultural Science 70: 909-918. https://doi.org/10.1080/14620316.1995.11515366.

Hogewoning, S.W., E. Wientjes, P. Douwstra, G. Trouwborst, W. Van Ieperen, R. Croce and J. Harbinson, 2012. Photosynthetic quantum yield dynamics: from photosystems to leaves. Plant Cell 24(5): 1921-1935. https://doi.org/10.1105/tpc.112.097972.

Hogewoning, S.W., G. Trouwborst, H. Maljaars, H. Poorter, W. Van Ieperen and J. Harbinson, 2010b. Blue light dose-responses of leaf photosynthesis, morphology, and chemical composition of cucumis sativus grown under different combinations of red and blue light. Journal of Experimental Botany 61(11): 3107-3117. https://doi.org/10.1093/jxb/erq132.

Hogewoning, S.W., H. Maljaars and J. Harbinson, 2007. The acclimation of photosynthesis in cucumber leaves to different ratios of red and blue light. Photosynthesis Research 91: 287-288. 32.

Hogewoning, S.W., P. Douwstra, G. Trouwborst, W. Van Ieperen and J. Harbinson, 2010a. An artificial solar spectrum substantially alters plant development compared with usual climate room irradiance spectra. Journal of Experimental Botany 61(5): 1267-1276. https://doi.org/10.1093/jxb/erq005.

Holder, R. and K.E. Cockshull, 1988. The effect of humidity and nutrition on the development of calcium deficiency symptoms in tomato leaves. The effects of high humidity on plant growth in energy saving greenhouses. Office for Official Publications of the European Communities, Luxembourg, pp. 53-60.

Hovi-Pekkanen, T. and R. Tahvonen, 2008. Effects of interlighting on yield and external fruit quality in year-round cultivated cucumber. Scientia Horticulturae 116(2): 152-161. https://doi.org/10.1016/J.SCIENTA.2007.11.010.

Hsiao, T.C., P. Steduto and E. Fereres, 2007. A systematic and quantitative approach to improve water use efficiency in agriculture. Irrigation Science 25(3): 209-231.

Hunt, R., D.R. Causton, B. Shipley and A.P. Askew, 2002. A modern tool for classical plant growth analysis. Annals of Botany 90(4): 485-488. https://doi.org/10.1093/aob/mcf214.

Incrocci, L., 2012. Nutrient solution calculator. EU-EUPHOROS project, calculation tools. Available at: https://tinyurl.com/yambd4gf.

Incrocci, L., C. Stanghellini, B. Dimauro and A. Pardossi, 2008. Rese maggiori a costi contenuti con la concimazione carbonica: risultati di studi in serre nel sud Italia e in Spagna. L'Informatore Agrario 21: 57-59.

Incrocci, L., G. Incrocci, A. Pardossi, G. Lock, C. Nicholl and J. Balendonck, 2009. The calibration of wet-sensor for volumetric water content and pore water electrical conductivity in different horticultural substrates. Acta Horticulturae 807: 289-294. https://doi.org/10.17660/ActaHortic.2009.807.39.

Incrocci, L., P. Marzialetti, G. Incrocci, A. Di Vita, J. Balendonck, C. Bibbiani, S. Spagnol and A. Pardossi, 2014. Substrate water status and evapotranspiration irrigation scheduling in heterogenous container nursery crops. Agricultural Water Management 131: 30-40. https://doi.org/10.1016/j.agwat.2013.09.004.

Incrocci, L., P. Marzialetti, G. Incrocci, A. Di Vita, J. Balendonck, C. Bibbiani, S. Spagnol and A. Pardossi, 2019. Sensor-based management of container nursery crops irrigated with fresh or saline water. Agricultural Water Management 213: 49-61. https://doi.org/10.1016/j.agwat.2018.09.054.

Incrocci, L., P. Marzialetti, G. Incrocci, J. Balendonck, S. Spagnol and A. Pardossi, 2010. Application of WET sensor for management of reclaimed wastewater irrigation in container-grown ornamentals (Prunus laurocerasus L.). The 3rd International Symposium on Soil Water Measurement Using Capacitance, Impedance and TDT. April 7-9, 2010. Murcia, Spain. Available at: http://edepot.wur.nl/139556.

Janssen, E.G.O.N., J. Ruigrok, A. van 't Ooster and J. de Wit, 2006. Buffering of geothermal heat and other renewable energy sources. TNO-report 2006-D-R0644, 56pp [In Dutch]. Available at: http://edepot.wur.nl/22107.

Jarvis, P.G., 1976. The interpretation of the variations in leaf water potential and stomatal conductance found in canopies in the field. Philosophical Transactions of the Royal Society of London B 273: 593-610. https://doi.org/10.1098/rstb.1976.0035.

Jones, H.G., N. Archer, E. Rotenberg and R. Casa, 2003. Radiation measurement for plant ecophysiology. Journal of Experimental Botany 54(384): 879-889. https://doi.org/10.1093/jxb/erg116.

Jongschaap, R.E.E., A. De Gelder, E. Heuvelink, F.L.K. Kempkes and C. Stanghellini, 2009. Nieuwe vormen van verwarming van gewas(delen). Wageningen UR Glastuinbouw, Wageningen, the Netherlands.

Kacira, M., S. Sase and L. Okushima, 2004. Effects of side Vents and span numbers on wind-induced natural ventilation of a gothic multi-span greenhouse. JARQ 38(4): 227-233.

Karlsson, M.G., R.D. Heins, J.O. Gerberick and M.E. Hackmann, 1991. Temperature driven leaf unfolding rate in hibiscus rosa-sinensis. Scientia Horticulturae 45(3-4): 323-331. https://doi.org/10.1016/0304-4238(91)90078-D.

Katsoulas, N., A. Sapounas, H.F. De Zwart, J.A. Dieleman and C. Stanghellini, 2015. Reducing ventilation requirements in semi-closed greenhouses increases water use efficiency. Agricultural Water Management 156: 90-99.

Katsoulas, N., E. Kitta, C. Kittas, I.L. Tsirogiannis, E. Stamati and D. Sayvas, 2006. Greenhouse cooling by a fog system: effects on microclimate and on production and quality of a soilless pepper crop. Acta Horticulturae 719: 455-462.

Kaya, C., D. Higgs, K. Saltali and O. Gezerel, 2007. Response of strawberry grown at high salinity and alkalinity to supplementary potassium. Journal of Plant Nutrition 25(7): 1415-1427. https://doi.org/10.1081/PLN-120005399.

Kempkes, F., H.F. De Zwart, P. Munoz, J.I. Montero, F.J. Baptista, F. Giuffrida, C. Gilli, A. Stepowska and C. Stanghellini, 2017b. Heating and dehumidification in production greenhouses at northern latitudes: energy use. Acta Horticulturae 1164: 445-452. https://doi.org/10.17660/ActaHortic.2017.1164.58.

Kempkes, F.L.K., J. Janse and S. Hemming, 2017a. Greenhouse concept with high insulating cover by combination of glass and film: design and first experimental results. Acta Horticulturae 1170: 469-486. https://doi.org/10.17660/ActaHortic.2017.1170.58.

Kim, H.J., M.Y. Lin, C.A. Mitchell, 2019. Light spectral and thermal properties govern biomass allocation in tomato through morphological and physiological changes. Environmental en Experimental Botany 157: 228-240. https://doi.org/10.1016/j.envexpbot.2018.10.019.

Kiyriacou, P.A., 2010. Biomedical sensors: temperature sensor technology. In: Biomedical sensors, Momentum Press, New York, NY, USA, pp. 1-38.

Kool, M.T.N., 1996. System development of glasshouse roses. PhD-thesis, Wageningen University, Wageningen, the Netherlands, 143 pp. Available at: https://edepot.wur.nl/210516.

Körner, O. and H. Challa, 2003. Design for an improved temperature integration concept in greenhouse cultivation. Computers and Electronics in Agriculture 39: 39-59. https://doi.org/10.1016/S0168-1699(03)00006-1.

Körner, Ch., J.A. Scheel and H. Bauer, 1979. Maximum leaf diffusive conductance in vascular plants. Photosynthetica 13(1): 45-82.

Körner, O., E. Heuvelink and Q. Niu, 2009. Quantification of temperature, CO_2, and light effects on crop photosynthesis as a basis for model-based greenhouse climate control. Journal of Horticultural Science and Biotechnology 84(2): 233-239. https://doi.org/10.1080/14620316.2009.11512510.

Kozai, T., 2013. Plant factory in Japan – current situation and perspectives. Chronica Horticulturae 53: 8-11.

Kozai, T., C. Kubota, M. Takagaki and T. Maruo, 2015. Greenhouse environment control technologies for improving the sustainability of food production. Acta Horticulturae 1107: 1-13.

Kozai, T., G. Niu and M. Takagaki, 2015. Plant factory: an indoor vertical farming system for efficient quality food production. Academic Press, Cambridge, MA, USA, 432 pp.

Krishna Bahadur, K.C., G.M. Dias, A. Veeramani, C.J. Swanton, D. Fraser, D. Steinke, E. Lee, H. Wittman, J.M. Farber, K. Dunfield, K. McCann, M. Anand, M. Campbell, N. Rooney, N.E. Raine, R. Van Acker, R. Hanner, S. Pascoal, S. Sharif, T.G. Benton and E.D.G. Fraser, 2018. When too much isn't enough: does current food production meet global nutritional needs? PLoS ONE 13(10): e0205683.

Kromdijk, J., F. Van Noort, S. Driever and T. Dueck, 2012. An enlightened view on protected cultivation of shade-tolerant pot-plants: benefits of higher light levels. Acta Horticulturae 956: 381-388. https://doi.org/10.17660/ActaHortic.2012.956.44.

KWIN, 2017. Quantitative Information on Dutch greenhouse horticulture 2016-2017. Report 5154, Wageningen University and Research Unit Greenhouse Horticulture, Wageningen, the Netherlands, 334 pp. Available at: https://www.wur.nl/en/newsarticle/Quantitative-Information-for-Greenhouse-Horticulture-KWIN-2016-2017.htm.

Langhans, R.W. and T.W. Tibbitts, 1997. Plant growth chamber handbook. Iowa Agricultural and Home Economics Experiment Station, Ames, IA, USA.

Larsen, R.U. and L. Persson, 1999. Modelling flower development in greenhouse chrysanthemum cultivars in relation to temperature and response group. Scientia Horticulturae 80(1-2): 73-89. https://doi.org/10.1016/S0304-4238(98)00219-2.

Lee, J.H., 2002. Analysis and simulation of growth and yield of cut chrysanthemum. PhD-thesis, Wageningen University, Wageningen, the Netherlands, 120 pp. Available at: https://wur.on.worldcat.org/oclc/907106667.

Lee, J.H., E. Heuvelink and H. Challa, 2002. A simulation study on the interactive effects of radiation and plant density on growth of cut chrysanthemum. Acta Horticulturae 593: 151-157. https://doi.org/10.17660/ActaHortic.2002.593.19.

Lee, J.H., J. Goudriaan and H. Challa, 2003. Using the expolinear growth equation for modelling crop growth in year-round cut chrysanthemum. Annals of Botany 92(5): 697-708. https://doi.org/10.1093/aob/mcg195.

Leonardi, C., S. Guichard and N. Bertin, 2000. High vapour pressure deficit influences growth, transpiration and quality of tomato fruits. Scientia Horticulturae 84(3-4): 285-296. https://doi.org/10.1016/S0304-4238(99)00127-2.

Li, Q. and C. Kubota, 2009. Effects of supplemental light quality on growth and phytochemicals of baby leaf lettuce. Environmental and Experimental Botany 67(1): 59-64. https://doi.org/10.1016/J.ENVEXPBOT.2009.06.011.

Li, S. and Willits, D., 2008. Comparing low-pressure and high-pressure fogging systems in naturally ventilated greenhouses. Biosystems Engineering 101(1): 69-77.

Li, T., E. Heuvelink and L.F.M. Marcelis, 2015. Quantifying the source-sink balance and carbohydrate content in three tomato cultivars. Frontiers in Plant Science 6: 416. https://doi.org/10.3389/fpls.2015.00416.

Li, T., E. Heuvelink, T.A. Dueck, J. Janse, G. Gort and L.F.M. Marcelis, 2014. Enhancement of crop photosynthesis by diffuse light: quantifying the contributing factors. Annals of Botany 114(1): 145-156. https://doi.org/10.1093/aob/mcu071.

Li, Y.L., C. Stanghellini and H. Challa, 2001. Effect of electrical conductivity and transpiration on production of greenhouse tomato. Scientia Horticulturae 88: 11-29. https://doi.org/10.1016/S0304-4238(00)00190-4.

Li, Y.L., L.F.M. Marcelis and C. Stanghellini, 2004. Plant water relations as affected by osmotic potential of the nutrient solution and potential transpiration in tomato (Lycopersicon esculentum L.). Journal of Horticultural Science & Biotechnology 79(2): 211-218. https://doi.org/10.1080/14620316.2004.11511750.

Limberger, J., T. Boxem, M. Pluymaekers, D. Bruhn, A. Manzella, P. Calcagno, F. Beekman, S. Cloetingh, and J.-D. Van Wees, 2018. Geothermal energy in deep aquifers: a global assessment of the resource base for direct heat utilization. Renewable and Sustainable Energy Reviews 82: 961-975.

López, J.C., C. Pérez, E. Baeza, J.J. Pérez-Parra, J.C. Garrido, G. Acien and S. Bonachela, 2008. Evaluation of the use of combustion gases originating from industry (SO2) as a contribution of CO_2 for greenhouse. Acta Horticulturae 797: 361-365. https://doi.org/10.17660/ActaHortic.2008.797.51.

Lu, N., T. Nukaya, T. Kamimura, D. Zhang, I. Kurimoto, M. Takagaki, T. Maruo, T. Kozai and W. Yamori, 2015. Control of vapor pressure deficit (VPD) in greenhouse enhanced tomato growth and productivity during the winter season. Scientia Horticulturae 197: 17-23. https://doi.org/10.1016/J.SCIENTA.2015.11.001.

Lund, J.B., T.J. Blom and J.M. Aaslyng, 2007. End-of-day lighting with different red/far-red ratios using light-emitting diodes affects plant growth of chrysanthemum. HortScience 42(7): 1609-1611. Available at: http://hortsci.ashspublications.org/content/42/7/1609.full.pdf.

Maas, E.V. and G.J. Hoffman, 1977. Crop salt tolerance – current assessment. Journal of the Irrigation and Drainage Division 103(2): 115-134.

MacNeal, B.L., J.D. Oster, R.J. Hatcher, 1970. Calculation of the electrical conductivity from solution composition data as an aid to *in situ* estimation of soil salinity. Soil Science 110: 405-414.

Marcelis, L.F.M., 1993. Effect of assimilate supply on the growth of individual cucumber fruits. Physiologia Plantarum 87(3): 313-320. https://doi.org/10.1111/j.1399-3054.1993.tb01736.x.

Marcelis, L.F.M., 1994. Fruit growth and dry matter partitioning in cucumber. PhD-thesis, Wageningen University, Wageningen, the Netherlands, 173 pp. Available at: https://wur.on.worldcat.org/oclc/906727182.

Marcelis, L.F.M., 1996. Sink strength as a determinant of dry matter partitioning in the whole plant. Journal of Experimental Botany 47: 1281-1291. https://doi.org/10.1093/jxb/47.Special_Issue.1281.

Marcelis, L.F.M. and J. Van Hooijdonk, 1999. Effect of salinity on growth, water use and nutrient use in radish (Raphanus Sativus L.). Plant and Soil 215(1): 57-64. https://doi.org/10.1023/A:1004742713538.

Marcelis, L.F.M., A.G.M. Broekhuijsen, E.M.F.M. Nijs, M.G.M. Raaphorst, 2006. Quantification of the growth response of light quantity of greenhouse-grown crops. Acta Horticulturae 711: 97-103. https://doi.org/10.17660/ActaHortic.2006.711.9.

Marcelis, L.F.M., A.G.M. Broekhuijsen, E.M.F.M. Nijs, M.G.M. Raaphorst and E. Meinen, 2006. Quantification of the growth response to light quantity of greenhouse grown crops. Acta Horticulturae 711: 97-104.

Marcelis, L.F.M., E. Heuvelink and J. Goudriaan, 1998. Modelling biomass production and yield of horticultural crops: a review. Scientia Horticulturae 74(1-2): 83-111. https://doi.org/10.1016/S0304-4238(98)00083-1.

Marshall, C., 2017. In Switzerland, a giant new machine is sucking carbon directly from the air. Available at: https://tinyurl.com/y6wzpb7o.

Martin, G.C. and G.E. Wilcox, 1963. Critical soil temperature for tomato plant growth. Soil Science Society of America Journal 27(5): 565. https://doi.org/10.2136/sssaj1963.03615995002700050028x.

Massa, D., L. Incrocci, R. Maggini, G. Carmassi, C.A. Campiotti and A. Pardossi, 2010. Strategies to decrease water drainage and nitrate emission from soilless cultures of greenhouse tomato. Agricultural Water Management 97: 971-980. https://doi.org/10.1016/j.agwat.2010.01.029.

Medrano, E., P. Lorenzo, M.C. Sanchez-Guerrero and J.I. Montero, 2005. Evaluation and modelling of greenhouse cucumber-crop transpiration under high and low radiation conditions. Scientia Horticulturae 105: 163-175. https://doi.org/10.1016/j.scienta.2005.01.024.

Meinen, E., T. Dueck, F. Kempkes and C. Stanghellini, 2018. Growing fresh food on future space missions: environmental conditions and crop management. Scientia Horticulturae 235: 270-278. https://doi.org/10.1016/j.scienta.2018.03.002.

Messelink, G., 2013. How to create a standing army of natural enemies in ornamental crops. Nursery/Floriculture Insect symposium. December 12, 2013. Watson Ville, FL, USA. Available at: https://ucanr.edu/sites/UCNFA/files/181226.pdf.

Milani, D., A. Qadir, A. Vassallo, M. Chiesa and A. Abbas, 2014. Experimentally validated model for atmospheric water generation using a solar assisted desiccant dehumidification system. Energy and Buildings 77: 236-246.

Misra, D. and S. Ghosh, 2018. Evaporative cooling technologies for greenhouses: a comprehensive review. Agricultural Engineering International: CIGR Journal 20(1): 1-15.

Mitchell, C.A., M.P. Dzakovich, C. Gomez, R. Lopez, J.F. Burr, R. Hernández, C. Kubota, C.J. Currey, Q. Meng and E.S. Runkle, 2015. Light-emitting diodes in horticulture. Horticultural Reviews 43: 1-87.

Moerkens, R., W. Van Lommel, R. Vanderbruggen and T. Van Delm, 2016. The added value of LED assimilation light in combination with high pressure sodium lamps in protected tomato crops in Belgium. Acta Horticulturae 1134: 119-124. https://doi.org/10.17660/ActaHortic.2016.1134.16.

Monteith, J.L., 1965. Evaporation and the environment. State and movement of water in living organisms. 19th Symposium of the Society for Experimental Biology, pp. 205-234. Available at: https://www.unc.edu/courses/2007fall/geog/801/001/www/ET/Monteith65.pdf.

Monteith, J.L. and M.H. Unsworth, 2014. Principles of environmental physics, 4th edition. Academic Press, Cambridge, MA, USA, 401 pp. https://doi.org/10.1016/C2010-0-66393-0.

Montero, J.I., A.A. Anton, P. Muñoz and P. Lorenzo, 2001. Transpiration from geraniums grown under high temperatures and low humidities in greenhouses. Agricultural and Forest Meteorology 107: 323-332. https://doi.org/10.1016/S0168-1923(01)00215-5.

Montero, J.I., P. Muñoz, E. Baeza and C. Stanghellini, 2017. Ongoing developments in greenhouse climate control. Acta Horticulturae 1182: 1-14. https://doi.org/10.17660/ActaHortic.2017.1182.1.

Morrow, R.C., 2008. LED in horticulture. HortScience 43: 1947-1950.

Mortensen, L.M. and H.R. Gislerød, 1999. Influence of air humidity and lighting period on growth, vase life and water relations of 14 rose cultivars. Scientia Horticulturae 82(3-4): 289-298. https://doi.org/10.1016/S0304-4238(99)00062-X.

Mortensen, L.M., R.I. Pettersen and H.R. Gislerød, 2007. Air humidity variation and control of vase life and powdery mil-dew in cut roses under continuous lighting. European Journal of Horticultural Science 6: 1611-4426. Available at: http://www.pubhort.org/ejhs/2007/file_491532.pdf.

Mulholland, B.J., M. Fussell, R.N. Edmondson, A.J. Taylor, J. Basham, J.M.T. Mckee and N. Parsons, 2002. The effect of split-root salinity stress on tomato leaf expansion, fruit yield and quality. Journal of Horticultural Science and Biotechnology 77(5): 509-519. https://doi.org/10.1080/14620316.2002.11511531.

Munns, R., 2002. Comparative physiology of salt and water stress. Plant, Cell and Environment 25(2): 239-250. https://doi.org/10.1046/j.0016-8025.2001.00808.x.

Muñoz, P., J.I. Montero, D. Piscia and A. Antón, 2011. Air exchange monitor. Euphoros calculation tools. Available at: https://www.wur.nl/en/Research-Results/Projects-and-programmes/Euphoros/Calculation-tools/Air-exchange-monitor.htm.

Munters, 2018. Evaporative cooling solutions, the Desicool. Available at: https://www.munters.com/en/solutions/cooling/.

Murage, E.N. and M. Masuda, 1997. Response of pepper and eggplant to continuous light in relation to leaf chlorosis and activities of antioxidative enzymes. Scientia Horticulturae 70(4): 269-279. https://doi.org/10.1016/S0304-4238(97)00078-2.

Nederhoff, E.M., 1994. Effects of CO_2 concentration on photosynthesis, transpiration and production of greenhouse fruit vegetable crops. PhD-thesis, Wageningen University, Wageningen, the Netherlands, 227 pp. Available at: https://edepot.wur.nl/206000.

Nederhoff, E.M., 2004. Carbon dioxide enrichment. Practical Hydroponics and Greenhouses Magazine. May/June edition. Casper Publications Pty Ltd, Narrabee, Australia, pp. 50-59.

Nederhoff, E.M., M. Warmenhoven and T. Dueck. 2009. Lichtmeetprotocol – afspraken voor lichtmetingen in proef met LED en SON-T belichting in kassen bij WUR in Bleiswijk in 2009/2010. Wageningen UR Greenhouse Horticulture, Wageningen, the Netherlands.

Nelson, J.A. and B. Bugbee, 2014. Economic analysis of greenhouse lighting: light emitting diodes vs high intensity discharge fixtures. PLoS ONE 9(6): e99010. https://doi.org/10.1371/journal.pone.0099010.

Ouzounis, T., E. Rosenqvist and C.-O. Ottosen, 2015. Spectral effects of artificial light on plant physiology and secondary metabolism: a review. HortScience 50(8): 1128-1135. Available at: http://www.photobiology.info/.

Öztürk, H.H., 2003. Evaporative cooling efficiency of a fogging system for greenhouses. Turkish Journal of Agriculture and Forestry 27(1): 49-57.

Papadakis, G., A. Frangoudakis and S. Krytsis, 1992. Mixed, forced and free convection heat transfer at the greenhouse cover. Journal of Agricultural Engineering Research 51: 191-205. https://doi.org/10.1016/0021-8634(92)80037-S.

Paradiso, R., E. Meinen, J.F.H. Snel, P. De Visser, W. Van Ieperen, S.W. Hogewoning and L.F.M. Marcelis, 2011. Spectral dependence of photosynthesis and light absorptance in single leaves and canopy in rose. Scientia Horticulturae 127(4): 548-554. https://doi.org/10.1016/J.SCIENTA.2010.11.017.

Pardossi, A., G. Carmassi, C. Diara, L. Incrocci, R. Maggini and D. Massa, 2011. Fertigation and substrate management in closed soilless culture. EU-FP7 EUPHOROS, Grant 211457, deliverable 15. Available at: www.euphoros.wur.nl/Reports.

Pardossi, A., L. Incrocci, G. Incrocci, F. Malorgio, P. Battista, L. Bacci, B. Rapi, P. Marzialetti, J. Hemming and J. Balendonck, 2009. Root zone sensors for irrigation management in intensive agriculture. Sensors 9(4): 2809-2835. https://doi.org/10.3390/s90402809.

Park, Y. and E.S. Runkle, 2017. Far-red radiation promotes growth of seedlings by increasing leaf expansion and whole-plant net assimilation. Environmental and Experimental Botany 136: 41-49. https://doi.org/10.1016/J.ENVEXPBOT.2016.12.013.

Paton, C. and P. Davies, 1996. The seawater greenhouse for arid lands. In: Mediterranean Conference on Renewable Energy Sources for water Production. Santorini, Greece.

Paul, O.U., I.H. John, I. Ndubuisi, A. Peter and O. Godspower, 2015. Calorific value of palm oil residues for energy utilisation. International Journal of Engineering Innovation and Research 4(4): 664-667.

Penman, H.L., 1948. Natural evaporation from open water, bare soil and grass. Proceedings of the Royal Society A: 193(1032): 120-145. https://doi.org/10.1098/rspa.1948.0037 and available at: http://rspa.royalsocietypublishing.org/content/193/1032/120.

Pérez Parra, J., E. Baeza, J.I. Montero and B.J. Bailey, 2004. Natural ventilation of parral greenhouses. Biosystems Engineering 87(3): 355-366. https://doi.org/10.1016/j.biosystemseng.2003.12.004.

Persoon, S. and S.W. Hogewoning, 2014. Onderzoek naar de fundamenten van energiebesparing in de belichte teelt. Projectnummer 14764.04. Productschap Tuinbouw. Available at: https://tinyurl.com/yas7y553.

Petersen, K.K., J. Willumsen and K. Kaack, 1998. Composition and taste of tomatoes as affected by increased salinity and different salinity sources. Journal of Horticultural Science and Biotechnology 73(2): 205-215. https://doi.org/10.1080/14620316.1998.11510966.

Pierik, R. and M. De Wit, 2014. Shade avoidance: phytochrome signalling and other aboveground neighbour detection cues. Journal of Experimental Botany 65(11): 2815-2824. https://doi.org/10.1093/jxb/ert389.

Pierik, R., M.L.C. Cuppens, L.A.C.J. Voesenek and E.J.W. Visser, 2004. Interactions between ethylene and gibberellins in phytochrome-mediated shade avoidance responses in tobacco. Plant Physiology 136(2): 2928-2936. https://doi.org/10.1104/pp.104.045120.

Piscia, D., J.I. Montero, B. Bailey, P. Muñoz and A. Oliva, 2013. A new optimisation methodology used to study the effect of cover properties on night-time greenhouse climate. Biosystems Engineering 116(2): 130-143. https://doi.org/10.1016/j.biosystemseng.2013.07.005.

Poorter, H. and M. Pérez-Soba, 2001. The growth response of plants to elevated CO_2 under non-optimal environmental conditions. Oecologia 129(1): 1-20. https://doi.org/10.1007/s004420100736.

Portree, J., 1996. Greenhouse vegetable production guide. British Columbia Ministry of Agriculture, Fisheries and Food, Abbotsford, British Columbia, Canada, 117 pp.

Qian, T., 2017. Crop growth and development in closed and semi-closed greenhouses. Wageningen University, Wageningen, the Netherlands.

Qian, T., J.A. Dieleman, A. Elings and L.F.M. Marcelis, 2012. Leaf photosynthetic and morphological responses to elevated CO_2 concentration and altered fruit number in the semi-closed greenhouse. Scientia Horticulturae 145: 1-9. https://doi.org/10.1016/J.SCIENTA.2012.07.015.

Qian, T., J.A. Dieleman, A. Elings, A. De Gelder, L.F.M. Marcelis and O. Van Kooten, 2011. Comparison of climate and production in closed, semi-closed and open greenhouses. Acta Horticulturae 893: 807-814. https://doi.org/10.17660/ActaHortic.2011.893.88.

Raaphorst, M., 2013. Praktijkexperiment ontvochtigen met zouten: gebruik en regeneratie van hygroscopisch zout in een kasproef bij Lans Zeeland. Wageningen UR Glastuinbouw, Wageningen, the Netherlands.

Rabobank International, 2000. The Mexican cut flower market. Ministry of Agriculture, Environment and Fisheries, The Hague, the Netherlands, 74 pp. Available at: http://edepot.wur.nl/118595.

Rabobank, 2018a. World vegetable map 2018. RaboResearch Food & Agribusiness. Available at: https://research.rabobank.com/far/en/sectors/regional-food-agri/world_vegetable_map_2018.html.

Rabobank, 2018b. Vertical farming in the Netherlands. Lambert van Horen, lecture 27 Jun 2018, Venlo, the Netherlands.

Raupach, M.R. and J.J. Finnigan, 1988. Single-layer models of evaporation from plant canopies are incorrect but useful, whereas multilayer models are correct but useless. Functional Plant Biology 15: 705-716. https://doi.org/10.1071/PP9880705.

Raviv, M. and J. Heinrich Lieth, 2008. Soilless culture: theory and practice. Elsevier, Amsterdam, the Netherlands.

Rijsdijk, A.A. and J.V.M. Vogelezang, 2000. Temperature integration on a 24-hour base: a more efficient climate control strategy. Acta Horticulturae 519: 163-170. https://doi.org/10.17660/ActaHortic.2000.519.16.

Roderick, M.L., 1999. Estimating the diffuse component from daily and monthly measurements of global radiation. Agricultural and Forest Meteorology 95: 169-185. https://doi.org/10.1016/S0168-1923(99)00028-3.

Ross, J., 1975. Radiative transfer in plant communities. In: Monteith, J.L. (ed.) Vegetation and the atmosphere. Academic Press, London, UK, pp. 13-55.

Roveti, D.K., 2001. Choosing a humidity sensor: a review of three technologies. Sensors Magazine: SensorsOnline. Available at: https://www.sensorsmag.com/components/choosing-a-humidity-sensor-a-review-three-technologies.

Sablani, S., M. Goosen, C. Paton, W. Shayya and H. Al-Hinai, 2003. Simulation of fresh water production using a humidification-dehumidification seawater greenhouse. Desalination 159(3): 283-288.

Sánchez-Guerrero, M.C., J.F. Alonso, P. Lorenzo and E. Medrano, 2010. Manejo Del Clima En El Invernadero Mediterráneo. Instituto de Investigación y Formación Agraria y Pesquera (IFAPA), Consejería de Agricultura y Pesca. Andalucía, Spain, 130 pp.

Sato, S., M.M. Peet and J.F. Thomas, 2002. Determining critical pre- and post-anthesis periods and physiological processes in Lycopersicon Esculentum mill. Exposed to moderately elevated temperatures. Journal of Experimental Botany 53(371): 1187-1195. https://doi.org/10.1093/jexbot/53.371.1187.

Savvides, A., W. Van Ieperen, J.A. Dieleman and L.F.M. Marcelis, 2013. Meristem temperature substantially deviates from air temperature even in moderate environments: is the magnitude of this deviation species-specific? Plant, Cell & Environment 36(11): 1950-1960. https://doi.org/10.1111/pce.12101.

Schapendonk, A.H.C.M., H. Challa, P.W. Broekharst and A.J. Udink Ten Cate, 1984. Dynamic climate control; an optimization study for earliness of cucumber production. Scientia Horticulturae 23(2): 137-150. https://doi.org/10.1016/0304-4238(84)90017-7.

Schoenmakers, H., 2018. Opbrengst berekenen. Kenniscentrum zonnepanelen. Available at: https://www.bespaarbazaar.nl/kenniscentrum/zonnepanelen/financieel/zonnepanelen-opbrengst/

Seginer, I., R. Linker, F. Buwalda, G. Van Straten and P. Bleyaert, 2004. The NICOLET lettuce model: a theme with variations. Acta Horticulturae 654: 71-78. https://doi.org/10.17660/ActaHortic.2004.654.7.

Shannon, M.C. and C.M. Grieve, 1998. Tolerance of vegetable crops to salinity. Scientia Horticulturae 78(1-4): 5-38. https://doi.org/10.1016/S0304-4238(98)00189-7.

Shibuya, T., R. Endo, Y. Kitaya and S. Hayashi, 2016. Growth analysis and photosynthesis measurements of cucumber seedlings grown under light with different red to far-red ratios. HortScience 51(&): 843-846.

Simha, R., 2012. Willis H carrier. Resonance 17(2): 117-138.

Singh, D., C. Basu, M. Meinhardt-Wollweber and B. Roth, 2015. LEDs for energy efficient greenhouse lighting. Renewable and Sustainable Energy Reviews 49: 139-147.

Singh, R.P. and D.R. Heldman, 2014. Psychrometrics. In: Singh, R.P. and Heldman D.R. (eds.) Introduction to food engineering, 5th edition. Academic Press, San Diego, CA, USA, pp. 593-616.

Sonneveld, C., 2000. Effects of salinity on substrate grown vegetables and ornamentals in greenhouse horticulture. PhD-thesis, Wageningen University, Wageningen, the Netherlands, 151 pp. Available at: https://wur.on.worldcat.org/oclc/906880136.

Sonneveld, C. and Voogt, W., 2009. Plant nutrition of greenhouse crops. Springer, Dordrecht, the Netherlands. https://doi.org/10.1007/978-90-481-2532-6.

Sonneveld, C., W. Voogt and L. Spaans, 1999. A universal algorithm for calculation of nutrient solutions. Acta Horticulturae 481: 331-340. https://doi.org/10.17660/ActaHortic.1999.481.38.

Sonneveld, P.J., G.L.A.M. Swinkels, F. Kempkes, J.B. Campen and G.P.A. Bot, 2006. Greenhouse with an integrated nir filter and a solar cooling system. Acta Horticulturae 719: 123-130.

Stanghellini, C., 1987. Transpiration of greenhouse crops: an aid to climate management. PhD-thesis, Wageningen University, Wageningen, the Netherlands, 150 pp. Available at: http://edepot.wur.nl/202121,

Stanghellini, C., 2014. Horticultural production in greenhouses: efficient use of water. Acta Horticulturae 1034: 25-32. https://doi.org/10.17660/ActaHortic.2014.1034.1.

Stanghellini, C. and F.L.K. Kempkes, 2004. Energiebesparing door verdampingsbeperking via klimaatregeling. Rapport 309, Wageningen UR Agrotechnology & Food Innovations, Wageningen, the Netherland, 31 pp. Available at: http://edepot.wur.nl/43211.

Stanghellini, C. and F.L.K. Kempkes, 2008. Steering of fogging: control of humidity, temperature or transpiration? Acta Horticulturae 797: 61-67. https://doi.org/10.17660/ActaHortic.2008.797.6.

Stanghellini, C. and H.F. De Zwart, 2015. Adaptive greenhouse design: what is best for growing tomatoes in northern Africa? International Symposium on New Technologies and Management for Greenhouses – Greensys 2015. July 19-23, 2015. Evora, Portugal. Available at: http://edepot.wur.nl/386128.

Stanghellini, C. and J.A. Bunce, 1993. Response of photosynthesis and conductance to light, CO_2, temperature and humidity in tomato plants acclimated to ambient and elevated CO_2. Photosynthetica 29(4): 487-497.

Stanghellini, C. and T. De Jong, 1995. A model of humidity and its application in a greenhouse. Agricultural and Forest Meteorology 76(2): 129-148. https://doi.org/10.1016/0168-1923(95)02220-R.

Stanghellini, C., A. Pardossi and N. Sigrimis, 2007. What limits the application of wastewater and/or closed cycle in horticulture? Acta Horticulturae 747: 323-330. https://doi.org/10.17660/ActaHortic.2007.747.39.

Stanghellini, C., A. Pardossi, F.L.K. Kempkes and L. Incrocci, 2005. Closed water loops in greenhouses: effect of water quality and value of produce. Acta Horticulturae 691: 233-241. https://doi.org/10.17660/ActaHortic.2005.691.27.

Stanghellini, C., D. Jianfeng and F.L.K. Kempkes, 2011. Effect of near-infrared-radiation reflective screen materials on ventilation requirement, crop transpiration and water use efficiency of a greenhouse rose crop. Biosystems Engineering 110(3): 261-271. https://doi.org/10.1016/j.biosystemseng.2011.08.002.

Stanghellini, C., F.L.K. Kempkes and L. Incrocci, 2009. Carbon dioxide fertilization in Mediterranean greenhouses: when and how is it economical? Acta Horticulturae 807: 135-142. https://doi.org/10.17660/ActaHortic.2009.807.16.

Stanghellini, C., F.L.K. Kempkes and P. Knies, 2003b. Enhancing environmental quality in agricultural systems. Acta Horticulturae 609: 277-283.

Stanghellini, C., F.L.K. Kempkes, E. Heuvelink, A. Bonasia and A. Karas, 2003a. Water and nutrient uptake of sweet pepper and tomato as (un)affected by watering regime and salinity. Acta Horticulturae 614: 591-598. https://doi.org/10.17660/ActaHortic.2003.614.88.

Stanghellini, C., J. Bontsema, A. De Koning and E.J. Baeza, 2012b. An algorithm for optimal fertilization with pure carbon dioxide in greenhouses. Acta Horticulturae 952: 119-124. https://doi.org/10.17660/ActaHortic.2012.952.13.

Stanghellini, C., L. Incrocci, J.C. Gazquez and B. Dimauro, 2008. Carbon dioxide concentration in Mediterranean greenhouses: how much lost production? Acta Horticulturae 801: 1541-1549. https://doi.org/10.17660/ActaHortic.2008.801.190.

Stanghellini, C., M.A. Bruins, V. Mohammadkhani, G.L.A.M. Swinkels, P.J. Sonneveld, 2012a. Effect of condensation on light transmission and energy budget of seven greenhouse cover materials. Acta Horticulturae 952: 249-254. https://doi.org/10.17660/ActaHortic.2012.952.30.

Sullivan, J.L., C.E. Clark, J. Han and M. Wang, 2010. Life-cycle analysis results of geothermal systems in comparison to other power systems. Center for Transportation Research, Energy Systems Division, Argonnen National Laboratory, Lemont, IL, USA.

Sutar, R.F. and G.N. Tiwari, 1995. Analytical and numerical study of a controlled-environment agricultural system for hot and dry climatic conditions. Energy and Buildings 23(1): 9-18.

Taiz, L. and E. Zeiger, 2010. Plant physiology, 5th edition. Sinauer Associates, Sunderland, MA, USA.

Takagi, M., H.A. El-Shemy, S. Sasaki, S. Toyama, S. Kanai, H. Saneoka and K. Fujita, 2009. Elevated CO_2 concentration alleviates salinity stress in tomato plant. Acta Agriculturae Scandinavica, Section B – Plant Soil Science 59(1): 87-96. https://doi.org/10.1080/09064710801932425.

Tarnavas, D., A.N.M. de Koning, I. Tsafaras, C. Stanghellini and J.A. Gonzalez, 2020. Practical implementation and evaluation of optimal carbon dioxide supply control. Acta Horticulturae, 1271: 193-197.

Thimijan, R.W. and R.D. Heins, 1983. Photometric, radiometric, and quantum light units of measure: a review of procedures for interconversion. HortScience 18: 818-822.

Thompson, R.B., C. Martínez, M. Gallardo, J.R. Lopez-Toral, M.D. Fernandez and C. Gimenez, 2006. Management factors contributing to nitrate leaching loss from a greenhouse-based intensive vegetable production system. Acta Horticulturae 700: 179-184. https://doi.org/10.17660/ActaHortic.2006.700.29.

Thompson, R.B., I. Delcour, E. Berkmoes and E. Stavridou, 2018. The fertigation bible. Available at: http://www.fertinnowa.com/the-fertigation-bible/.

Thornley, J.H.M., 1976. Mathematical models in plant physiology: a quantitative approach to problems in plant and crop physiology. Academic Press (Inc.), London, UK, 318 pp. Available at: https://wur.on.worldcat.org/oclc/2129632.

Tiwari, A., H. Dassen and E. Heuvelink, 2007. Selection of sweet pepper (Capsicum Annuum L.) genotypes for parthenocarpic fruit growth. Acta Horticulturae 761: 135-140. https://doi.org/10.17660/ActaHortic.2007.761.16.

Trouwborst, G., S.W. Hogewoning, J. Harbinson and W. Van Ieperen, 2011. The influence of light intensity and leaf age on the photosynthetic capacity of leaves within a tomato canopy the influence of light intensity and leaf age on the photosynthetic capacity of leaves within a tomato canopy. Journal of Horticultural Science and Biotechnology 86(4): 403-407. https://doi.org/10.1080/14620316.2011.11512781.

Tsafaras, I. and De Koning, A.N.M., 2017. Real-time application of crop transpiration and photosynthesis models in greenhouse process control. Acta Horticulturae 1154: 65-72.

Tüzel, I.H., Y. Tüzel, M.K. Meric, R. Whalley and G. Lock, 2009. Response of cucumber to deficit irrigation. Acta Horticulturae 807: 259-264. https://doi.org/10.17660/ActaHortic.2009.807.34.

US Standard Atmosphere, 1976. US standard atmosphere 1976. US Government Printing Office, Washington, DC, USA.

Van Beveren, P.J.M., J. Bontsema, G. Van Straten and E.J. Van Henten, 2015. Minimal heating and cooling in a modern rose greenhouse. Applied Energy 137: 97-109. https://doi.org/10.1016/j.apenergy.2014.09.083.

Van Ieperen, W., 1996. Consequences of diurnal variation in salinity on water relations and yield of tomato. PhD-thesis, Wageningen University, Wageningen, the Netherlands, 176 pp.

Van Ieperen, W., 2016. Plant growth control by light spectrum: fact or fiction? Acta Horticulturae 1134: 19-24. https://doi.org/10.17660/ActaHortic.2016.1134.3.

Van Kooten, O., E. Heuvelink C. Stanghellini, 2008. New development in greenhouse technology can mitigate the water shortage problem of the 21st century. Acta Horticulturae 767: 45-52.

Van Noort, F., W. Kromwijk, J. Snel, M. Warmenhoven, E. Meinen, T. Li, F. Kempkes and L. Marcelis, 2013. 'Grip op licht' bij potanthurium en bromelia: meer energie besparing bij het nieuwe telen potplanten met diffuus licht en verbeterde monitoring meer natuurlijk. Wageningen UR Glastuinbouw, Rapport 1287. Wageningen UR Glastuinbouw, Wageningen, the Netherlands. Available at: http://edepot.wur.nl/295313.

Van Os, E.A. and C. Blok, 2016. Disinfection in hydroponic systems and hygiene. Technical Information sheet No.6. Available at: http://edepot.wur.nl/403803.

Van 't Ooster, A., E.J. Van Henten, E.G.O.N. Janssen and H.R.M. Bongaerts, 2008. Use of supplementary lighting top screens and effects on greenhouse climate and return on investment. Acta Horticulturae 801: 645-652. https://doi.org/10.17660/ActaHortic.2008.801.74.

Vanthoor, B.H.E., 2011. A model-based greenhouse design method. PhD-thesis, Wageningen University, Wageningen, the Netherlands, 307 pp. Available at: http://edepot.wur.nl/170301.

Vanthoor, B.H.E., J.C. Gázquez, J.J. Magán, M.N.A. Ruijs, E. Baeza, C. Stanghellini, E.J. Van Henten and P.H.B. De Visser, 2012. A methodology for model-based greenhouse design: part 4, economic evaluation of different greenhouse designs: a Spanish case. Biosystems Engineering 111: 336-349. https://doi.org/10.1016/j.biosystemseng.2011.12.008.

Velez-Ramirez, A.I., W. Van Ieperen, D. Vreugdenhil, P.M.J.A. Van Poppel, E. Heuvelink and F.F. Millenaar, 2014. A single locus confers tolerance to continuous light and allows substantial yield increase in tomato. Nature Communications 5(1): 4549. https://doi.org/10.1038/ncomms5549.

Verkerke, W., C. De Kreij, J. Janse, 1993. Keukenzout maakt zacht, maar lekker. Groenten en Fruit/Glasgroenten 51: 14-15.

Vermeulen, P. and F. Van Wijmeren, 2013. Innovation network energy saving systems East Brabant (INES OB): collaborative exploration of the energy landscape. Wageningen UR Glastuinbouw, Bleiswijk the Netherlands. [in Dutch]

Von Zabeltitz, C., 1986. Gewächshäuser, planung und bau. Verlag Eugen-Ulmer, Germany, 284 pp.

Voogt, W., 2014. Soil fertility management in organic greenhouse crops; a case study on fruit vegetables. Acta Horticulturae 1041: 21-35. https://doi.org/10.17660/ActaHortic.2014.1041.1.

Voogt, W. and E.A. Van Os, 2012. Strategies to manage chemical water quality related problems in closed hydroponic systems. Acta Horticulturae 927: 949-955. https://doi.org/10.17660/ActaHortic.2012.927.117.

Voogt, W., G.-J. Swinkels and E. Van Os, 2012. 'Waterstreams': a model for estimation of crop water demand, water supply, salt accumulation and discharge for soilless crops. Acta Horticulturae 957: 123-130. https://doi.org/10.17660/ActaHortic.2012.957.13.

Voogt, W., J.A. Kipp, R. De Graaf and L. Spaans, 2000. A fertigation model for glasshouse crops grown in soil. Acta Horticulturae 537: 495-502.

Wollaeger, A.M. and E.S. Runkle, 2014. Growing seedlings under LEDs. Greenhouse Grower: 80-85. Available at: https://tinyurl.com/yawuylsy.

Wubs, A.M., 2010. Towards stochastic simulation of crop yield: a case study of fruit set in sweet pepper. PhD-thesis, Wageningen University, Wageningen, the Netherlands. Available at: http://edepot.wur.nl/150654.

Xu, C. and B. Mou, 2015. Evaluation of lettuce genotypes for salinity tolerance. HortScience 50(10): 1441-1446. Available at: http://hortsci.ashspublications.org/content/50/10/1441.full.

Yang, C., L. Jia, C. Chen, G. Liu and W. Fang, 2011. Bio-oil from hydro-liquefaction of Dunaliella salina over Ni/REHY catalyst. Bioresource Technology 102(6): 4580-4584.

About the authors

Cecilia Stanghellini
Senior Scientist at Wageningen University & Research

Cecilia Stanghellini, an Italian physicist with a Dutch PhD in Agricultural and Environmental Sciences, has been working in Wageningen for over 30 years, on greenhouse technology. Main topics are: greenhouse climate simulation and management; crop yield and resource use efficiency; economy and environmental impact of greenhouse crops. She has lead several large international projects, on efficient greenhouse vegetable production and capacity building. She teaches 'Greenhouse Technology' at Wageningen University and is the scientific coordinator of the yearly 'Wageningen Summerschool on Greenhouse Horticulture'. She is [co]author of some 200 publications, is fluent in 5 languages and has been invited as lecturer in 5 continents.

Email: cecilia.stanghellini@wur.nl
Internet (url): www.wur.eu/greenhousehorticulture

Bert van 't Ooster
Lecturer at Wageningen University & Research

Bert van 't Ooster has a long record of experience in education and scientific research in Biosystems Engineering at Wageningen University. He is involved in many courses and supervised over 150 MSc thesis students. Currently he teaches Greenhouse Technology, Biosystems Design, Building Physics and Climate Engineering, Livestock Technology, Engineering problem Solving, and Research methods. His main topics in research are modelling and design of agricultural systems, systems operations, continuous time and discrete event systems. In greenhouses technology he (co)developed simulation models that are used in research and in education. He is (co-)author of more than 50 publications in refereed scientific journals, conference proceedings, books, lecture books and professional magazines.

Email: bert.vantooster@wur.nl
Internet (url): www.fte.wur.nl

Ep Heuvelink

Associate Professor at Wageningen University & Research

Ep Heuvelink has over 30 years of experience in scientific research and education in greenhouse crops. He is associate professor in the Horticulture and Product Physiology group of Wageningen University & Research. His expertise is greenhouse crop physiology and crop simulation. He supervised over 120 MSc thesis students and is co-promotor of 22 PhD students. Ep is frequently invited as a keynote speaker at international scientific symposia and teaches advanced intensive courses on greenhouse production, crop physiology and crop modelling all over the world. He is a teacher in the 'Wageningen Summerschool on Greenhouse Horticulture' and (co-)authored 94 papers in refereed scientific journals, over 150 papers in professional journals and 5 books.

Email: ep.heuvelink@wur.nl
Internet (url): www.hpp.wur.nl

List of boxes

Chapter 1 – Introduction

Box 1.1. Various types of crop protection and what they do.   21

Chapter 2 – Crop as a production machine

Box 2.1. Example 1: the time from anthesis to a harvest-ready tomato.   30

Box 2.2. Example 2: radish, time from emergence to harvest.   31

Box 2.3. Think like a detective.   37

Box 2.4. Formulae for the photosynthesis-light response.   42

Box 2.5. Photosynthesis in analogy to Ohm's law.   43

Box 2.6. From CO_2 assimilation to fresh yield.   47

Box 2.7. Light use efficiency: how many g per mol light?   48

Box 2.8. Simulation of LAI based on daily crop growth rate.   50

Box 2.9. Example of calculating partitioning.   52

Chapter 3 – Radiation, greenhouse cover and temperature

Box 3.1. The energy of radiation: the link between Joules and micromoles.   56

Box 3.2. The human eye versus the plant.   57

Box 3.3. How do we measure radiation?   58

Box 3.4. Light-diffusing covers: what scattering does and what material
technology can do.   63

Box 3.5. NIR filters: absorption or reflection?   64

Box 3.6. Linearization.   66

Box 3.7. How do we measure net radiation?   67

Box 3.8. A traditional Chinese solar greenhouse.   71

Box 3.9. Energy saving: balancing insulation and light transmission.   72

Chapter 4 – Properties of humid air and physics of air treatment
Box 4.1. Measuring temperature and humidity in greenhouses.   76
Box 4.2. Molar mass of dry air and vapour.   79
Box 4.3. Example – calculation of enthalpy.   83
Box 4.4. Finding specific humidity at saturation.   84
Box 4.5. Physical basis of the energy balance based psychrometric equation.   90

Chapter 5 – Ventilation and mass balance
Box 5.1. Definitions of the ventilation rate.   102
Box 5.2. Local climate and ventilation capacity.   104
Box 5.3. Shading and ventilation.   108
Box 5.4. Ventilation rate and carbon dioxide balance.   110
Box 5.5. How to 'measure' ventilation.   111

Chapter 6 – Crop transpiration and humidity in greenhouses
Box 6.1. Boundary layer resistance.   116
Box 6.2. Stomatal resistance.   117
Box 6.3. The slope of saturated vapour functions.   118
Box 6.4. Net radiation of a crop.   120
Box 6.5. How can we measure crop transpiration?   123
Box 6.6. Is ventilation always the best humidity control?   128
Box 6.7. Air circulation.   129

Chapter 7 – Crop response to environmental factors
Box 7.1. Potted plants love light too.   134
Box 7.2. Light sum determines the production time of young plants.   135
Box 7.3. Contribution of supplementary light to total light integral. 136
Box 7.4. Supplementary light shortens the production time of young plants.   138
Box 7.5. Poor nutrient distribution in root substrate is not always a disaster.   159

Chapter 8 – Heating in climate-controlled greenhouses

Box 8.1. Basics of climate control.   166
Box 8.2. Calculation of the U-value by summing thermal resistances.   167
Box 8.3. Alternative for calculating the heat loss of dry and thin construction layers.   168
Box 8.4. Example of calculating heat capacity.   171
Box 8.5. Heating circuits within the greenhouse.   178
Box 8.6. The heat exchanger.   183
Box 8.7. Co-generator more efficient than central power plant.   185
Box 8.8 Buffer capacity and energy storage.   191

Chapter 9 – Cooling and dehumidification

Box 9.1. Compression heat pump.   203
Box 9.2. Absorption heat pump.   204
Box 9.3. Primary energy ratio.   207
Box 9.4. The benefit of reducing ventilation.   209

Chapter 10 – Supplementary lighting

Box 10.1. Advantages of LED over HPS lamps.   214
Box 10.2. Photometric units – to be used for conversion purposes only.   215
Box 10.3. Conversion factors.   217
Box 10.4. Calculation of light intensity at plant level.   217
Box 10.5. An example of light distribution in a greenhouse.   219
Box 10.6. Efficiency of current lighting systems.   221

Chapter 11 – Carbon dioxide supply

Box 11.1. CO_2 concentration in flue gas: how much air do we need to displace for
CO_2 supply?   230
Box 11.2. What can a 100 m^3 ha^{-1} buffer do in terms of separation of CO_2 and
heat production?   232
Box 11.3. How much CO_2 is needed to increase greenhouse air concentration?   233

Chapter 12 – Managing the shoot environment

Box 12.1. The basics of a standard climate control program.   241
Box 12.2. The typical cropping cycle of 'dark winter' regions.   244
Box 12.3. Old and new heating and ventilation management in Dutch glasshouses.   245
Box 12.4. Optimal lighting in 'discrete' production systems.   249
Box 12.5. Optimal light intensity.   250
Box 12.6. Management options.   253

Chapter 13 – Root zone management: how to limit emissions
Box 13.1. Electrical conductivity and salt concentration. 259
Box 13.2. Mineral composition of the 'average' plant. 260
Box 13.3. How much water is in the root system? 262
Box 13.4. Water-holding capacity and available water. 263
Box 13.5. Sensors for steering irrigation and nutrition. 265
Box 13.6. What is soil-less? 267
Box 13.7. Crops absorbing non-nutrient salts. 272
Box 13.8. How to calculate emissions. 273

Chapter 14 – Vertical farms
Box 14.1. Is 0 km always better? 284

Printed in the United States
by Baker & Taylor Publisher Services